# Telecommunications
# Transmission
# Engineering

# Telecommunications Transmission Engineering

## Volume 1 — Principles

### Second Edition

**AT&T**

This book was written for the practicing Transmission Engineer and for the student of Transmission Engineering in an undergraduate curriculum. A major part of this audience was the technical community of the Bell System. The authors were also Bell System people. This accounts for the references to Bell System facilities, products and services. Although subsequent events changed the Bell System and the national telecommunications network, this book continues to serve the original purpose. The book is not for sale through ordinary commercial channels, but inquiries may be directed to:

AT&T Technologies
Commercial Sales Clerk
Select Code 350-053
P. O. Box 19901
Indianapolis, Indiana 46219
1-800-432-6600

ISBN 0-932764-13-4

Printed in the United States of America

Telecommunications
Transmission
Engineering

## Introduction

Communication Engineering is concerned with the planning, design, implementation, and operation of the network of channels, switching machines, and user terminals required to provide communication between distant points. Transmission Engineering is the part of Communication Engineering which deals with the channels, the transmission systems which carry the channels, and the combinations of the many types of channels and systems which form the network of facilities. It is a discipline which combines many skills from science and technology with an understanding of economics, human factors, and system operations.

This three-volume book is written for the practicing Transmission Engineer and for the student of transmission engineering in an undergraduate curriculum. The material was planned and organized to make it useful to anyone concerned with the many facets of Communication Engineering. Of necessity, it represents a view of the status of communications technology at a specific time. The reader should be constantly aware of the dynamic nature of the subject.

Volume 1, *Principles*, covers the transmission engineering principles that apply to communication systems. It defines the characteristics of various types of signals, describes signal impairments arising in practical channels, provides the basis for understanding the relationships between a communication network and its components, and provides an appreciation of how transmission objectives and achievable performance are interrelated.

Volume 2, *Facilities*, emphasizes the application of the principles of Volume 1 to the design, implementation, and operation of transmission systems and facilities which form the telecommunications network. The descriptions are illustrated by examples taken from modern types of facilities most of which represent equipment of Bell Laboratories design and Western Electric manufacture; these examples are used because they are familiar to the authors.

Volume 3, *Networks and Services,* shows how the principles of Volume 1 are applied to the facilities described in Volume 2 to provide a variety of public and private telecommunication services. This volume reflects a strong Bell System operations viewpoint in its consideration of the problems of providing suitable facilities to meet customer needs and expectations at reasonable cost.

The material has been prepared and reviewed by a large number of technical personnel of the American Telephone and Telegraph Company, Bell Telephone Companies, and Bell Telephone Laboratories. Editorial support has been provided by the Technical Publications Organization of the Western Electric Company. Thus, the book represents the cooperative efforts of many people in every major organization of the Bell System and it is difficult to recognize individual contributions. One exception must be made, however. The material in Volume 1 and most of Volume 2 has been prepared by Mr. Robert H. Klie of the Bell Telephone Laboratories, who was associated in this endeavor with the Bell System Center for Technical Education. Mr. Klie also coordinated the preparation of Volume 3.

C. H. Elmendorf, III
Assistant Vice President —
Transmission Division.
American Telephone and Telegraph Company

# Volume 1 — Principles

# *Preface*

This volume, comprised of five sections, covers the basic principles involved in transmitting communication signals over Bell System facilities. Section 1 provides a broad description of the transmission environment and an overview of how transmission parameters affect the performance of the network. The second section consists of a review of most of the mathematical relationships involved in transmission engineering. A wide range of subjects is discussed, from an explanation of and justification for the use of logarithmic units (decibels) to a summary of information theory concepts.

The third section is devoted to the characterization of the principal types of signals transmitted over Bell System facilities. Speech, television, PICTUREPHONE®, digital and analog data, address, and supervisory signals are described. Multiplexed combinations of signal types are also characterized. The fourth section describes a variety of impairments suffered by signals transmitted over practical channels, which have imperfections and distortions. Also discussed are the units in which impairments are expressed and the methods by which they are measured. The fifth section discusses the derivation of transmission objectives, gives many established values of these objectives, and relates them to requirements applied to system design and operation. The section concludes with a chapter on international communications and internationally applied transmission objectives.

# Contents

# Contents

# Contents

# Contents

# Contents

# Contents

# Contents

# Telecommunications Transmission Engineering

## Section 1

## Background

The Bell System should be regarded as a single, huge, and far-flung telecommunications system made up of station sets, cables, switching systems, transmission systems, wires, and a conglomeration of other hardware of all sorts and sizes. This telecommunications system has grown rapidly and is still growing at a rapid rate. It has within it a large number of interconnected and interrelated systems and subsystems, each of which was designed with an approach that provided for successful development and overall Bell System evolution. This relationship between the parts and the total has permitted the orderly growth of a giant and the rendering of telecommunications services throughout the United States, Canada, and indeed the world.

Historically, the first telephone systems consisted of two remote station sets interconnected by wires normally used for telegraph communications. As interest in telephone communication built up, the transmission capabilities of the station sets and the interconnecting wires were gradually improved. Soon, manually operated switching systems were introduced in local communities to provide flexible interconnections among people living close together and sharing a high community of interest. These switching systems and the surrounding station sets and interconnecting wires have become known as the *local plant*.

The expanding local areas, the increasing demands for a wider range of services, and improvements in technology soon permitted the interconnection of one central office with another. As these interconnections increased in numbers and distances over which service had to be provided grew larger, the evolving long distance network became a separate entity known as the *toll plant*. Larger and more complex switching and transmission systems were designed to meet the unique needs of this part of the overall system.

Chapter 1 provides an overview of the operating Bell System plant with emphasis on the transmission and switching facilities that provide nationwide telephone service. Equipment used for other services that share the message network facilities is also briefly discussed.

An introduction to transmission concepts is given in Chapter 2. Brief descriptions of telephone, program, television, and data signals are presented, transmission terminology is defined, and basic techniques and modes of transmission are explained. Some specialized equipment, used to improve plant performance, is described to illustrate the interactions of various parts of the network.

Chapter 1

# The Transmission Environment

The Bell System provides a variety of communications services to large numbers of people over a very wide geographical area. To accomplish this task, a vast and complex physical plant has evolved. This plant is by no means static; it is highly dynamic in terms of growth, change, and the manner in which it is used for providing customer services.

The services provided by the Bell System are not readily categorized. The basic service of voice communications is handled by what is known as the switched message network; however, some services such as telegraph, facsimile, and voiceband data also utilize this network. In addition, a growing list of other services (e.g., point-to-point private line, television network service, wideband data, etc.) are provided, some of which require special switching arrangements and some of which require no switching at all.

The provision of transmission paths, or channels, and the flexible interconnection of these paths by switching are the two principal functions performed by the switched message network, the largest of the service categories that use the plant. The facilities involved are shared by many other services provided by the Bell System. The network transmission paths, highly variable in length, are of two major types, customer loops and interoffice trunks. The switching arrangements are also of two major types, local and toll. The design, operation, and maintenance of this huge network is further complicated by the multiplicity of its parts.

## 1-1 TRANSMISSION PATHS

Transmission paths are designed to provide economic and reliable transmission of signals between terminals. The designs must accom-

3

modate a wide range of applied signal amplitudes and must guarantee that impairments are held to acceptably low values so that received signals can be recovered to satisfy the needs of the recipient, whether a person or a machine.

Many transmission paths are designed as two-wire circuits; that is, transmission may occur simultaneously in both directions over a single pair of wires. Other paths, voice-frequency or carrier, are designed as four-wire; each of the two directions of transmission is carried on a separate wire pair.

The four major elements in transmission paths are station equipment (telephones, data sets, etc.), customer loops (cables and wires that connect station equipment to central offices), local and toll trunks (interconnections between central offices, consisting of cables or transmission systems and the transmission media they use), and the switching equipment (found primarily in the central offices). In its simplest form, a transmission path might consist of two station sets interconnected by a pair of wires.

## The Station Set

The station set accepts a signal from a source and converts it to an electrical form suitable for transmission to a receiver which reverses the process at a distant point. In most cases the station set is a telephone; however, many other types of station sets are used. Examples include facsimile sets, which operate to convert modulated light beams to modulated electrical analog signals and back to light at the receiving station, and voiceband data sets, which translate the signal format used by a computer to an electrical representation suitable for transmission over the telephone network and then back to the appropriate computer signal format. Many of these types of sets must meet transmission requirements for voice communications.

## Customer Loops

The station set is connected to the central office by the customer loop. This connection is most commonly made through a pair of insulated wires bundled together with many other wire pairs into a cable which may be carried overhead on poles, underground in ducts, or buried directly in the ground. For urban mobile service, however, the loop consists of a radio connection between the station set and the central office. The design of the customer loop must satisfy transmission requirements for all types of signals to be carried, e.g., speech, data, dial pulsing, TOUCH-TONE®, ringing, or supervision.

Loops are busy (i.e., connected to trunks or other loops) only a small percentage of the time — in some cases, less than 1 percent of the time. Where suitable calling patterns exist, this has led to the consideration of line concentrators for introduction between the station sets and the central office. A concentrator is a small switching machine which allows a number of loops to be connected to the central office over a smaller number of shared lines which are, in effect, trunks.

For some services, the loop plant is frequently reconfigured. In providing private branch exchange (PBX) services, for example, the loop plant provides PBX trunks connecting the customer's switching arrangement to the local or serving central office. In other services, loop plant may be used to form a part of a loop to be intermixed with trunks to provide an extended loop, or it may be used as a part of a channel between various customer locations for the transmission of wideband signals.

### Switching Machines

For switched message telephone service, the loop connects the station set to a switching machine in the local central office, which enables connections to be established directly to other local station sets or, through trunks and other switching offices, to any other station set on the switched network. The various types of switching offices which house this equipment are illustrated in Figure 1-1.

The principal switching machines in use today are electromechanical, e.g., the step-by-step and the crossbar types. Coming into increasing use, however, are electronic switching systems, which provide improvements in flexibility, versatility, and ease of maintenance, along with a considerable reduction of space requirements.

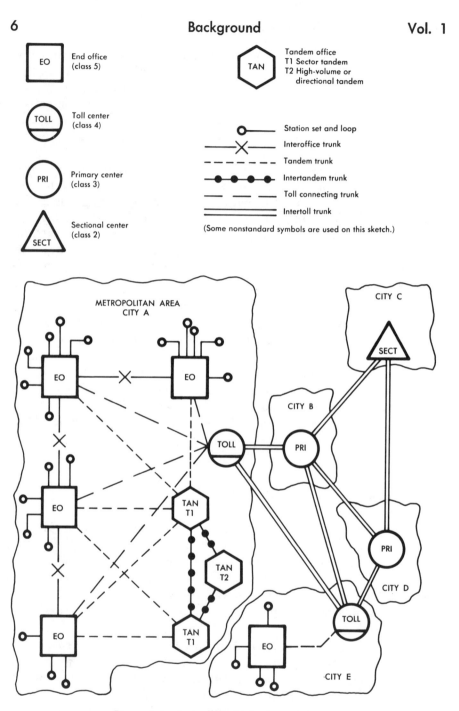

Figure 1-1. A simplified telephone system.

Transmission paths are provided through switching machines in a variety of ways and by a number of different mechanisms, including step-by-step switches, crossbar switches, and ferreed switching networks. These all have one thing in common, the ability to connect any one of a set of several thousand terminals to any other in the set. By design, this is accomplished with only a minimum of blocking during the busiest hour; i.e., only a very small percentage of calls is not completed as a result of all paths being busy. Each of the many paths is designed to provide satisfactory transmission quality through the central office.

As mentioned earlier, many transmission circuits are operated on a two-wire basis and, as a result, are also switched on a two-wire basis. Thus, especially in the local area, most switching machines provide two-wire paths. In the toll network, most of the transmission paths are four-wire; as a result, many toll switching machines must provide four-wire switching and transmission paths.

## Trunks

The transmission paths which interconnect switching machines are called trunks. One essential difference between a loop and a trunk is that a loop is permanently associated with a station set, whereas a trunk is a common connection shared by many users. There are several classes and types of trunks depending on signalling features, operating functions, classes of switching offices interconnected, transmission bandwidth, etc.

There are three principal types of interoffice trunks: local (interoffice, tandem, and intertandem), toll connecting, and intertoll. These trunk types and the switching offices that they interconnect are illustrated in Figure 1-1 which shows a representative metropolitan area and typical connections to the toll portion of the network.

All trunks must provide transmission and supervision in both directions simultaneously. However, trunks are designated *one-way* or *two-way* according to whether signalling is provided in both directions or only one. Two-way signalling is usually provided on intertoll trunks; calls can be originated on the trunk from the switching machine at either end. One-way signalling is the usual method of operating local and toll connecting trunks; therefore, separate trunk groups are provided for the two directions of originating traffic between the two offices involved.

Any trunk may use carrier transmission systems. However, local and toll connecting trunks rely heavily on voice-frequency cable media, although short-haul analog and digital carrier systems are becoming more widely used, especially in large metropolitan areas. The intertoll trunks, for the most part, utilize long-haul analog carrier systems and microwave radio relay systems.

## 1-2  SWITCHING ARRANGEMENTS

The service offered by the Bell System consists fundamentally of providing transmission capability upon demand between two or more points. Implied by "upon demand" is a switching arrangement capable of finding the distant end or ends of a desired connection and completing the connection between the originating and distant ends promptly and accurately. This is accomplished by a large number of switching machines connected together and organized around considerations of geography, concentrations of population, communities of interest, and diversity of facilities. These switching arrangements are illustrated in Figure 1-1 and may be broadly classified as either the local switching hierarchy (utilized for local transmission) or the toll switching hierarchy (utilized for transmission outside the local area). The switching equipment of either arrangement, however, is not totally divorced from that of the other. For example, tandem offices, operated by an associated company, are frequently used to switch toll traffic. Two methods are used. One is to segregate trunks between interlocal and toll use by maintaining separate groups. The second is to use a common tandem trunk group for both toll and interlocal. When trunks are so shared, the more severe transmission requirements for either use must be applied to the common group.

### The Local Switching Hierarchy

Figure 1-2 illustrates the various degrees of complexity that may be involved in switching within a local area. The simplest connection

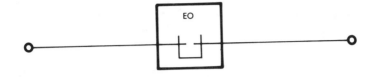

(a) Station-to-station connection; same end office

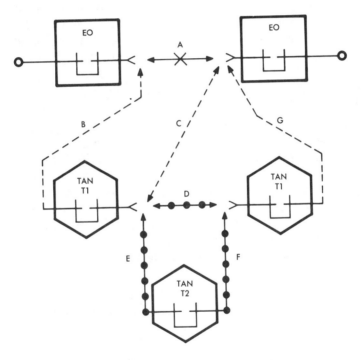

(b) Station-to-station connections using local trunks

Note :

Symbols are same as in Figure 1-1.

Figure 1-2. Illustrative telephone connections.

in the switched network is from one station set to another through a single local central office. Transmission over such a connection involves only the two station sets, their loops, and the transmission path through the switching machine as shown in Figure 1-2(a).

A connection in a multioffice area might be set up between two local, or class 5, offices in a number of ways as shown in Figure 1-2(b). Within the metropolitan area, it can be seen that trunks might interconnect two offices directly, using trunk A. Alternately, one, two, or three tandem switching machines might be used; with one machine, trunks B and C are used; with two machines, trunks B, D, and G are used. Finally, if three machines are involved, trunks B, E, F, and G are all used. These tandem machines, used in large metropolitan areas, provide economies through switching versus trunk facility costs and also provide alternate routing of traffic.

The complexity of the transmission network is obviously increased by this multitrunk local area switching arrangement, which is quite separate from the toll switching hierarchy discussed below. Since a connection might use just one interoffice trunk between the two end offices or as many as four tandem and intertandem trunks interconnecting the end offices and the tandem offices, the network arrangement must be designed and built according to objectives that take into account the number of trunks that might be connected together in tandem to complete a connection from one station to another. While local trunks are usually short, their numbers comprise the largest segment of trunks in the Bell System.

## The Toll Switching Hierarchy

The hierarchy of toll switching offices, developed to facilitate the transmission of signals beyond a local area, is illustrated in Figure 1-3. Working from the top down, it can be seen that the hierarchy consists of regional centers, sectional centers, primary centers, toll centers, and end offices. These centers and offices are also classified by a numbering system as shown in Figure 1-4. The figure also shows the quantity of each type of office operating in the Bell System in early 1970.

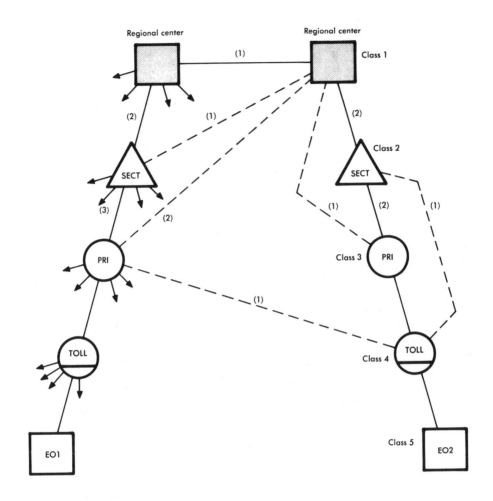

Notes:

1. Numbers in ( ) indicate order of choice of route at each center for calls originating at EO1.

2. Arrows from a center indicate trunk groups to other lower rank centers that home on it. (Omitted in right chain.)

3. Dashed lines indicate high-usage groups.

Figure 1-3. Choice of routes on assumed call.

| CLASS | DESIGNATION | APPROXIMATE NO. IN SERVICE, 1970 |
|-------|-------------|----------------------------------|
| 1 | Regional center | 10* |
| 2 | Sectional center | 50 |
| 3 | Primary center | 200 |
| 4 | Toll center | 1000 |
| 5 | End office | 10,000-15,000 |

*In addition to the ten regional centers in the U.S.A., there are two in Canada.

Figure 1-4. The toll switching hierarchy (Bell System only).

Access to the toll network is made through toll connecting trunks. In general, they are classified with local trunks since they are relatively short and intermixed on facilities with interoffice, tandem, and intertandem trunks. Generally, toll connecting trunks provide connections between class 5 and class 4 offices (end offices and toll centers). However, since class 5 offices may connect into the toll network at any level of the hierarchy, toll connecting trunks may also connect to class 3, class 2, or class 1 offices as well as to class 4 offices. In these cases, the higher offices also perform the functions of class 4 offices. The facilities used by toll connecting trunks may be voice-frequency or carrier. The termination at the class 5 office is two-wire; at the higher class offices it may be two-wire or four-wire, depending on the switching machine.

The toll switching network is provided with intertoll trunks between various combinations of office classes. One such combination is shown in Figure 1-3. Note that final trunk groups (i.e., those carrying traffic for which they are the only route and overflow traffic for which they are the "last choice" route) are provided between each lower ranking office and the higher ranking office on which it homes. All regional centers are interconnected by final trunk groups. High-usage trunk groups, which provide for alternate routing, are installed between any two offices that have sufficient community of interest. Automatic switching of toll circuits facilitates the use of alternate routing, so that a number of small loads may be concentrated into large trunk groups, resulting in higher efficiencies and attendant economies.

The order of choice of trunks for a call originating in end office 1 and terminating in end office 2 is indicated in Figure 1-3 by the numbers in parentheses. In the example there are ten possible routes for the call. Note that the first choice route involves two intermediate links. In many cases a single direct link, which would be first choice, exists between the two toll centers. Only one route requires seven intermediate links (intertoll trunks in tandem), the maximum permitted in the design of the network.

The probability that a call will require more than $n$ links in tandem to reach its destination decreases rapidly as $n$ increases from 2 to 7. First, a large majority of toll calls are between end offices associated with the same regional center. The maximum number of toll trunks in these connections is therefore less than seven. Second, even a call between telephones associated with different regional centers is routed over the maximum of seven intermediate toll links only when all of the normally available high-usage trunk groups are busy. The probability of this happening in the case illustrated in Figure 1-3 is only $p^5$, where $p$ is the probability that all trunks in any one high-usage group are busy. Finally, many calls originate above the base of the hierarchy since each higher class of office incorporates the functions of lower class toll offices and usually has some class 5 offices homing on it. Figure 1-5 makes these points more specific. The middle column of this table shows, for the hypothetical system

| NUMBER OF INTERMEDIATE LINKS, $n$ | PROBABILITY | |
|---|---|---|
| | FIGURE 1-3 | 1961 STUDY |
| Exactly 1 | 0.0 | 0.50 |
| 2 or more | 1.0 | 0.50 |
| Exactly 2 | 0.9 | 0.30 |
| 3 or more | 0.1 | 0.20 |
| 4 or more | 0.1 | 0.06 |
| 5 or more | 0.0109 | 0.01 |
| 6 or more | 0.00109 | 0 |
| Exactly 7 | 0.00001 | 0 |

Figure 1-5. Probability that $n$ or more links will be required to complete a toll call.

of Figure 1-3, the probability that the completion of a toll call will require $n$ or more links between toll centers, for values of $n$ from 1 to 7. In computing probabilities for this illustration, the assumptions are: (1) the chance that all trunks in any one high-usage group are simultaneously busy is 0.1; (2) the solid line routes are always available; and (3) of the available routes the one with the fewest links will always be selected. The values in Figure 1-5 illustrate that connections requiring more and more links become increasingly unlikely. These numbers are, of course, highly idealized and simplified.

Actual figures from a Bell System study made in 1961 are shown in the last column of the table of Figure 1-5. These numbers represent the probability of encountering $n$ links in a completed toll call between an office near White Plains, New York, and an office in the Sacramento, California, region. The assumption was made that all traffic had alternate routing available and that blocking due to final groups was negligible. Note that at that time 50 percent of the calls were completed over only one intermediate link. This is not possible in the layout shown in Figure 1-3, where it may be assumed that traffic volume does not yet justify a direct trunk between toll centers. The maximum number of links involved in the 1961 study was five; this number was required on only 1 percent of the calls.

More recent studies, reported informally, indicate that the trend continues in the direction of involving fewer trunk links in toll calls. In 1970, approximately 75 percent of all toll calls were completed over only one intertoll trunk; 20 percent required two intertoll trunks in tandem; about 4 percent required three trunks; the remaining 1 percent required four or more intertoll trunks in tandem. This trend is a result of increasing connectivity between offices by providing increased numbers of high-usage trunk groups (direct connections) between lower classes of offices in the hierarchy.

## 1-3  IMPACT OF SYSTEM MULTIPLICITY ON NETWORK PERFORMANCE

The provision of customer-to-customer communications channels can involve a multiplicity of instrumentalities, facilities, and systems interconnected in many ways. Station sets, loops, and end offices are particularly important, especially in the switched message network, since they are used in every connection. Toll connecting and intertoll trunks, toll transmission systems, and toll switching machines are also important when communications beyond the local area are considered. The overall comprises a complex configuration of plant items whose interactions give rise to several broad problems in the total network design and operation.

The first problem is that the accumulation of performance imperfections (such as loss, noise, and impedance irregularities) from a large number of systems leads to severe requirements on individual units and to great concern with the mechanisms causing imperfections and with the ways in which imperfections accumulate.

The second problem is that the variable complement of systems forming overall connections makes quite complex the problem of economically allocating tolerable imperfections among these systems. Deriving objectives for a connection of fixed length and composition is a problem involving customer reactions and economics. However, when these objectives must be met for connections of widely varying length and composition, the problem of deriving objectives for a particular system requires an even more complex statistical study involving considerable knowledge of plant layout, operating procedures, and the performance of other systems.

A third problem involves the satisfactory operation of each part with nearly all other parts. Compatibility is particularly important when new equipment and new systems are being developed, because the existing plant and the new interact importantly in many ways and also because plant growth must take place by gradual additions rather than by massive junking and replacement.

A fourth problem is that of reliability. Only small percentages of outage time are acceptable for the communications services provided by the Bell System, and these must account for all causes of failure—equipment failure, natural or man-made disaster, operating errors, etc.

Finally, to be complete, any discussion of the environment must recognize that telephone plant and power transmission and distribution systems share the same geography, either aerially or underground. This fact is important from a safety standpoint and from the standpoint of quality of transmission on telephone facilities. Power systems may come in contact with telephone plant as a result of storms, plant failures, or induction, endangering customers, employees, and property unless protective measures are applied. The presence of power systems in proximity to the telephone plant can also be damaging from the standpoint of quality of telephone transmission since noise induction is a distinct possibility.

## 1-4  MAINTENANCE AND MAINTENANCE SUPPORT

The switching patterns that have been described impose strict requirements on all transmission circuits. For example, up to seven intertoll trunks may be connected in tandem, and successive calls between the same two telephones may take different routes which involve different numbers and kinds of circuits. The losses encountered on calls routed over different numbers of links must not vary excessively, nor may the transmission quality vary significantly. If unsatisfactory transmission occurs, it cannot be observed by an operator as in the past, and the customer's attempt to report the trouble disconnects the impaired circuit, making difficult the identification of the source of trouble.

To cope with this situation, many central offices have extensive test facilities associated with them. Some of these facilities are test switchboards which have access to the lines and trunks in the office by manual patch or cross-connecting means. New automatic test facilities are also now available and are used extensively to test interoffice trunks by way of special trunk circuits and access arrangements provided in the switching machines. In addition, many central offices are equipped with voiceband data test centers for both DATA-PHONE® and private line service.

A great variety of portable, special purpose, and general purpose test equipment is also usually available in most central offices. This equipment, fixed and portable, manual and automatic, is described in greater detail later.

Extensive test equipment is also available for special services. For example, test equipment for television and wideband data services is located at the Television Operating Centers and the Wideband Data Test and Service Bays.

In addition to equipment that is directly involved in maintenance, there is an extensive list of equipment and transmission system features that may be classified as maintenance support. These equipment and service features are designed to facilitate trouble identification, isolation, and repair, to prevent extensive proliferation of trouble conditions, to provide for emergency restoration of broadband facilities on a temporary basis, to provide for remote telemetering and remote control of maintenance equipment and alarms, and to provide special communications channels (order wires) for maintenance personnel. These also are described in greater detail in a later chapter.

#### REFERENCES

1. Technical Staff of Bell Telephone Laboratories. *Transmission Systems for Communications,* Fourth Edition (Winston-Salem, N. C.: Western Electric Company, Inc., 1970).

2. Bell System Technical Reference PUB41005, *Data Communications Using the Switched Telecommunications Network* (American Telephone and Telegraph Company, August 1970).

3. *Switching Systems* (American Telephone and Telegraph Company, 1961).

Chapter 2

# Introduction To Transmission

The movement of intelligence from one point to another is the basic task of the Bell System. The intelligence to be moved can be called a *message,* regardless of the form it takes or its purpose. The most common form, of course, is speech, and the telephone system was initially developed around the need for voice communications. Over the years, however, many other types of messages (such as facsimile, program, video, and data) have evolved.

In general, transmission technology has advanced in parallel with this evolution, providing a means of translating these messages into electrical signals and developing the communications channels that make it possible to transmit the messages in reversible form via existing transmission media. Extension of the capabilities of the existing multiple-link plant and the development of new plant compatible with the old and capable of fulfilling transmission requirements are among the problems confronting the transmission engineer.

The variety of message signals and types of channels interact in many ways. Different types of message signals require channels of various bandwidths and operating characteristics. These channels utilize voice-frequency and carrier facilities which must meet stringent requirements if they are to provide satisfactory service economically. To meet these requirements, it is sometimes necessary to use specially designed ancillary equipment on the channels or systems.

## 2-1 MESSAGE SIGNALS

The characterization of transmitted message signals is essential to an understanding of how such signals interact with the channels over which they are transmitted. The *message signal* is defined as

18

an electrical representation of a message, which can be transmitted in its electrical form from source to destination. Qualitative descriptions of the more common signals found in the Bell System are given here. The signals described in this chapter include voice, program, video, data and facsimile, and control signals. The latter, usually classified as signalling and supervision, are transmitted in order to activate switching operations and to perform other subsidiary functions. Variations of these signal types are used to transmit all messages presently offered as Bell System communication services. Any of the signals may be transmitted in either digital or analog form; the choice is dependent in some cases on the transmission facilities available. More detailed quantitative characterizations of all these signal types are given in Chapters 12 through 16.

## Speech

The most common signal transmitted over Bell System facilities is the speech signal, an electrical signal generated in the telephone station set as an analog of the acoustical speech wave generated in the voice box, or larynx, of the speaking telephone user. This signal carries most of its information in a band of frequencies between 200 Hz and 3500 Hz. Most of the energy is peaked near 800 Hz; most of the articulation is above 800 Hz. It has higher frequency and lower frequency components, but these are not normally transmitted. It is an extremely complex signal, not only because of the large number of frequency components it contains, but also because of the wide range of amplitudes that any component may have and because of the rapidity with which the frequency and amplitude of its components may change.

Another complexity is the time relationships inherent in the speech signal. By one definition or criterion, the signal duration might be measured from the time the connection is established until it is broken. By another criterion, the signal duration might be defined as the speaking interval—during a typical telephone connection, each party speaks about half the time and listens the other half. But the situation is even more complex. There are short intervals, sometimes only milliseconds in length, during which a speaker pauses for breath or for other reasons. Signal duration could be defined as covering the time between those pauses. So, it is a matter of definition; care must be taken to define the signal precisely when circumstances demand it, for example, when considering system loading or crosstalk effects.

## Program

Program signals are those associated with the distribution of radio program material, the audio portion of television program material, or "wire music" systems. These signals are usually transmitted over one-way channels having a somewhat wider bandwidth than the standard voice-frequency message channel. The signals may include speech and a wide range of musical material. The signal energy is usually maintained at a higher average level than that of switched telephone voice-frequency signals and may be transmitted continuously for hours; however, since there is such a small percentage of circuits assigned to this type of service, little effect is felt in system loading.

## Video

There are three types of video signals commonly transmitted over Bell System facilities—television, PICTUREPHONE®, and multilevel facsimile. Multilevel facsimile represents a very small percentage of transmitted signals and is described as a data signal. Brief descriptions of television and PICTUREPHONE signals are given here.

A television signal contains information in electrical form from which a picture can be re-created with fidelity. A still monochrome picture may be expressed as a variation in luminance over a two-dimensional field. In a moving picture, however, the luminance function also varies with time. The moving picture, therefore, is a function of three independent variables: luminance, position, and time.

The electrical signal (characterized in Chapter 15) consists of a current or voltage amplitude which is a function of time. At any instant, the signal can represent the value of luminance at only one point in the picture. It is necessary, therefore, in the translation of a complete picture into an electrical signal, that the picture be scanned in a systematic manner. If the scanning pattern is sufficiently detailed and conducted rapidly, a satisfactory reproduction of picture detail and motion is obtained. The basic system consists of a series of scans in nearly horizontal lines from left to right, starting at the top of the image field. When the bottom of the field is reached, the process is started again from the top with alternate fields interlaced to form a frame.

For the successful decoding of the signal into a picture at the receiver, it is necessary to transmit a key to the scanning pattern. In the standard signal, this consists of frequent short-duration synchronizing pulses indicating characteristic points in the course of the scanning pattern, such as the beginning of scanning lines and fields. This is coupled with the condition that the motion of the scanning spot between pulses is uniform with time in the field of view. The picture signal is interrupted during retrace time and replaced by a black signal known as a blanking pulse. Because of this, the return trace is not visible in the picture.

The PICTUREPHONE signal is conceptually similar to the television signal. Both signals use a frame rate of 30 per second and a field rate of 60 per second. Lines from alternate fields are interlaced. The following tabulation compares the two signals in other important respects:

|  | SCAN RATE | LINES/FIELD | BANDWIDTH |
|---|---|---|---|
| TELEVISION | 15.75 kHz | 525 | 4.3 MHz |
| PICTUREPHONE | 8.0 kHz | 250 | 1.0 MHz |

The signal duration in television operation is long, an hour or more, with only short breaks for commercial and station-break announcements. These hardly qualify as signal terminations for most situations. For PICTUREPHONE signal transmission, the signal duration is the full period of the call since picture information is transmitted in both directions during this entire period.

## Data and Facsimile

The basic data signal usually consists of a train of pulses which represent, in coded form, the information to be transmitted. Such signals are processed in many ways to make them suitable for transmission over Bell System facilities. To represent coded values of the signal, the amplitude may be shifted, the frequency may be shifted, or the phase of a carrier may be shifted. In addition, the relative positions of pulses or the duration of pulses may be changed.

The speed with which changes are made, no matter which parameter is changed, determines to a large extent the bandwidth required to transmit data signals. Transmission speeds used in the Bell System vary from a few bits (binary digits) per second for supervisory control channels, to a few hundred bits per second for teletype or telegraph signals, to over one megabit per second for use on digital carrier systems.

Some two-valued facsimile signals (black and white facsimile) closely resemble binary data signals and may be compared with them in many ways. Multivalued facsimile signals are more like video signals, as mentioned earlier; they produce pictures at slow speeds with gradations of grey between black and white. Such facsimile signals are often regarded as special forms of data because channel requirements for facsimile transmission are quite similar to those for data transmission. The latter signals, together with other forms of data (such as the electrocardiogram signal), may be regarded as analog data signals.

Data signal durations are highly variable. Some data messages tend to be very short while others can last for hours. Facsimile messages last several minutes typically.

## Control Signals

In order to implement the functions of any switched network, it is necessary to transmit three types of control, or signalling, information. These are (1) alerting signals, (2) address signals, and (3) supervisory signals. These signals are usually transmitted over the loops or trunks directly involved in an overall connection. However, they may also be transmitted over a separate, dedicated signalling channel used as a common signalling facility for many message channels. Such a common channel system is under study and development.

Alerting signals include the ringing signal, which is supplied to a loop to alert the customer to an incoming call on his line, and a variety of signals that are used to alert operators to a need for assistance on a call. Addressing signals provide information, transmitted over loops and trunks, concerning the desired destination of the call (the called number) and, sometimes, the identification of the calling number. Supervisory signals are used to indicate a demand for service, the termination of a call, and the busy/idle status of each loop or trunk.

Many forms of addressing and supervisory signals are employed. These include pulsing of the direct current supplied on loops for talking purposes or on voice-frequency trunks for supervision, changes in state of direct current supplied on voice-frequency trunks, and single-frequency or multiple-frequency alternating current signals which may be transmitted within or outside the voiceband of a carrier or voice-frequency circuit. The most important of these signals are described and characterized in Chapter 13.

Many other types of signalling information are transmitted for subsidiary functions. These include dial tone, audible ringing tone, coin signals (deposit, return, and collect), busy and reorder tones, and recorded announcements such as time and weather information. None of these relates importantly to transmission work, however, and so they are not described in detail.

The duration of information signals varies widely. Addressing signals last for only a short time, one to several seconds. Supervisory signals, on the other hand, may be present for minutes, hours, or even days when a trunk, for example, is not called into use. It is interesting to note that address signals may be regarded as transient by nature, and supervisory signals are steady state.

## 2-2   CHANNELS

A *channel* is defined as a frequency band, or its equivalent in the time domain, established in order to provide a communications path between a message source and its destination. The characteristics of the signal derived from the message source determine the requirements imposed upon the channel in respect to bandwidth, signal-to-noise performance, etc.

In the switched telephone message network, a variety of channels are provided on a full-time, dedicated basis in the form of loops, local trunks, toll connecting trunks, and intertoll trunks. Each such channel is a well defined entity between its terminals for long periods of time. Changes in the channel makeup or configuration can be made only by changing soldered connections or by patching within a jack field.

The end-to-end frequency band established between station sets in a built-up telephone call is also a channel. This frequency band is dedicated and maintained only for the duration of the call. In this case, the channel is made up of other tandem-connected channels—the interconnected loops and trunks used to establish the connection.

With the advent of time division switching and its integration with time division transmission systems, the end-to-end concept of a channel in a built-up telephone connection may have to be modified. Present systems maintain the integrity of the channel in the time domain equivalent of the analog channel, but it is theoretically possible to change channel assignments during a call. Analog channel assignments are changed during a call in the TASI system, described later in this chapter.

Thus, channels in the switched telephone message network may be regarded as fixed, changeable, or switchable. In any case, each type of channel must be designed to have a transmission response that will satisfy the objectives set for the type of service to be provided. That is, they must be of sufficient bandwidth, must have gain/frequency and phase/frequency characteristics that are well controlled, and must not be contaminated by excessive noise or other interference. These parameters will be discussed more quantitatively in later chapters.

Channels may also be regarded as one-way or two-way. Carrier systems are usually operated on a four-wire basis, a separate path for each direction of transmission. On one such path the dedicated band of frequencies (i.e., the channel) carries signal energy in one direction only, and so each path represents a one-way channel. Voice-frequency circuits (loops and trunks), on the other hand, are frequently operated so that both directions of transmission are carried on the same wire pair—a two-way channel. In any case, in the switched message network, loops and trunks must be capable of full duplex, i.e., two-way simultaneous usage.

In this discussion, definitions involving channels in the switched telephone message network have been stressed. It must be recognized that many other types of channels are provided in the Bell System. These include very wideband channels for high-speed data or video signal transmission, channels somewhat wider than speech channels for radio and television program or sound signals, voiceband channels that are specially treated to meet data or facsimile transmission objectives, and very narrowband channels for telegraph and low-speed data signal transmssion.

## 2-3　VOICE-FREQUENCY (VF) TRANSMISSION

To a large extent, the line facilities and apparatus that are applied in practice to the local telephone plant operate at voice frequency. The loop plant employs a two-wire mode of operation almost exclusively, and the local network trunk plant is operated in both the two-wire and the four-wire modes. In either mode, the transmission medium introduces signal loss which must be controlled within established limits in order to provide satisfactory service. When the losses exceed the established limits, compensation must be made by means of voice-frequency repeaters (amplifiers and associated circuit features) whose gains are designed to restore signal amplitudes. For economical circuit design, then, proper choice must be made of the minimum wire size compatible with circuit length, as well as the appropriate repeater type relative to mode of operation, wire size, and circuit length.

### Modes of Voice-Frequency Transmission

The telephone station set is basically a four-wire instrument, one that requires two wires for the transmitter and two wires for the receiver. If the four-wire nature of the set were extended into the entire local plant including both loops and trunks, four wires would have to be provided for every connection including the transmission paths through the switching machines. Such an arrangement, illustrated in Figure 2-1(a), would offer some transmission advantages, but it would be inordinately expensive since it would nearly double the amount of copper required for cables and other types of conductors needed to provide transmission paths and would impose a burden on

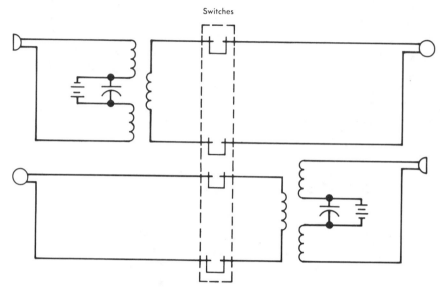

(a) Four-wire connection

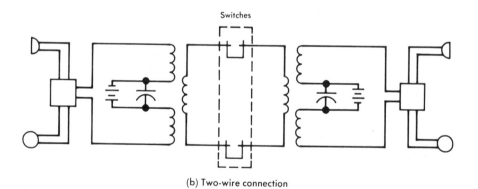

(b) Two-wire connection

Figure 2-1. Voice-frequency modes of transmission.

local switching machines, nearly all of which provide two-wire transmission paths only. To avoid this expense, the station set is provided with circuitry that combines the transmitter and receiver conductors so that only one pair of wires is needed for transmission in both directions. This arrangement, called two-wire transmission, is illustrated in Figure 2-1 (b).

Two-wire transmission is used almost exclusively in loops and commonly in short trunks between local central offices. However, when trunks are long or when the bandwidth is significantly greater than the 4 kHz used for speech transmission, the technical problems are such that four-wire transmission is necessary. Net losses can be held at lower values, and there are fewer echo and singing paths. Therefore, there are applications for four-wire voice-frequency circuits even before carrier applications become economical.

### Voice-Frequency Repeaters

The selection of repeater type in solving voice-frequency application problems depends on the required gain and on the mode of transmission, two-wire or four-wire. The E-type repeater, shown schematically in Figure 2-2, is used in many two-wire trunks and some loops to provide the necessary gain and equalization. Its unique shunt and series negative impedance characteristics provide gain in both directions of transmission in a two-wire circuit.

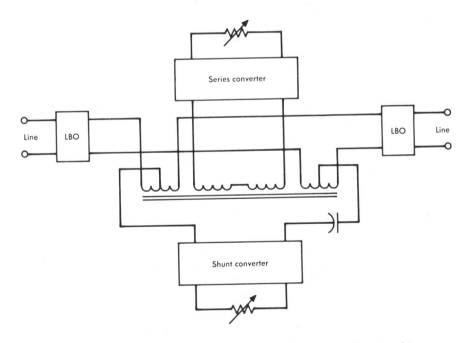

Figure 2-2. Negative impedance repeater for a two-wire repeatered line.

The repeaters used in four-wire lines use separate amplifiers for each direction of transmission. Figure 2-3 shows how four-wire repeaters are used in a four-wire trunk. Notice that repeaters used along the line connect four-wire to four-wire, while at the ends of the trunk the arrangement provides interconnection between four-wire and two-wire facilities.

New designs, designated *facility terminals,* are now available to provide either two-wire or four-wire gain. In addition, this equipment provides flexibility in interconnecting equipment needed for other circuit functions such as signalling and equalization as required.

## 2-4 CARRIER SYSTEMS

Since the per-channel copper costs for VF transmission are often prohibitive, carrier systems have been developed to reduce overall costs by the substitution of electronics for copper. Carrier systems (which for the purpose of this discussion include microwave radio systems) are broadband, multichannel, four-wire facilities. The carrier principle proves to be economical because its broadband, multichannel features allow one carrier channel to be used for a multiplicity of narrower band channels (for speech, data, or other signal transmission). These individual channels operate, of course, in the four-wire mode also, since there is a separate path for each direction of transmission.

Systems designed for submarine cable operation and some short-haul carrier systems use a mode of transmission called equivalent four-wire. In this mode, the two directions of transmission are separated in frequency on a single pair of wires, rather than in space on separate wire pairs. Two circuit arrangements that are commonly used are illustrated in the block diagrams of Figure 2-4 (a) and (b). The advantage of this mode, of course, is that only one pair is required for both directions of transmission.

A carrier system may be regarded as consisting functionally of three major parts: (1) high-frequency line or radio relay equipment which, with the transmission medium, provides a broadband channel of specified characteristics to permit simultaneous bidirectional transmission of a wide range of communications signals; (2) modulating equipment to process signals from one form to another more suitable for transmission in each direction of transmission from the terminal;

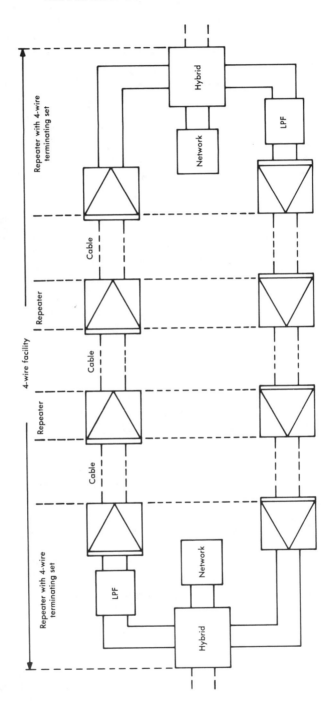

Figure 2-3. Four-wire repeatered line.

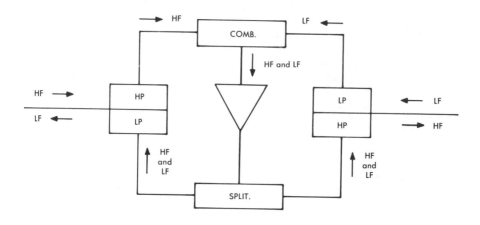

(a) Single amplifier connection

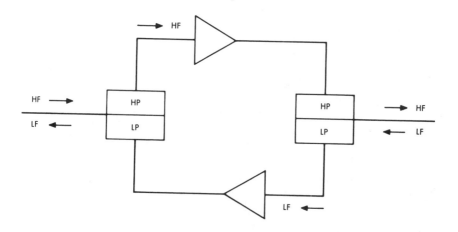

(b) Dual amplifier connection

Figure 2-4. Equivalent four-wire repeater configurations.

and (3) multiplexing equipment which combines, at the system input, and separates, at the system output, the various signals sent through the system. To achieve efficient equipment packaging, the modulating and multiplexing equipment are usually combined. Together these three major parts provide transmission channels having fixed gain, acceptably low noise and distortion, and high velocity of propagation.

## High-Frequency Line Equipment

To provide a broadband channel, high-frequency line equipment must perform a number of functions which differ depending on the type of system. The basic function found in all systems is that of amplification to compensate for losses in the medium between repeater points. Such compensation may require the gain to be a nonflat function of frequency and, as such, it is often considered as the first step in equalization.

In *analog coaxial cable systems,* amplification is the only primary function of the high-frequency line equipment. Other functions, such as regulation to maintain constant the overall system gain in the presence of temperature changes, and equalization to compensate for small deviations in the transmission response, are also provided, but for present purposes they may be regarded as secondary. Another secondary function is that of protection line switching; in the event of failure or for line maintenance, service may be switched automatically or manually to a spare line.

Another important function of the line equipment of some short-haul cable systems is frequency frogging, whereby the signal is modulated alternately between two frequency bands at each repeater. This process is required to control systematic impairments (equalization) that arise due to the transmission response of the medium and to limit unwanted crosstalk paths.

The repeaters in the high-frequency line equipment in *microwave radio systems* also provide gain as their primary function. Secondary functions include modulation to translate the signal from high radio frequencies to intermediate or baseband frequencies (where designs of amplifiers are more tractable), frequency frogging between radio frequency bands, and high-speed switching to provide alternate paths for the signals in order to overcome the effects of fading in the medium (the atmosphere) or to overcome the effects of equipment failure.

In the high-frequency line equipment of *digital systems* there are two primary functions. In addition to providing gain, a digital repeater regenerates the transmitted pulses and reshapes them for transmission to the next repeater. The regeneration process must also include a timing function so that the pulses are transmitted in correct time relationship to one another.

## Modulating Equipment

Input signals to a carrier system must be processed to make them suitable for transmission over the line equipment. The processing is usually referred to as modulation. There are several forms of modulation, and they may be used singly or in combination according to the needs of the system.

The process involves modifying the signal in some reversible manner to prepare it for combining with other signals or for transmission over the high-frequency line or both. This may be accomplished by varying (modulating) a carrier in amplitude or frequency in accordance with the amplitude and frequency variations of the input signal (sometimes called the baseband signal); or the input signal, regarded as a continuous wave, may be sampled in time and then coded into a stream of pulses as is done in digital transmission systems.

## Multiplex Equipment

Multiplexing means the combining of multiple signals for simultaneous transmission over a common medium. The simplest form of multiplexing might be called *space division multiplexing*. It occurs when many signals are transmitted over separate pairs of wires all in the same cable. The term is usually applied, however, to the two categories called frequency division multiplex and time division multiplex.

If in a *frequency division multiplex* (FDM) system a number of signals modulate carriers at some high frequencies, they may be transmitted simultaneously over a common medium provided (1) the band of each signal covers a part of the broadband spectrum of the high-frequency line equipment different from all other modulated signals and (2) the total bandwidth does not exceed that of the high-frequency line equipment. Such signals are combined in electrical networks in the transmitting terminal of the system and are separated by frequency-selective networks at the receiving terminal.

In a *time division multiplex* (TDM) system, the pulses, which are formed for different signals in the modulating equipment, are interspersed in a regular time relationship at the transmitting terminal. Timing pulses, transmitted with the signal information, permit the operation and control of gate circuits at the receiving terminal. These circuits separate the signals from one another so that they may be processed, or demodulated, individually at the receiving terminal of the system.

## 2-5  ANCILLARY EQUIPMENT AND FUNCTIONS

Included in the transmission plant are a large number of equipment items and operating techniques that have been developed so that transmission and operating requirements may be met more economically. Among these are circuits such as compandors and echo suppressors, operating techniques such as frogging, and complete switching/transmission systems that employ time assignment speech interpolation techniques.

The word *compandor* is made up of syllables taken from the words *COMpressor* and *exPANDOR*. The performance of some carrier systems is improved, especially for speech signals, by the use of these devices. At the transmitting terminals of a telephone circuit, speech signal amplitudes are compressed into a narrower than normal range and then restored by the expandor at the receiving terminal. The result is a significant reduction of noise during periods of small signal transmission and during quiet intervals. These are the periods when noise is most objectionable to the telephone user.

In addition to compensating for losses incurred in the medium, the design of telephone trunks involves dealing with and overcoming various other impairments suffered in the course of transmission. One such impairment is echo. If a trunk is so designed that the likelihood of a disturbing echo is high, it is often equipped with an echo suppressor. This device acts as a pair of voice-operated switches; while one subscriber is talking, the echo suppressor inserts high loss in the opposite direction of transmission to attenuate the echo before it is returned to the speaker.

Two types are used, full and split. In the full echo suppressor, the voice-operated switches and echo attenuation circuits for both directions of transmission are located at one end of the trunk. In the split echo suppressor, the circuitry is split between the two ends of the four-wire trunk.

The performance of transmission systems is often improved by some kind of frogging, a term adopted from the railroad industry where a frog is a special section of rail used to cross one track over another. In transmission, some impairments may accumulate due to channels being in close relationship to one another. These relationships and the impairments can be altered significantly by frogging in space (by changing the medium or by reversing or transposing wire positions) or by frequency frogging (changing the relative positions of channels in a common spectrum). Both space and frequency frogging techniques are used in telephone practice.

A Time Assignment Speech Interpolation (TASI) system is used to increase the efficiency of bandwidth utilization on some transmission systems. It operates as a high-speed switching system to allow a number of talkers to share a smaller number of trunks on the high-frequency line. The switches are voice-operated and allow a channel in the transmission system to be taken from a speaker during breaks in his conversation. These breaks occur during periods that a user is listening, rather than talking, and during other pauses in normal conversation.

### REFERENCES

1. Technical Staff of Bell Telephone Laboratories. *Transmission Systems for Communications, Fourth Edition* (Winston-Salem, N. C.: Western Electric Company, Inc., 1970), Chapter 29.

## Section 2

## Elements of Transmission Analysis

In this section of the book a number of technical subjects are treated in a manner designed to acquaint the reader with fundamental principles of transmission analysis, which have application to transmission system design, operation, and planning. The subjects are covered in a series of nine chapters.

While the book has been written for persons with an electrical engineering background, it must be appreciated that each of the subjects covered in this section has been worthy of entire textbooks at both the undergraduate and graduate levels of study. It has been impossible, therefore, to discuss most of these subjects without using a higher level of mathematics than can normally be assumed for second or third year undergraduate students. For the most part, the mathematics used are presented without apology, without proof, and without thorough mathematical development that might satisfy a mathematician. For additional background information, the inquisitive reader is referred to the literature listed at the end of each chapter.

Chapter 3 provides a transition from the "Background" section of the book to the more theoretical subjects to follow. Some terminology is defined, and justification for the use of logarithmic units in transmission work is presented. The concept of transmission level points is discussed, and measurements of certain types of signals and interferences are described.

Chapters 4 and 5 cover the related subjects of "Four-Terminal Linear Networks" and "Transmission Line Theory." The material in these chapters includes discussions of the basic Ohm's and Kirchoff's laws and their application; the analysis of networks and their interactions, impedance relationships, return loss and reflections; and transformer and hybrid coil theory and applications. Transmission lines are treated in terms of equivalent circuits, characteristic impedance, primary electrical constants, velocity of propagation, and loading.

Chapter 6, on wave analysis, is presented in order to increase the reader's understanding of the Fourier series and Fourier transform. This permits a more general understanding of time-domain and frequency-domain relationships between signals and transmission channels.

Chapter 7 covers negative feedback amplifiers from the points-of-view of how design limitations and compromises are made to accomplish design objectives and how these objectives are related to the performance of transmission systems in the field. The principal benefits of feedback are discussed and means for providing feedback, as well as the manner in which feedback mechanisms interact with each other, are described.

In Chapter 8, a number of methods of signal processing are described in order to show how signals are modified for more efficient transmission over existing media and then restored to their original form for final transmission to the receiver. Various forms of amplitude, angle, and pulse modulation processes are covered.

"Probability and Statistics" is the subject of Chapter 9. The application of this branch of science to transmission system design and operation is among the most important aspects of transmssion work. Without the application of probability and statistics, the Bell System could not operate economically and perhaps could not operate at all. The terminology and symbology of this branch of mathematics are first described. Examples of statistical and probabilistic analyses are given to illustrate how such techniques may be used to solve transmission problems.

Chapter 10 covers a brief history and description of information theory and its application in transmission engineering. Mathematical expressions are presented to show the theoretically maximum channel capacity for both ideal (distortion-free) channels and for typical noisy channels. While the subject is of most concern to development and research workers, an understanding of the principles should enhance the work of the transmission engineer in the field.

Chapter 11, the last chapter of this section, consists of a presentation of the more important aspects of conducting engineering economy studies. Transmission problems usually have more than one technically sound solution; the selection of one of several alternative lines of action can often be best made on the basis of economic comparisons of the alternatives.

Chapter 3

# Fundamentals of Transmission Theory

Transmission systems for communications are made up of a large number of tandem-connected two-port (four-terminal) discrete networks and distributed networks such as transmission lines. In the analysis of transmission systems, the properties of these networks must be defined mathematically. Logarithmic units are commonly used because the ratios of currents, voltages, and powers found in these networks are large and awkard to manipulate. If the input-output relations, or *transfer characteristics*, of the individual two-port networks are determined, the transfer characteristics of the tandem connection of several such networks can be found by taking a product of the appropriate network transfer characteristics.

Transmission parameters of communication systems are measured in a manner consistent with mathematical analysis techniques. Thus, many types of test equipment are designed to measure signal and interference amplitudes in logarithmic units (decibels). Other test equipment types measure more conventional parameters, such as volts, amperes, or milliwatts. Some test instruments are arranged to display signals or interferences as functions of either time or frequency.

## 3-1 POWER AND VOLTAGE RELATIONS IN LINEAR CIRCUITS

Some of the mathematical relations necessary for the evaluation of system performance can be explained in terms of the simple circuit diagram of Figure 3-1. The transducer in this circuit is assumed to be linear; i.e., the relation of the output signal to the input signal can be described by a set of linear differential equations with constant coefficients.

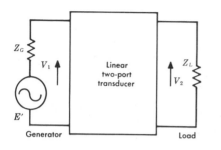

Figure 3-1. Terminated two-port circuit.

Energy is transferred from generator to load via the linear two-port transducer. The transducer may take on a wide variety of forms, ranging from a simple pair of wires to a complex assortment of cables and amplifiers, modulators, filters, and other circuits. The four terminals are associated in pairs; the pair connected to the generator is commonly called the input port and the pair connected to the load as the output port. If the energy at the output is greater than at the input, the transducer is said to have gain. If the energy is less at the output than at the input, the transducer is said to have loss.

As illustrated in Figure 3-1, a generator may be characterized by its open-circuit voltage, $E'$, and its internal impedance, $Z_G$, and a load by its impedance, $Z_L$. If the signal produced by the generator is periodic, it may be represented by a Fourier series, $V = V_1 + V_2 \ldots + V_k + \ldots V_n$, each term of which has the form

$$V_k = E_k \cos(\omega t + \phi_k), \qquad (3\text{-}1)$$

or, more conveniently,

$$V_k = E_k\, e^{j(\omega t + \phi_k)}. \qquad (3\text{-}2)$$

In Equations (3-1) and (3-2), the subscript $k$ represents the $k$th term of the Fourier series, $V_k$ represents its instantaneous voltage, and $E_k$ represents its peak voltage. The input-output relations of the transducer are not dependent on the presence or absence of other similar terms in the series nor of their magnitudes. The Fourier series signal representation is thus convenient for this type of analysis.

For example, if the generator voltage of Figure 3-1 is a single-frequency signal represented by

$$E' = E_s\, e^{j(\omega t + \phi_s)}, \qquad (3\text{-}3)$$

the ratio of $V_1$ to $V_2$ is given by

$$V_1/V_2 = [E_1 \, e^{j(\omega_1 t + \phi_1)}] / [E_2 \, e^{j(\omega_2 t + \phi_2)}]$$

$$= (E_1/E_2) \, e^{j(\omega_1 t - \omega_2 t + \phi_1 - \phi_2)}. \tag{3-4a}$$

Where the radian frequency, $\omega$, is the same for $E_1$ and $E_2$ (generally true except for modulators), this equation may be written

$$V_1/V_2 = (E_1/E_2) \, e^{j(\phi_1 - \phi_2)}. \tag{3-4b}$$

As mentioned above, the ratios encountered in telephone transmission are often very large, and the numerical values involved are awkward. Moreover, it is frequently necessary to form the products of several ratios in order to express the gain or loss of a network or a tandem connection of networks. The expression and manipulation of voltage or power ratios are simplified by the use of logarithmic units. The natural logarithm (ln) of the ratio of Equation (3-4b) is a complex number.

$$\ln(V_1/V_2) = \theta = \alpha + j\beta = \ln(E_1/E_2) + j(\phi_1 - \phi_2). \tag{3-5}$$

The real and imaginary parts of Equation (3-5) are uniquely identifiable, which is to say

$$\alpha = \ln(E_1/E_2), \text{ the attenuation constant,}$$

and

$$\beta = \phi_1 - \phi_2, \text{ the phase constant.} \tag{3-6}$$

When this measure of voltage (or current) ratio is used, $\alpha$ is said to be expressed in nepers and $\beta$ in radians. The term *neper* is an adaptation of Napier, the name of the Scottish mathematician credited with the invention of natural logarithms.

## The Decibel

The logarithmic unit of signal ratio which now finds wide acceptance is the decibel (dB). The decibel is equal to 0.1 bel, a unit named for Alexander Graham Bell whose investigations of the human ear revealed its logarithmic response. Strictly speaking, the decibel is defined only for power ratios; however, as a matter of common usage, voltage or current ratios also are expressed in decibels. The precautions required to avoid misunderstanding of such usage are developed in the following.

If two powers, $p_1$ and $p_2$, are expressed in the same units (watts, microwatts, etc.), then their ratio is a dimensionless quantity, and as a matter of definition

$$D = 10 \log \ (p_1/p_2) \qquad \text{dB} \qquad\qquad (3\text{-}7)$$

where log denotes logarithm to the base 10, and $D$ expresses the relative magnitude of the two powers in decibels. If an arbitrary power is represented by $p_0$, then

$$D = 10 \log \ (p_1/p_0) - 10 \log \ (p_2/p_0) \qquad \text{dB.} \qquad (3\text{-}8)$$

Each of the terms on the right of Equation (3-8) represents a power ratio expressed in dB, and their difference is a measure of the relative magnitudes of $p_1$ and $p_2$. Thus, the value of this difference is independent of the value assigned to $p_0$. However, it is often convenient to use a value of one milliwatt for $p_0$. The terms 10 log $(p_1/p_0)$ and 10 log $(p_2/p_0)$ are then expressions of power ($p_1$ or $p_2$) relative to one milliwatt, abbreviated dBm. Note, however, that their difference is in dB, not dBm. In short, Equation (3-7) is a measure of the difference in dB between $p_1$ and $p_2$. Note that nepers and decibels may be related by the expression nepers/dB = 20 log 2.718 = 8.686. This relationship is derived from the definitions of Naperian and common logarithms.

As mentioned above, voltage and current ratios are also often expressed in decibels as a matter of common usage. Such relationships are simple and direct when the impedances are equal at the points where the voltages or currents are measured. If the impedances are not equal, errors may be introduced unless care is taken to use appropriate correction factors as explained below.

If there is an rms drop of $e$ volts across a complex impedance ($Z = R + jX$ ohms) as a result of an rms current of $i$ amperes flowing through the impedance, the power dissipated in the impedance may be written

$$p = i^2 R \qquad \text{watts.} \qquad (3\text{-}9)$$

The rms voltage and rms current are related by Ohm's law, discussed in Chapter 4, in such a way that $i = e/|Z|$. By substituting this value in Equation (3-9) and expanding $|Z|^2$, it can be shown that the power dissipated may also be written

$$p = \frac{e^2}{R(1 + X^2/R^2)} \qquad \text{watts.} \qquad (3\text{-}10)$$

Using appropriate subscripts to indicate two different measurements, the value of $p$ from Equation (3-9) may be substituted in Equation (3-7) to yield

$$D = 10 \log (p_1/p_2) = 10 \log (i_1/i_2)^2 + 10 \log (R_1/R_2)$$

$$= 20 \log (i_1/i_2) + 10 \log (R_1/R_2) \qquad \text{dB.} \qquad (3\text{-}11)$$

Similarly, the value of $p$ from Equation (3-10) may be substituted in Equation (3-7). This gives

$$D = 10 \log (p_1/p_2) = 10 \log \frac{e_1^2/R_1 \, (1 + X_1^2/R_1^2)}{e_2^2/R_2 \, (1 + X_2^2/R_2^2)}$$

$$= 20 \log (e_1/e_2) - 10 \log (R_1/R_2) - 10 \log \frac{(1 + X_1^2/R_1^2)}{(1 + X_2^2/R_2^2)} \qquad \text{dB.}$$

$$(3\text{-}12)$$

The terms beyond $20 \log (i_1/i_2)$ and $20 \log (e_1/e_2)$ in Equations (3-11) and (3-12) give rise to serious error unless they are included when expressing voltage and current ratios in decibels except when the impedances $Z_1$ and $Z_2$ are equal. The extent of these errors may best be illustrated by some simple examples.

*Example 3-1:*

Let

$$p_1 = 2 \text{ mW} = 0.002 \text{ watt,}$$

$$p_2 = 1 \text{ mW} = 0.001 \text{ watt.}$$

Then, from Equation (3-7)

$$D = 10 \log (p_1/p_2) = 10 \log 2 = 3 \text{ dB.}$$

Let

$$R_1 = 10 \text{ ohms}, R_2 = 10 \text{ ohms}, X_1 = 10 \text{ ohms}, X_2 = 10 \text{ ohms.}$$

Then, from Equation (3-9)

$$i_1 = (p_1/R_1)^{1/2} = (0.002/10)^{1/2} = 0.014 \text{ Amp}$$

$$i_2 = (p_2/R_2)^{1/2} = (0.001/10)^{1/2} = 0.01 \text{ Amp}$$

and from Ohm's law

$$e_1 = i_1 \, | \, Z_1 \, | = 0.014 \sqrt{10^2 + 10^2} = 0.2 \text{ volt}$$

$$e_2 = i_2 \, | \, Z_2 \, | = 0.01 \sqrt{10^2 + 10^2} = 0.14 \text{ volt.}$$

From Equation (3-11)

$$D = 20 \log (i_1/i_2) + 10 \log (R_1/R_2)$$

$$= 20 \log (0.014/0.01) + 10 \log 1$$

$$= 3 \text{ dB.}$$

From Equation (3-12)

$$D = 20 \log (e_1/e_2) - 10 \log (R_1/R_2) - 10 \log \frac{(1 + X_1^2/R_1^2)}{(1 + X_2^2/R_2^2)}$$

$$= 20 \log (0.2/0.14) - 10 \log 1 - 10 \log 1$$

$$= 3 \text{ dB.}$$

Thus, no error results from computing the current or voltage differences in dB simply by taking 20 log of the current or voltage ratios. This is because the impedances $Z_1$ and $Z_2$ are equal and all terms after the first in Equations (3-11) and (3-12) reduce to zero.

*Example 3-2:*

In this example, the assumption is again

$$p_1 = 2 \text{ mW} = 0.002 \text{ watt}$$
$$p_2 = 1 \text{ mW} = 0.001 \text{ watt.}$$

Then

$$D = 10 \log (p_1/p_2) = 10 \log 2 = 3 \text{ dB.}$$

Now, assume $R_1 = 10$ ohms, $R_2 = 20$ ohms, and $X_1 = X_2 = 0$.

From Equation (3-9)

$$i_1 = (p_1/R_1)^{1/2} = (0.002/10)^{1/2} = 0.014 \text{ Amp}$$
$$i_2 = (p_2/R_2)^{1/2} = (0.001/20)^{1/2} = 0.007 \text{ Amp.}$$

From Ohm's law

$$e_1 = i_1 |Z_1| = 0.014 \times 10 = 0.14 \text{ volt}$$
$$e_2 = i_2 |Z_2| = 0.007 \times 20 = 0.14 \text{ volt.}$$

From Equation (3-11)

$$D = 20 \log (i_1/i_2) + 10 \log (R_1/R_2)$$
$$= 20 \log 2 + 10 \log (1/2)$$
$$= 6 - 3 = 3 \text{ dB.}$$

From Equation (3-12)

$$D = 20 \log (e_1/e_2) - 10 \log (R_1/R_2) - 10 \log \frac{(1 + X_1^2/R_1^2)}{(1 + X_2^2/R_2^2)}$$
$$= 20 \log 1 - 10 \log (1/2) - 10 \log 1$$
$$= 0 + 3 + 0 = 3 \text{ dB.}$$

Once again the three expressions for $D$ give the same answer, 3 dB. Note, however, that in this example significant errors would occur if $D$ were computed for current or voltage ratios without concern for the impedance of the circuits. In the case of the current ratio, the answer would have been 6 dB; in the case of the voltage ratio, the answer would have been 0 dB.

*Example 3-3:*

Once again, assume

$$p_1 = 2 \text{ mW} = 0.002 \text{ watt}$$

$$p_2 = 1 \text{ mW} = 0.001 \text{ watt.}$$

Then

$$D = 10 \log (p_1/p_2) = 10 \log 2 = 3 \text{ dB.}$$

Now, assume $R_1 = 10$ ohms, $R_2 = 20$ ohms, $X_1 = 20$ ohms, and $X_2 = 10$ ohms.

From Equation (3-9)

$$i_1 = (p_1/R_1)^{1/2} = (0.002/10)^{1/2} = 0.014 \text{ Amp}$$

$$i_2 = (p_2/R_2)^{1/2} = (0.001/20)^{1/2} = 0.007 \text{ Amp.}$$

From Ohm's law

$$e_1 = i_1 \, | \, Z_1 \, | = 0.014 \sqrt{10^2 + 20^2} = 0.31 \text{ volt}$$

$$e_2 = i_2 \, | \, Z_2 \, | = 0.007 \sqrt{20^2 + 10^2} = 0.16 \text{ volt.}$$

Then, from Equation (3-11)

$$D = 20 \log (i_1/i_2) + 10 \log (R_1/R_2)$$

$$= 20 \log 2 + 10 \log (1/2)$$

$$= 6 - 3 = 3 \text{ dB.}$$

From Equation (3-12)

$$D = 20 \log (e_1/e_2) - 10 \log (R_1/R_2) - 10 \log \frac{(1 + X_1^2/R_1^2)}{(1 + X_2^2/R_2^2)}$$

$$= 20 \log 2 - 10 \log (1/2) - 10 \log (5/1.25)$$

$$= 6 + 3 - 6 = 3 \text{ dB.}$$

Again, as in Example 3-2, the value of $D$ is 3 dB no matter how computed. However, the importance of including the impedance factors in the computation is demonstrated.

## Loss, Delay, and Gain

There are several different methods of describing the transfer characteristics of a two-port network. Such characteristics require specification of four complex quantities representing input and output relationships. However, in many cases where the network environment (such as source and load impedances) is controlled, the transfer can often be characterized more readily by one frequency-dependent complex number describing the loss (or gain) and phase shift through the network. Several different means of expressing the transfer characteristic have come into use, each having merit for a particular set of circumstances and each depending in part on the definition of the network parameters involved.

**Insertion Loss and Phase Shift.** In the circuit of Figure 3-1, assume that it has been determined that power, $p_2$, is delivered to the load, $Z_L$, when the open-circuit voltage $E'$, is applied. Next assume that the two-port network is removed, the generator is connected directly to the load, and the power delivered to $Z_L$ is $p_0$. The difference in dB between $p_0$ and $p_2$ is called the insertion loss of the two-port network; i.e.,

$$\text{Insertion loss} = 10 \log (p_0/p_2) \quad \text{dB.} \quad (3\text{-}13)$$

If the impedances are matched throughout, there is no ambiguity in expressing insertion loss as a voltage or current ratio. The instantaneous voltages, $V_0$ and $V_2$, corresponding respectively to $p_0$ and $p_2$, may be expressed in terms of peak values, $E_0$ and $E_2$. By proceeding as in the development of Equation (3-6), the insertion loss and a definition of the *insertion phase shift* may be written:

$$\text{Insertion loss} = 20 \log (E_0/E_2) = 20 \log (I_0/I_2) \quad \text{dB;} \quad (3\text{-}14)$$

$$\text{Insertion phase shift} = 57.3 (\phi_0 - \phi_2) \quad \text{degrees} \quad (3\text{-}15)$$

where $\phi_0$ and $\phi_2$ are given in radians.

If the transducer of Figure 3-1 furnishes gain, then $E_2 > E_0$, and the insertion loss values are negative. In order to avoid talking about negative loss, it is customary to write

$$\text{Insertion gain} = 20 \log (E_2/E_0) \quad \text{dB.} \tag{3-16}$$

If complex gain is expressed in the form of Equation (3-5), the phase shift will be the negative of the value found in Equation (3-15). Unfortunately, there is no standard name which clearly distinguishes between the phase shift calculated from a loss ratio and that calculated from a gain ratio. The ambiguity is entirely a matter of algebraic sign and can always be resolved by observing the effect of substituting a shunt capacitor for the transducer. This gives a negative sign to the value of $\phi_2$ and a positive change in the phase of Equation (3-15).

**Phase and Envelope Delay.** The phase delay and envelope delay of a circuit are defined as

$$\text{Phase delay} = \beta/\omega$$

$$\text{Envelope delay} = d\beta/d\omega$$

where $\beta$ is in radians, $\omega$ is in radians per second, and delay is therefore expressed in seconds. In accordance with the sign convention adopted previously, both the phase and the envelope delay of an "all-pass" network are positive at all finite frequencies. The above expressions show that the envelope delay is the rate of change, or slope, of the phase delay curve. If the phase delay is linear over the frequency band of interest, the envelope delay is a constant over that band.

For cables or similar transmission media, the phase shift is usually quoted in radians per mile. In this case, phase delays and envelope delays are expressed in seconds per mile. Their reciprocals are called *phase velocity* and *group velocity*, respectively, and the units are miles per second.

**Available Gain.** The maximum power available from a source of internal impedance, $Z_G$, is obtained when the load connected to its terminals is equal to its conjugate, $Z_G^*$, i.e., if

and

$$\left.\begin{array}{l} Z_G = R_G + jX_G \\[2mm] Z_G^* = R_G - jX_G. \end{array}\right\} \tag{3-17}$$

It should be noted that maximizing available power is not necessarily the optimum relationship because, when conjugate impedances are interconnected, large reflections occur. As a result, other impedance relationships are preferable. For an open-circuit generator voltage having an rms value, $e$, the maximum available power is

$$p_{aG} = e^2/4R_G. \qquad (3\text{-}18)$$

The power actually delivered to $Z_L$ in Figure 3-1 is also maximized if the output impedance of the transducer is conjugate to $Z_L$. Designating this power as $p_{a2}$ leads to a definition of *available gain*, $g_a$, as

$$g_a = 10 \log (p_{a2}/p_{aG}). \qquad (3\text{-}19)$$

**Transducer Gain.** Ordinarily the impedances do not meet the conjugacy requirements, and it is necessary to define the *transducer gain*, $g_t$, of the two-port circuit as

$$g_t = 10 \log (p_L/p_{aG}) \qquad (3\text{-}20)$$

where $p_L$ is the power actually delivered to the load. Transducer gain is dependent on load impedance and can never exceed available gain. Transducer gain is equal to available gain only when the load impedance is equal to the conjugate of the network output impedance.

**Power Gain.** Finally, *power gain*, $g_p$, is defined as

$$g_p = 10 \log (p_L/p_1) \qquad (3\text{-}21)$$

where $p_1$ is the power actually delivered to the input port of the transducer. The power gain is equal to the transducer gain of a network when the input impedance of the network is equal to the conjugate of the source impedance. The power gain is equal to the insertion gain of the network when the input impedance of the network is equal to the load impedance connected to the output of the network.

## 3-2  TRANSMISSION LEVEL POINT

In designing transmission circuits and laying them out for operation and maintenance, it is necessary to know the signal amplitude at various points in the system. These values can be determined conveniently by use of the transmission level point concept.

> *The transmission level at any point in a transmission circuit or system is the ratio, expressed in decibels, of the power of a signal at that point to the power of the same signal at a reference point called the zero transmission level point (0 TLP).*

Thus, any point in a transmission circuit or system may be referred to as a *transmission level point*. Such a point is usually designated as a $-x$ dB TLP, where $x$ is the designed loss from the 0 TLP to that point. Since the losses of transmission facilities and circuits tend to vary with frequency, the TLP is specified for designated frequencies. For voiceband circuits, this frequency is usually 1000 Hz. For analog carrier systems, the frequency in the carrier band must be specified.

The TLP concept is convenient because it enables circuit losses or gains to be quickly and accurately determined by finding the difference between the transmission level point values at the points of interest. This principle may be extended from relatively simple circuits, such as message trunks, to very complex broadband transmission systems where the TLP values often vary with frequency across the carrier band.

The transmission level point concept is also a convenience in that signals and various forms of interference can easily be expressed in values referred to the same transmission level point. This facilitates the addition of interference amplitudes, the expression of signal-to-noise ratios, and the relation of performance to objectives in system evaluation. These important advantages are apparent where various types of signals and interferences are involved.

Transmission level points are applied within the switched message network and special services networks. Similar concepts are applied to wideband services such as PICTUREPHONE, television, and wideband data signal transmission. The channels used for these services are given specially-defined transmission reference points.

Confusion often arises because the word *level* is used (properly and improperly) in so many ways. Frequent references may be found to such things as power level, voltage level, signal level, or speech level. To add to the confusion, the word *level* is often used interchangeably with the word power. Here, *level* is generally used only as a part of the phrase *transmission level point*. Signal power and voltage are referred to in appropriate units such as watts, milliwatts, dBm, volts, or dBV.

A troublesome correlation exists between transmission level point and power. When a test signal of the correct frequency is applied to a properly adjusted circuit at a power in dBm that corresponds numerically with the TLP at which it is applied, the test signal power measured at any other TLP in the circuit corresponds numerically with the designated TLP value. Careless use of terminology often leads to referring to a TLP as the $x$ dBm level point. It cannot be stressed too strongly that *this is improper terminology even if it happens that a test signal of $x$ dBm is measured at the $x$ dB TLP*. This correlation is unfortunate in that it has led to some confusion. On the other hand, when properly used, TLPs simplify loss computations.

While *0 TLP* is used in this book as the abbreviation for the reference transmission level point, it should be pointed out that several other forms of terminology are sometimes used elsewhere. These include zero level, zero-level point, 0-dB point, 0-dB TL, and 0 SL (SL for system level).


## Commonly Used TLPs

Application of the transmission level point concept must begin with the choice of a common datum or reference point and the arbitrary assignment of 0-dB transmission level to that point. Other transmission level points in the trunk or system are then related to the reference point by the number of dB of gain from the reference point to the point of interest. If (in a properly adjusted circuit) a signal of $x$ dBm is applied or measured at the reference point and if that signal is measured as $y$ dBm at the point of interest, the point of interest is designated as the $(y\text{-}x)$ dB transmission level point.

The 0-dB transmission level point (0 TLP) is so defined as a matter of convenience and uniformity. It would be convenient also to have the 0 TLP available as an access point for connecting probes and measuring equipment. However, there is no requirement that such an access point be available and, in fact, as a result of changes in circuit arrangements resulting from changing objectives, the 0 TLP is seldom available physically in the toll plant.

Originally, the 0 TLP was conveniently defined at the transmitting jack of a toll switchboard. Intertoll trunks were equipped at each end with 4-dB pads which could be switched in or out of the circuit to suit best the needs of a particular application. As technology improved and the need for better performance increased, these pads were reduced to 2 dB; later, under the via net loss (VNL) design plan, they were eliminated entirely from the intertoll trunks. The loss corresponding to that of the pads is now assigned to the toll connecting trunks, two of which must be used in each toll connection.

With these changes in intertoll trunk designs, it would have been possible to redefine the reference transmission level point. However, this would have resulted in changing all transmission level point values. It was instead deemed desirable to maintain the original 0 TLP concept as well as other important transmission level points. As a result, the outgoing side of the switch to which an intertoll trunk is connected is designated a −2 dB TLP and the outgoing side of the switch at which a local area trunk is terminated is defined as 0 TLP.

In the layout of four-wire trunks, a patch bay, called the four-wire patch bay, is usually provided to facilitate test, maintenance, and circuit rearrangements between the trunks and the switching machine terminations. Transmission level points at these four-wire patch bays have been standardized for all four-wire trunks. On the transmitting side the TLP is −16 dB, and on the receiving side the TLP is +7 dB. Thus, a four-wire trunk, whether derived from voice-frequency or carrier facilities, must be designed to have 23-dB gain between four-wire patch bays. These standard transmission level points are necessary to permit flexible telephone plant administration.

In four-wire circuits, the TLP concept is easily understood and applied because each transmission path has only one direction of transmission. In two-wire circuits, however, confusion or ambiguity may be introduced by the fact that a single point may be properly designated as two different TLPs, each depending on the assumed direction of transmission.

## Illustrative Applications of TLP

The transmission level point concept is applied to an individual trunk as illustrated in Figure 3-2. The circuit elements within each trunk are interconnected by design to produce predetermined gains and losses so that each point in the trunk may be assigned a transmission level value. Some of the important transmission level points discussed earlier and the assignment of transmission level point values within a toll trunk are illustrated in the figure.

Starting in the upper left corner of the diagram, the outgoing side of the switch is designated as the −2 dB TLP. As the circuit is followed from left to right, the office equipment transforms the circuit from two-wire to four-wire. The diagram shows an office loss of 14 dB so that, at the input to the four-wire trunk (MOD IN), the TLP value is −16 dB; i.e., the input to the four-wire trunk is a −16 dB TLP. As the connection is traced toward the right, the trunk between office A and office B provides +23 dB of gain so that the output of the trunk (DEMOD OUT) is a +7 dB TLP. The office equipment has 11 dB of loss to effect a −4 dB TLP at the office side of the first switch encountered in office B.

If the circuit is followed from right to left, similar losses are observed and appropriate transmission level points are shown along the circuit. Note that at closely related points (four-wire trunk input and output), the transmission level points are quite different for the two directions of transmission. In the two-wire circuits, the same point (electrically), e.g., the switch at the end of the trunk, has two values of TLP, −2 dB for one direction and −4 dB for the other.

Figure 3-3 shows a built-up connection of three toll trunks and illustrates the fact that the transmission level point concept is applied to an individual trunk and not to the built-up connection. If the circuit is traced from left to right, the TLP is shown as −2 dB at office A and −4 dB at office B. Each of the interconnected trunks from A to D is shown with specific TLPs, −2 dB at the left and

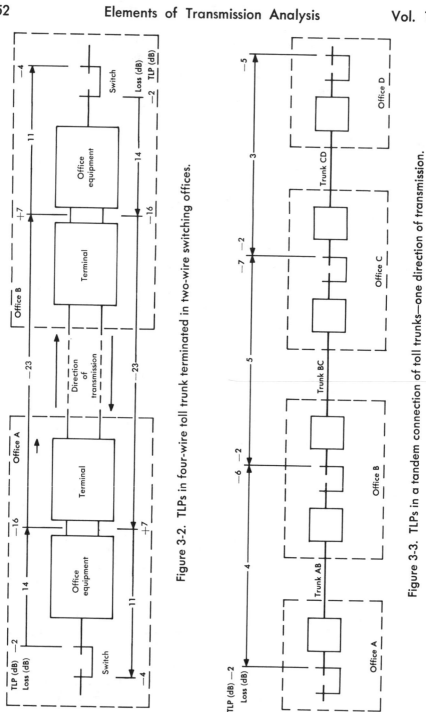

Figure 3-2. TLPs in four-wire toll trunk terminated in two-wire switching offices.

Figure 3-3. TLPs in a tandem connection of toll trunks—one direction of transmission.

— (2 + L) dB at the right, where L is the trunk loss in dB. That is, transmission level points are redefined for each trunk in the built-up connection. *Thus, there is no one unique 0 TLP for the Bell System.* Each trunk has a defined 0 TLP which often is a reference that does not exist in fact.

By common consent and usage, transmission level point values are found by determining the gain or loss between TLPs at 1000 Hz in the circuit of interest. The use of modulators in the terminal equipment of frequency division multiplex carrier systems produces a shift of frequency from 1000 Hz in the original circuit to some higher frequency in the carrier system. The TLP value can be determined at the higher frequency and related to the 1000-Hz value in the original circuit to obtain the TLP in the carrier system.

If the value of a transmission level point is not known, it can be determined by measurement. The process depends on the direction of transmission and on having proper values of pad losses and amplifier gains in the circuit between a known TLP, $A$, and the TLP to be determined, $X$. If the unknown is to be established by transmitting from $A$ to $X$, a 1000-Hz signal (or equivalent in a carrier channel) may be applied at $A$ and measured at $A$ and $X$. The TLP at $X$ is determined by subtracting from the TLP value at $A$ the loss (in dB) from $A$ to $X$. If the TLP at $X$ is to be determined by transmitting from $X$ to $A$, the value at $X$ is the value of the TLP at $A$ plus the loss (in dB) from $X$ to $A$.

In order to avoid overloading transmission systems, the applied test signal power (in dBm) should be at least 10 dB below the TLP value at any point. Since signal power is often expressed in terms of its value at 0 TLP, the unit dBm0 is used as an abbreviation for "dBm at 0 TLP."

## 3-3  SIGNAL AND NOISE MEASUREMENT

The TLP concept is valuable in system design, operation, and maintenance in that it provides a means of calculating signal and interference amplitudes at given points in a system as well as the gains or losses between TLPs; nevertheless, operating systems must be checked at times by actual measurement to see that signal or interference amplitudes are being maintained at the expected, or calculated, values. When there is excessive gain or loss in a system, measurement is also a means of locating trouble.

In the telephone system there are complex signals and noises to be measured. Simple instruments are inadequate, particularly since they do not take into account any of the subjective factors which determine the final evaluation of a telephone circuit. Both the instruments and units of measure used in telephony for signal and noise measurements must be adapted to the special needs involved.

Since telephone circuits operate with signal and interference powers which rarely are as large as 0.1 watt and which may be lower than $10^{-12}$ watt, the use of the watt as a unit of measurement is awkward. A more convenient unit is the milliwatt, or $10^{-3}$ watt. An exception is in radio transmitter work, where output power is frequently measured in watts.

Many other types of equipment are used for evaluating transmission quality and facilitating maintenance procedures. These include oscillators, ammeters, voltmeters, and transmission measuring sets. The parameters measured, the units of measurement, and the techniques involved are all important aspects of transmission engineering. The cathode ray oscilloscope is one of the most powerful of these specialized instruments in that the parameters of interest can be displayed for study and analysis.

## Volume

The amplitude of a *periodic* signal can be characterized by any of four related values: the rms, the peak, the peak-to-peak, or the average. The choice depends upon the particular purpose for which the information is required. It is more difficult to deal with *nonperiodic* signals such as the speech signals transmitted over telephone circuits where the rms, peak, peak-to-peak, and average values and the ratio of one to another are all irregular functions of time, so that one number cannot easily specify any of them.

Regardless of the difficulty of the problem, the amplitude of the telephone signal must be measured and characterized in some fashion that will be useful in designing and operating systems involving electronic equipment and transmission media of various kinds. Signal amplitudes must be adjusted to avoid overload and distortion, and gain and loss must be measured. If none of the simple characterizations is adequate, a new one must be invented. The unit used for expressing speech signal amplitude is called the volume unit (vu). It is an empirical kind of measure evolved initially to meet the needs of AM radio broadcasting and is not definable by any precise mathematical formula. The volume is determined by reading a volume indicator, called the vu meter, in a carefully specified fashion.*

The development of the vu meter was a joint project of the Bell System and two large broadcasting networks. Its principal functions are measuring signal amplitude to enable the user to avoid overload and distortion, checking transmission gain and loss for the complex signal, and indicating the relative loudness of the signal when converted to sound.

The vu meter can be used equally well for all speech, whether male or female. There is some difference between music and speech in this respect, and so a different reading technique is used for each.

The meter scale is logarithmic, and the readings bear the same relationship to each other as do decibels; however, the scale units are in vu, not in dB. The transient response (damping characteristic) of the meter movement prevents the meter needle from registering very short high-amplitude impulses such as those created by percussive sounds in speech. A correlation between talker volume and long-term average power and peak power can be established. Also, a vu meter reading for a sinusodial signal delivered to a 600-ohm resistive termination is numerically equal to the power in dBm delivered to the termination. Such correlations are valuable, but the fact that they exist should not be allowed to confuse the real definition of volume and vu. Putting it as simply as possible, a —10 vu talker is one whose signal is read on a calibrated volume indicator (by someone who knows how) as —10 vu. It should be noted that the vu meter has a flat frequency response over the audible range, and it is not frequency weighted in any fashion. Some, but not all, meters calibrated in dB can be used to read vu; however, the transient response of the meter movement must meet certain carefully defined specifications as in the vu meter.

---

*Objective measurements of speech signals are now possible [3, 4].

## Noise

The measurement of telephone channel noise, like the measurement of volume, is an effort to characterize a complex signal. The measurement is further complicated by an interest in how much it annoys the telephone user rather than in the absolute power. Consider the requirements of a meter which can measure the subjective effects of noise:

(1) The readings should take into consideration the fact that the interfering effect of noise is a function of frequency as well as of amplitude.

(2) When dissimilar noise components are present simultaneously, the meter should combine them in the same manner as do the ear and brain to measure the overall interfering effect.

(3) When different types of noise cause equal interference as determined in subjective tests, use of the meter should give equal readings.

The 3-type noise measuring set is essentially an electronic voltmeter which meets these requirements, respectively, by incorporating (1) frequency weighting, (2) a detector approximating an rms detector, and (3) a transient response similar to that of the human ear.

The first of the requirements for noise measurement involves annoyance and the effect of noise on intelligibility. Since both are functions of frequency, frequency weighting is included in the set. To determine the weighting characteristic, annoyance was measured in the absence of speech by adjusting the amplitude of a tone until it was as annoying as a reference 1000-Hz tone. This was done for many tones and many observers, and the results are averaged and plotted. A similar experiment was performed in the presence of speech at average received volume to determine the effect of noise on articulation. The results of the two experiments were combined and smoothed, resulting in the C-message weighting curve shown in Figure 3-4. The experiments were made with a 500-type telephone; therefore, the weighting curve includes the frequency characteristic of this telephone as well as the hearing of the average subscriber. The remainder of the telephone plant is assumed to provide transmission which is essentially flat across the band of a voice channel. Therefore, the C-message

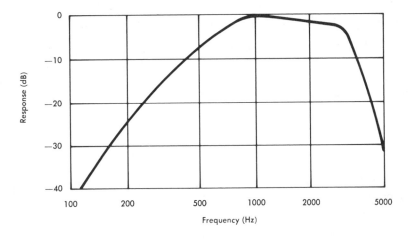

Figure 3-4. C-message frequency weighting.

weighting is applicable to measurements made almost anywhere except across the telephone receiver.

To illustrate the significance of the weighting curve of Figure 3-4, a 200-Hz tone of given power is found to be 25 dB less disturbing to a listener using a 500-type telephone than a 1000-Hz tone of the same power. Hence, the weighting network incorporated in the noise meter has 25 dB more loss at 200 Hz than at 1000 Hz.

Other weighting networks can be substituted in 3-type noise measuring sets. For example, the 3 KC FLAT network may be used to measure the power density of Gaussian noise. This network has a nominal low-pass response down 3 dB at 3 kHz and rolls off at 12 dB per octave. The response to Gaussian noise is almost identical to that of an ideal (sharp cutoff) 3-kHz low-pass filter.

The second factor affecting the measurement of the interfering effect of noise involves the evaluation of simultaneous occurrences of noise components at different frequencies and of different characteristics. Experimentally, narrow bands of noise were used in various combinations. It was found that the closest agreement between the judgment of the listener and the reading of the noise measuring set was obtained when the noises were added on an rms, or power, basis. Thus, for example, if two tones having equal interfering effect when applied individually are applied simultaneously, the effect when both are present is 3 dB worse than for each separately.

The third factor which affects the manner in which noise must be measured is the transient response of the human ear. It has been found that, for sounds shorter than 200 milliseconds, the human ear does not fully appreciate the true power in the sound. For this reason the meter on the noise measuring set (as well as the vu meter) is designed to give a full indication on bursts of noise longer than 200 milliseconds. For shorter bursts, the meter indication decreases.

These three characteristics of the 3-type noise measuring set— frequency weighting, power addition, and transient response—essentially prescribe the way message circuit noise is measured for speech signal transmission. This is not yet enough; a noise reference datum and a scale of measurement must also be provided.

The chosen reference is $10^{-12}$ watt, or $-90$ dBm. The scale marking is in decibels, and measurements are expressed in decibels above reference noise (dBrn). A 1000-Hz tone at a power of $-90$ dBm gives a 0-dBrn reading regardless of which weighting network is used. For all other measurements, the weighting must be specified. The unit dBrnc is commonly used when readings are made using the C-message weighting network.

As with dBm power readings, vu and dBrn readings may be taken at any transmission level point and referred to 0-dB TLP by subtracting the TLP value from the meter reading. Thus a typical noise reading might be 25 dBrn at 0-dB TLP, abbreviated 25 dBrn0. Similarly, values of dBrnc referred to 0 TLP are identified as dBrnc0.

Other noise measuring instruments have been designed to evaluate the effects of noise on other types of signals or in other types of channels.

## Display Techniques

Among the many specialized measurements that are needed in the evaluation of transmission circuits and signals are those taken from displays of signal or interference amplitudes on the tube face of a cathode ray oscilloscope. Two types of displays are commonly used. One type shows amplitude as a function of time and the other shows amplitude as a function of frequency. These are referred to as time-domain and frequency-domain displays.

## 3-4   ADDITION OF POWER IN DB

The merits of expressing power and power ratios in dBm and dB, respectively, have been demonstrated by the preceding discussions. A difficulty arises, principally in noise and interference studies, when it is necessary to find the sum of two or more powers that are given in dBm. Although the necessary steps are straightforward (the values must be converted to milliwatts, added, and then reconverted to dBm), they are time consuming. Specifically, suppose powers $P_1$ dBm and $P_2$ dBm are being dissipated in a circuit and it is desired to determine the sum, also in dBm.

The expression for the sum of the two powers is

$$p = p_1 + p_2 \text{ milliwatts.}$$

This may also be written

$$p = p_1 \left(1 + p_2/p_1\right). \tag{3-22}$$

The sum and each of the individual powers may be expressed as $P$, $P_1$, and $P_2$ dBm and the summing expression may be represented by the shorthand notation

$$P = P_1 \text{ "+" } P_2$$

where $P = 10 \log p$, $P_1 = \log p_1$, and $P_2 = 10 \log p_2$.

Equation (3-22) may be written in logarithmic form as

$$10 \log p = 10 \log p_1 + 10 \log \left(1 + p_2/p_1\right)$$

or

$$10 \log p = 10 \log p_1 + 10 \log s_p$$

where $s_p = 1 + p_2/p_1$. Then,

$$P = P_1 + S_p \qquad \text{dBm} \tag{3-23}$$

where $S_p = 10 \log s_p$   dB.

It is convenient to assign the symbol $p_1$ to the larger of the two powers to be added (to either, if they are equal) so that $s_p$ lies in the range of $1 \leqq s_p \leqq 2$ and $S_p$ is in the range of $0 \leqq S_p \leqq 3$ dB. The value of $S_p$ is shown in Figure 3-5 as a function of the difference between $P_1$ and $P_2$ in dB. Thus, the sum of two powers can be determined by first finding the value of $P_1 - P_2$, next estimating the value of $S_p$ from the figure, and then adding $S_p$ to $P_1$ as in Equation (3-23). It should be noted that this method may be applied

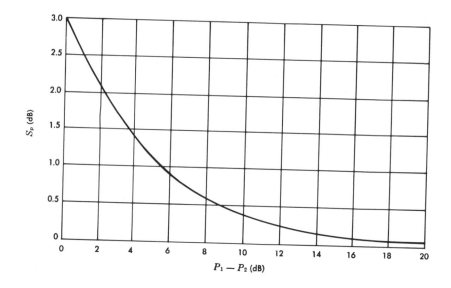

Figure 3-5. Power sum of two signals both expressed in dBm.

to any two powers, such as dBW or dBrnc, expressed in dB relative to an absolute value.

In the foregoing, it is assumed that the two powers to be added act independently and that the resultant is the linear sum of the two components. Such an assumption is valid, for example, when two sine waves of different frequencies or a sine wave and a band of random noise are to be added. It is not true when two sine waves of the same frequency are to be added. In this case, the two are said to be coherent, the resultant power depends on the phase relationship between them, and the summing process must be treated somewhat differently.

For two sine waves of the same frequency, the power sum may be written as

$$p = (\sqrt{p_1} + \sqrt{p_2} \cos \theta)^2 + (\sqrt{p_2} \sin \theta)^2 \text{ milliwatts}$$

where $\theta$ is the phase angle between the two sine waves. The above equation may also be written

$$p = p_1 + p_2 + 2 \sqrt{p_1 p_2} \cos \theta$$

or

$$p = p_1 (1 + p_2/p_1 + 2 \sqrt{p_2/p_1} \cos \theta). \tag{3-24}$$

This expression may be converted to a logarithmic form similar to Equation (3-23) and is then written

$$P = P_1 + S_v \qquad \text{dBm} \qquad (3\text{-}25)$$

where $S_v = 10 \log s_v$

$$= 10 \log (1 + p_2/p_1 + 2 \sqrt{p_2/p_1} \cos \theta) \qquad \text{dB.}$$

Of primary interest in noise and interference studies is the case in which $\theta = 0$, i.e., the case representing in-phase addition of interferences. As in the earlier analysis, it is convenient to assign the symbol $p_1$ to the larger of the two interference signals to be added so that $s_v$ lies in the range $1 \leq s_v \leq 4$ and $S_v$ is in the range $0 \leq S_v \leq 6$ dB. With this choice ($\theta = 0$), the value of $S_v$ is shown in Figure 3-6 as a function of the difference between $P_1$ and $P_2$ in dB. The sum for such in-phase addition may thus be found by determining $P_1 - P_2$, estimating the value of $S_v$ from the figure, and then adding $S_v$ to $P_1$ as in Equation (3-25).

The subscripts $p$ and $v$ applied to $S_p$ and $S_v$ are used to denote "power" and "voltage" addition as these processes are commonly called. The shorthand notation used to represent in-phase addition is usually written

$$P = P_1 \;\; ''+'' \;\; P_2 \qquad ,$$

a form analogous to that used earlier to represent "power" addition.

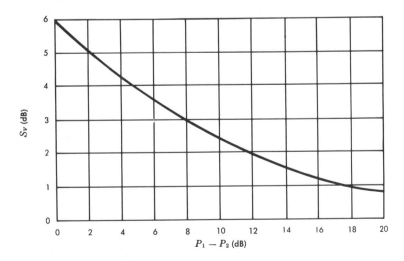

Figure 3-6. Power sum of two in-phase signals both expressed in dBm.

### REFERENCES

1. Technical Staff of Bell Telephone Laboratories. *Transmission Systems for Communications*, Fourth Edition (Winston-Salem, N. C.: Western Electric Company, Inc., 1970).

2. Cochran, W. T. and D. A. Lewinski. "A New Measuring Set for Message Circuit Noise," *Bell System Tech. J.*, Vol. 39 (July 1960).

3. Brady, P. T. "Statistical Basis for Objective Measurement of Speech Levels," *Bell System Tech. J.*, Vol. 44 (Sept. 1965).

4. Brady, P. T. "Equivalent Peak Level: A Threshold—Independent Speech-Level Measure," *Journal of the Acoustical Society of America*, Vol. 44 (Sept. 1968).

Chapter 4

# Four-Terminal Linear Networks

The transmission of an electrical signal from transmitting to receiving terminal is accomplished by transferring energy from one electrical network to the next until the receiving terminal is reached. An understanding of the complete transmission process requires an understanding of the general principles of linear alternating-current networks and of how they interact when they are tandem-connected to form a complete signal path from transmitter to receiver. Linear networks are those whose output voltages or currents are directly proportional to the input voltages or currents. The networks may or may not be bilateral.

An understanding of the mathematical properties of network impedances and their interactions is made easier by several basic theorems. The analysis of transformers, series and parallel resonant circuits, and electric wave filters are of special interest.

Network computations are approached differently depending on whether the problem is one of analysis or synthesis. In analysis the stimulus and the network are given, and the problem is to determine the response of the network to the stimulus; i.e., the problem is to determine the output given the input and the network configuration. In synthesis the stimulus and response (input and output) are given, and the problem is to determine the network configuration and component values that satisfy the given input-output relationships. Since the synthesis process is of interest only to the network designer and developer, the primary concern here is only with analysis; however, there are references at the end of the chapter which describe some

63

of the increasingly sophisticated methods of both analysis and synthesis that are now available.

In considering the layout and application of transmission circuits, it is essential that basic limitations be recognized. It is also important to recognize circumstances in which these limitations do or do not apply and the corrective measures that may be appropriate to overcome the limitations in specific situations. These situations may be as simple as connecting a telephone station set to its loop or as complex as changing the mode of operation of the message network so that a trunk is added to a built-up connection covering thousands of miles. In either case, judgements must be exercised as to the effects. Lengthy and clumsy calculations, previously avoided by the use of charts and nomographs, are now made simple and tractable by the use of high-speed digital computers.

For many purposes in telephone transmission analysis, the performance of a circuit, a piece of equipment, or even a complete system may be approximated over the voice range of frequencies by its performance at one frequency. Such approximations are often made by measuring performance at 1000 Hz. The procedure has the merit of simplicity, especially in measuring transmission line loss, but it neglects certain elements of the transmission process which must be measured over the whole band of frequencies. No single frequency can be fully representative of a complex electrical wave, nor can transmission through a complex network be fully represented by transmission at a single frequency.

## 4-1  THE BASIC LAWS

In the analysis of the usual electrical networks making up communications circuits. Ohm's and Kirchoff's laws are of fundamental importance. Certain other theorems which are of considerable assistance in analyzing and characterizing networks and their performance have been derived from these laws.

### Ohm's Law

*The current, I, which flows through an impedance, Z ohms, is equal to the voltage developed across the impedance divided by the value of the impedance, or*

$$I = E/Z \qquad \text{amperes.} \qquad (4\text{-}1)$$

This law is illustrated in Figure 4-1 where $E$ and $I$ may be direct or alternating voltages and currents and where $Z$ may be a simple resistor or a complex impedance involving resistance, inductance, and capacitance.

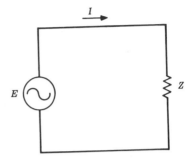

Figure 4-1. Simple series circuit containing an impedance, Z.

## Kirchoff's Laws

**Law 1:** *At any point in a circuit, there is as much current flowing to the point as there is flowing away from it.* For example, at point x in Figure 4-2,

$$I_1 = I_2 + I_3. \tag{4-2}$$

**Law 2:** *In any closed electrical circuit, the algebraic (or vector) sum of the electromotive forces (emf's) and the potential drops is equal to zero.* In Figure 4-2,

$$E - I_1 Z_A - I_3 Z_C = 0,$$
$$E - I_1 Z_A - I_2 Z_B = 0, \tag{4-3}$$
and
$$I_2 Z_B - I_3 Z_C = 0.$$

The arrows in Figure 4-2 indicate the assumed direction of current flow. A battery is assumed to produce a voltage rise from the negative to the positive terminal. A voltage due to current flowing through an impedance is assumed to be in the direction of positive to negative corresponding to the assumed direction of current flow. This accounts for the signs of the terms in Equations (4-3).

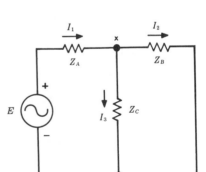

Figure 4-2. Simple series—parallel circuit.

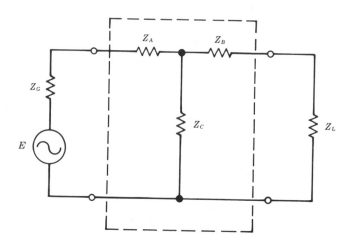

(a) T network

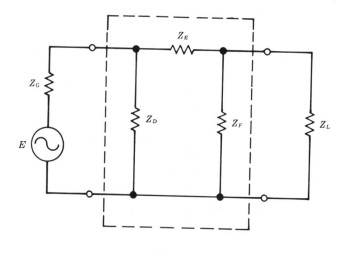

(b) π network

Figure 4-3. Equivalent networks.

## 4-2 APPLICATION AND THEOREMS

The application of Ohm's and Kirchoff's laws to more complicated circuits involves setting up simultaneous linear equations for solution. This can be very laborious, and several network theorems have been developed to expedite the process.

### Equivalent Networks

From their configurations, two important types of networks are called the T and $\pi$ electrical networks. A three-element T structure and a three-element $\pi$ structure can be interchanged provided certain relations exist between the elements of the two structures and provided the impedances can be realized.

Figure 4-3 represents two forms of a circuit connecting a generator of voltage $E$ and impedance $Z_G$ to a receiver having impedance $Z_L$. If the impedances enclosed in the boxes are related by the relationships shown in Figure 4-4, one box may be substituted for the other without affecting the voltages or currents in the circuit outside the boxes.

This property of networks permits any three-terminal structure, no matter how complex, to be reduced to a simple T. For example, a $\pi$ to T transformation permits converting the circuit in Figure 4-5(a) to that shown in Figure 4-5(b). By combining $Z_C$

| $\pi$ TO T | T TO $\pi$ |
|---|---|
| $Z_A = \dfrac{Z_D\,Z_E}{Z_D + Z_E + Z_F}$ | $Z_D = \dfrac{Z_A\,Z_B + Z_B\,Z_C + Z_C\,Z_A}{Z_B}$ |
| $Z_B = \dfrac{Z_E\,Z_F}{Z_D + Z_E + Z_F}$ | $Z_E = \dfrac{Z_A\,Z_B + Z_B\,Z_C + Z_C\,Z_A}{Z_C}$ |
| $Z_C = \dfrac{Z_F\,Z_D}{Z_D + Z_E + Z_F}$ | $Z_F = \dfrac{Z_A\,Z_B + Z_B\,Z_C + Z_C\,Z_A}{Z_A}$ |

Figure 4-4. Equivalent network relationships.

with $Z_5$ and $Z_B$ with $Z_6$ and making a second $\pi$ to T transformation,
Figure 4-5(b) can be reduced to the simple T shown in Figure 4-5(c).

These relationships apply only to networks having three terminals.
Similar relations can be developed for four-terminal networks.
Figure 4-6 is a typical four-terminal network. If only the voltages
measured across terminals 2-2 are significant, the five impedances
in Figure 4-6(a) can be replaced by the T structure in Figure 4-6(b).

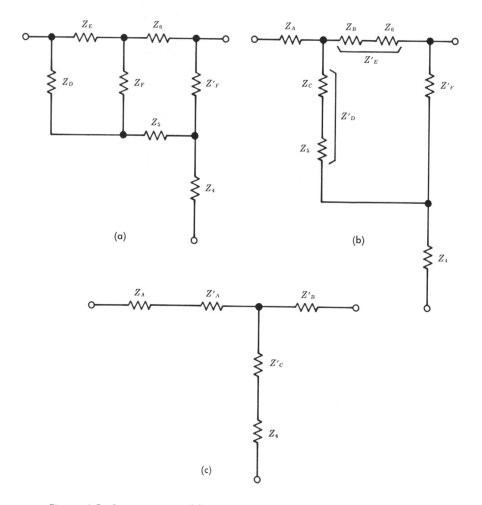

Figure 4-5. Successive simplification of networks by $\pi$ to T transformations.

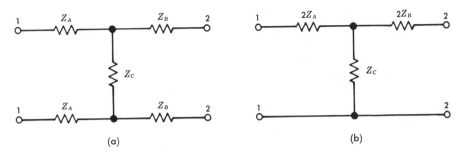

Figure 4-6. Equivalent four-terminal networks.

## Thevenin's, or Pollard's, Theorem

For the purpose of simplifying calculations, an arrangement such as that in Figure 4-7 (a) may be considered as two networks with one supplying energy to the other. The first of these networks is then replaced by an equivalent simplified circuit consisting of an emf and an impedance in series, as shown in Figure 4-7 (b).

Thevenin's theorem gives the rules required for this simplification as follows: *The current in any impedance, $Z_L$, connected to two terminals of a network is the same as that resulting from connecting $Z_L$ to a simple generator whose generated voltage is the open-circuit voltage at the original terminals to which $Z_L$ was connected and whose internal impedance is the impedance of the network looking*

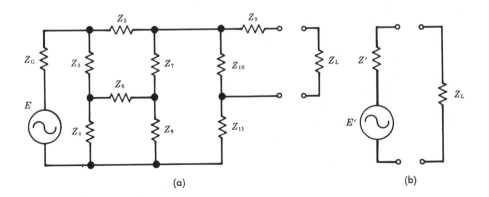

Figure 4-7. Application of Thevenin's theorem.

*back from those terminals with all generators in the original network replaced by their internal impedances.*

For example, if the equivalent emf, $E'$ in Figure 4-7 (b), is the open-circuit voltage at the terminals of Figure 4-7 (a) and if the equivalent impedance, $Z'$ of Figure 4-7 (b), is the impedance presented at the terminals of Figure 4-7 (a) when $E$ is made zero, the two circuits are equivalent. Another way to compute $Z'$ is to set it equal to the open-circuit voltage at the network terminals divided by the short-circuit current at the terminals. Under these conditions the load draws the same current as in the original connection.

## Superposition Theorem

*If a network has two or more generators, the current through any component impedance is the sum of the currents obtained by considering the generators one at a time, each of the generators other than the one under consideration being replaced by its internal impedance.*

Multigenerator networks can be solved by Kirchoff's laws, but their solution by superposition requires less complicated mathematics. Perhaps of even greater importance is the fact that this theorem is a useful tool for visualizing the currents in a circuit.

Before an example of the superposition theorem is given, it may be beneficial to review the concepts of the internal impedance of a generator. The open-circuit voltage of a battery is greater than the voltage across its terminals when supplying current to a load. The open-circuit voltage is a fixed value determined by the electrochemical properties of the materials from which the battery is made. Under load, the decrease in terminal voltage is due to the voltage drop across the internal resistance of the battery. If it were possible to construct a battery from materials that had no resistance, the battery would have no internal resistance and no internal voltage drop. Since there are no materials with infinite conductivity, every practical voltage source can be resolved into a voltage in series with an internal resistance or impedance.

Perhaps the superposition theorem can be most easily explained by working out a simple problem. In Figure 4-8 (a), which way does the current flow in the 10-ohm resistor?

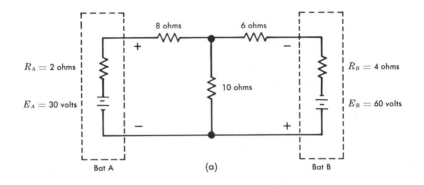

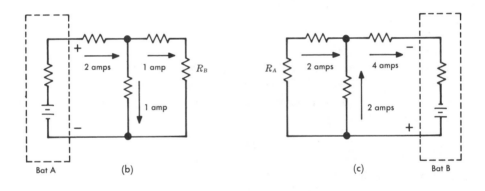

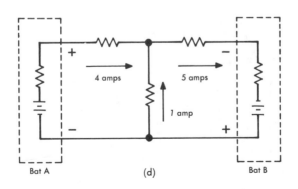

Figure 4-8. Superposition theorem.

According to the theorem the currents caused by each battery should be determined, in turn, with all other batteries replaced by their internal resistances. The currents indicated in Figures 4-8(b) and 4-8(c) are computed by Ohm's law. The currents flowing in the circuit with two batteries are the sum of these component currents; of course, sum means algebraic sum (or vector sum if the problem is ac). Currents flowing in opposite directions subtract. The resultant currents are shown in Figure 4-8(d), which shows that the 10-ohm resistor carries one ampere in the upward direction. The direction of the current in the 10-ohm resistor could have been estimated by inspection, since the resistances are symmetrical and the 60-volt battery produces the larger component of current. However, going through the arithmetic illustrates the application of the theorem.

### Compensation Theorem

*Any linear impedance in a network may be replaced by an ideal generator, one having zero internal impedance, whose generated voltage at every instant is equal in amplitude and phase to the instantaneous voltage drop caused by the current flowing through the replaced impedance.*

In Figure 4-9(a), the impedance has been separated from the rest of the network for consideration. The equations of Kirchoff's laws determine the currents and voltages in all parts of the network. According to the compensation theorem, these equations would not be altered if the network is changed to that of Figure 4-9(b) where the generator voltage is the product of current $I$ and impedance $Z$ from Figure 4-9(a).

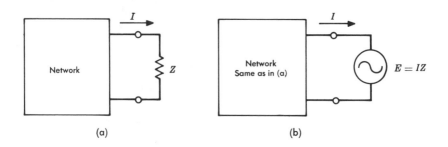

(a)                    (b)

Figure 4-9. Illustration of compensation theorem.

## 4-3   NETWORK IMPEDANCE RELATIONSHIPS

The analysis of a four-terminal network and its interactions in tandem connections is primarily related to the impedances of the network itself and of its terminations. Relationships among these impedances permit calculation of transmission effects (attenuation and phase shift), return loss, echo (magnitude and delay), power transfer, and stability. The networks may be relatively simple discrete components, such as transformers or attenuators, or they may be transmission lines, radio circuits, or carrier circuits, any of which may have gain or loss.

### Image Impedance

In a four-terminal network, such as that in Figure 4-10, impedances $Z_1$ and $Z_2$ may be found such that if a generator of impedance $Z_1$ is connected between terminals 1-1 and impedance $Z_2$ is connected as a load between terminals 2-2, the impedances looking in both directions at 1-1 are equal and the impedances looking in both directions at 2-2 are also equal. Impedances $Z_1$ and $Z_2$ are called the *image impedances* of the network.

The values of $Z_1$ and $Z_2$ may be determined from Ohm's law and the solution of two simultaneous equations. From inspection of Figure 4-10, the two equations may be written as

$$Z_1 = Z_A + \frac{(Z_B + Z_2)Z_C}{Z_B + Z_C + Z_2}$$

and

$$Z_2 = Z_B + \frac{(Z_A + Z_1)Z_C}{Z_A + Z_C + Z_1},$$

where $Z_A$, $Z_B$, and $Z_C$ are the impedances of the T-network equivalent to the four-terminal network.

Solving for $Z_1$ and $Z_2$ yields

$$Z_1 = \sqrt{\left(Z_A + Z_C\right)\left(Z_A + \frac{Z_B Z_C}{Z_B + Z_C}\right)}$$

and

$$Z_2 = \sqrt{\left(Z_B + Z_C\right)\left(Z_B + \frac{Z_A Z_C}{Z_A + Z_C}\right)}.$$

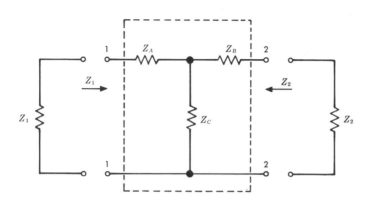

Figure 4-10. Image termination of a four-terminal network.

Examination of the latter equations and Figure 4-10 reveals some interesting relationships. The parenthetical expressions $(Z_A + Z_C)$ and $(Z_B + Z_C)$ are seen to be the impedances of the T network if the impedances are computed, respectively, from terminals 1-1 with terminals 2-2 open and from terminals 2-2 with terminals 1-1 open. Similarly, $Z_A + (Z_B Z_C)/(Z_B + Z_C)$ and $Z_B + (Z_A Z_C)/(Z_A + Z_C)$ are the impedances at terminals 1-1 and 2-2 if the opposite pair of terminals is short-circuited.

Thus, the image impedances of a four-terminal network are most easily determined by measuring the open-circuit and short-circuit impedances as above. Then,

$$Z_1 = \sqrt{Z_{oc} Z_{sc}} \tag{4-4}$$

and

$$Z_2 = \sqrt{Z'_{oc} Z'_{sc}} \;, \tag{4-5}$$

where

$Z_{oc}$ = impedance at 1-1 with 2-2 open

$Z_{sc}$ = impedance at 1-1 with 2-2 short-circuited

$Z'_{oc}$ = impedance at 2-2 with 1-1 open

$Z'_{sc}$ = impedance at 2-2 with 1-1 short-circuited.

As shown previously, the conversion of any complex network to an equivalent T network can be accomplished for any given frequency. Thus, the processes described above permit the determination of the image impedances of a network at any frequency.

Note that if the network is symmetrical, i.e., $Z_A = Z_B$, the image impedances are equal, $Z_1 = Z_2$.

## T-Network Equivalent

In the above determination of network image impedances as functions of open-circuit and short-circuit impedances measured (or computed) from the input and output terminals of the network, the assumed impedances of the T network were mathematically eliminated. Sometimes, however, it is also necessary to determine values of $Z_A$, $Z_B$, and $Z_C$ of Figure 4-10 in terms of the open-circuit and short-circuit measurements. For a four-terminal network containing only passive components or one in which the gains in the two directions of transmission are equal, this may again be accomplished by solving simultaneous equations.

In the discussion of image impedance, the following relationships among the impedances of Figure 4-10 are shown:

$$Z_{oc} = Z_A + Z_C, \text{ or } Z_A = Z_{oc} - Z_C \tag{4-6a}$$

$$Z'_{oc} = Z_B + Z_C, \text{ or } Z_B = Z'_{oc} - Z_C \tag{4-6b}$$

$$Z_{sc} = Z_A + \frac{Z_B Z_C}{Z_B + Z_C} \tag{4-7a}$$

and

$$Z'_{sc} = Z_B + \frac{Z_A Z_C}{Z_A + Z_C}. \tag{4-7b}$$

Into Equations (4-7a and b), substitute the value of $Z_A$ and $Z_B$ from Equations (4-6a and b) and solve for $Z_C$:

$$Z_C = \sqrt{(Z'_{oc} - Z'_{sc}) Z_{oc}} \tag{4-8}$$

and also

$$Z_C = \sqrt{(Z_{oc} - Z_{sc}) Z'_{oc}} \tag{4-9}$$

The values of $Z_C$ from Equations (4-8) and (4-9) may now be substituted directly in Equations (4-6a and b) to give expressions for $Z_A$ and $Z_B$ in terms of input and output open-circuit and short-circuit impedances. Thus, all legs of the equivalent T network may be determined from these measurements provided the network is bilateral, i.e., contains only passive components or has equal gain in the two directions of transmission.

If the network contains sources of amplification such that the gains in the two directions of transmission are not equal, the circuit cannot be reduced to a simple equivalent T network. Transfer effects, which account for the difference in gain in the two directions, must be taken into account.

### Transfer Effects

The determination of image impedances of a four-terminal network and the conversion of such a network to an equivalent T configuration permit input and output current and voltage relationships to be established directly from the application of Ohm's and Kirchoff's laws. However, these relationships may be applied directly only when the four-terminal network is bilateral. When it is not bilateral, these

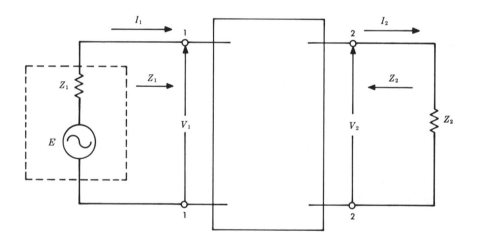

Figure 4-11. Image-terminated network.

relatively simple relationships do not apply directly because of transfer effects that occur as current flows through the network.

Consider the circuit of Figure 4-11. This circuit is similar to Figure 4-10 except that impedance $Z_1$ is replaced by a voltage generator having an internal impedance equal to $Z_1$. The circuit arrangements result in voltage $V_1$ across terminals 1-1 and voltage $V_2$ across terminals 2-2 when the network is terminated in its image impedances, $Z_1$ and $Z_2$. The input current is $I_1$, and the output current is $I_2$.

If the voltage source is connected at the 2-2 terminals, analogous voltage and current expressions may be written with the symbols $V$ and $I$ changed to $V'$ and $I'$.

The following relationships may then be written as definitions:

$$G_{1-2} = \frac{I_2 V_2}{I_1 V_1},$$
(4-10)

$$G_{2-1} = \frac{I'_1 V'_1}{I'_2 V'_2},$$
(4-11)

and

$$G_I = \sqrt{G_{1-2} \, G_{2-1}} \quad ,$$
(4-12)

where $G_I$ is sometimes called the *image transfer efficiency*.

In these equations, the currents and voltages are complex quantities. The current-voltage products in Equations (4-10) and (4-11) are quantities usually called volt-amperes, or *apparent power*. Thus, Equations (4-10) and (4-11) may be regarded as the gain, in the 1-2 or 2-1 direction, in apparent power resulting from transmission through the network. Equation (4-12) expresses the geometric mean of the apparent power gain in the two directions. In all cases, these definitions apply only when the network is image-terminated.

It is convenient to express the quantity $G_I$ in terms of open-circuit and short-circuit impedances. It can be shown that $G_I$ may take any of the following forms:

$$G_I = \frac{1 - \sqrt{\dfrac{Z_{sc}}{Z_{oc}}}}{1 + \sqrt{\dfrac{Z_{sc}}{Z_{oc}}}} = \frac{1 - \sqrt{\dfrac{Z'_{sc}}{Z'_{oc}}}}{1 + \sqrt{\dfrac{Z'_{sc}}{Z'_{oc}}}} ; \qquad (4\text{-}13)$$

$$G_I = \frac{\sqrt{Z_{oc}} - \sqrt{Z_{sc}}}{\sqrt{Z_{oc}} + \sqrt{Z_{sc}}} = \frac{\sqrt{Z'_{oc}} - \sqrt{Z'_{sc}}}{\sqrt{Z'_{oc}} + \sqrt{Z'_{sc}}} ; \qquad (4\text{-}14)$$

and

$$G_I = \frac{Z_1 - Z_{sc}}{Z_1 + Z_{sc}} = \frac{Z_2 - Z'_{sc}}{Z_2 + Z'_{sc}}. \qquad (4\text{-}15)$$

These equations for $G_I$ will be found useful in subsequent discussions of sending-end impedance, echo, and stability.

## Sending-End Impedance

The *sending-end impedance* of a four-terminal network is the impedance seen at the input of the network when the output is terminated in any impedance, $bZ_2$; $b$ is a factor used as a mathematical convenience to modify the terminating image impedance, $Z_2$. When $bZ_2$ is equal to the image impedance, $Z_2$ (i.e., b = 1), the sending-end impedance, $Z_S$, is equal to the image impedance, $Z_1$. It is important to consider the effects on the value of $Z_S$ of different impedance values for $bZ_2$, because these effects are related to phenomena such as return loss, singing, and talker echo, any or all of which may be important when a network is terminated in other than its image impedance, as in Figure 4-12.

The development of useful expressions for the analysis of the performance of a four-terminal network terminated in other than its image impedance can be demonstrated conveniently by starting with the image-terminated case as illustrated in Figure 4-13. The voltage $V_1$ is equal to $E/2$, as shown, because of the assumption of image terminations at both ends of the network. At the receiving terminals 2-2, the network is again assumed to be terminated in its

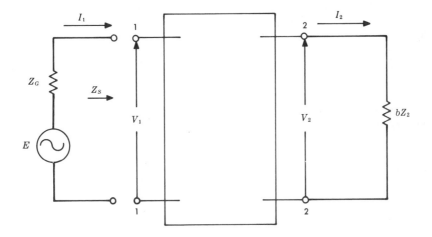

Figure 4-12. Sending-end impedance.

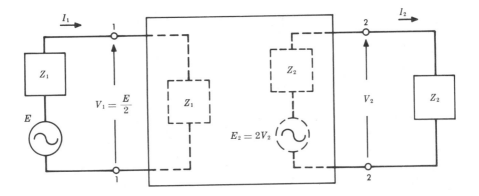

Figure 4-13. Image-terminated network.

image impedance, $Z_2$. The voltage appearing across terminals 2-2, $V_2$, may be defined in terms of Thevenin's theorem. The four-terminal network, which now is the driving point for the load, $Z_2$, is replaced by a simple impedance (by definition, equal to $Z_2$) and a generator whose open-circuit voltage is such as to produce $V_2$; i.e., $E_2 = 2V_2$.

By means of Equation (4-10), the input and output portions of the four-terminal network of Figure 4-13 may be related.

Thus,

$$G_{1-2} = \frac{I_2 V_2}{I_1 V_1} \qquad (4\text{-}16)$$

The input and output currents may be related to their corresponding voltage drops and impedances by

$$I_1 = V_1/Z_1$$

and

$$I_2 = V_2/Z_2.$$

Substituting these values of current in Equation (4-16),

$$G_{1-2} = \frac{V_2{}^2 Z_1}{V_1{}^2 Z_2}$$

from which

$$V_2 = V_1 \sqrt{G_{1-2}} \sqrt{Z_2/Z_1} \quad . \qquad (4\text{-}17)$$

It can also be shown that

$$I_2 = I_1 \sqrt{G_{1-2}} \sqrt{Z_1/Z_2} \quad . \qquad (4\text{-}18)$$

Thus, Equations (4-17) and (4-18) may be used to relate input and output voltages and currents in an image-terminated four-terminal network. If the network were driven from the right (generator impedance of $Z_2$), similar expressions could be derived. Then,

$$V_1 = V_2 \sqrt{G_{2-1}} \sqrt{Z_1/Z_2} \qquad (4\text{-}19)$$

and

$$I_1 = I_2 \sqrt{G_{2-1}} \sqrt{Z_2/Z_1} \quad . \qquad (4\text{-}20)$$

Now consider a termination having a value other than $Z_2$ at terminals 2-2 of the network. Its value can be expressed in terms of $Z_2$ and an incremental impedance, $Z_r$, in series with $Z_2$. Note that $Z_r$ is a complex impedance whose components may be positive, negative, or zero.

By use of the compensation theorem, $Z_r$ may be replaced by an ideal generator whose internal impedance is zero and whose generated voltage, $E_r$, is equal to the voltage drop across $Z_r$ caused by the current $I_R$ flowing through it.

The circuit may now be analyzed using the superposition theorem. The currents and voltages in the input and output circuits are shown in Figure 4-14. Symbology is the same as in Figure 4-13 for voltages

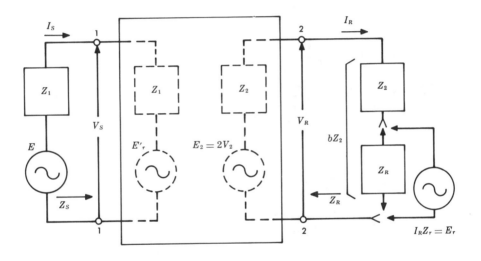

| Voltages | Currents |
|---|---|
| $V_S = V_1 + \dfrac{E'_r}{2} = \dfrac{E + E'_r}{2}$ | $I_S = I_1 + I'_r$ |
| $V_R = \dfrac{E_r}{2} + V_2 = \dfrac{E_r + E_2}{2}$ | $I_R = I_2 + I_r$ |
| $E\sqrt{G_{1\text{-}2}}\sqrt{Z_2/Z_1} = E_2 = 2V_2$ | $I_1\sqrt{G_{1\text{-}2}}\sqrt{Z_1/Z_2} = I_2$ |
| $E_r\sqrt{G_{2\text{-}1}}\sqrt{Z_1/Z_2} = E'_r$ | $I_r\sqrt{G_{2\text{-}1}}\sqrt{Z_2/Z_1} = I'_r$ |

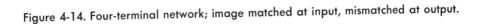

Figure 4-14. Four-terminal network; image matched at input, mismatched at output.

and currents analogous to the image-terminated case; other current and voltage components are shown to reflect the presence of the compensating voltage, $E_r$, substituted for $Z_r$.

Note that $E'_r/2$ (a component of $V_S$) and $I_r$ (a component of $I_R$) may be considered as reflected values of voltage and current at terminals 1-1 that would exist if the compensating voltage, $E_r/2$, at terminals 2-2 acted alone. Similarly, $E_r/2$ and $I_r$ may be considered as the reflected voltage and current at terminals 2-2 due to the compensating voltage.

Now, the sending-end impedance may be written

$$Z_S = \frac{V_S}{I_S} = \frac{(E + E'_r)/2}{I_1 + I'_r},\qquad (4\text{-}21)$$

where values of $V_S$ and $I_S$ are taken from Figure 4-14.

Equation (4-21) may be further developed. Note that in Figure 4-14 voltage $V_R$ may be written

$$V_R = \frac{E_r + E_2}{2} = I_R b Z_2$$

where $b$ is defined as

$$b = \frac{Z_2 + Z_r}{Z_2}.\qquad (4\text{-}22)$$

Then

$$E_r + E_2 = 2 I_R b Z_2.$$

Since

$$I_R = \frac{E_2}{Z_2(1 + b)},$$

$$E_r + E_2 = \frac{2E_2 b Z_2}{Z_2(1 + b)} = \frac{2b E_2}{1 + b},$$

and

$$\frac{E_r}{E_2} = -\left(\frac{1 - b}{1 + b}\right).$$

From Figure 4-14,

$$E'_r = E_r \sqrt{G_{2\text{-}1}} \sqrt{Z_1/Z_2} = - E_2 \left( \frac{1-b}{1+b} \right) \sqrt{G_{2\text{-}1}} \sqrt{Z_1/Z_2}$$

$$= - E \left( \frac{1-b}{1+b} \right) \sqrt{G_{2\text{-}1}} \sqrt{Z_1/Z_2} \sqrt{G_{1\text{-}2}} \sqrt{Z_2/Z_1} \ .$$

By substituting Equation (4-12),

$$E'_r = - G_I E \left( \frac{1-b}{1+b} \right). \tag{4-23}$$

Then,

$$V_S = \frac{E + E'_r}{2} = \frac{E}{2} \left[ 1 - G_I \left( \frac{1-b}{1+b} \right) \right]. \tag{4-24}$$

The current $I_S$ may be written

$$I_S = I_1 + I'_r = I_1 - \frac{E'_r}{2Z_1}$$

Substituting Equation (4-23) in the above gives

$$I_S = I_1 + G_I \frac{E}{2Z_1} \left( \frac{1-b}{1+b} \right) = I_1 \left[ 1 + G_I \left( \frac{1-b}{1+b} \right) \right]. \tag{4-25}$$

Equations (4-24) and (4-25) may now be substituted in Equation (4-21) to give

$$Z_S = \frac{V_S}{I_S} = \frac{\dfrac{E}{2} \left[ 1 - G_I \left( \dfrac{1-b}{1+b} \right) \right]}{I_1 \left[ 1 + G_I \left( \dfrac{1-b}{1+b} \right) \right]}.$$

The image impedance at the input is

$$Z_1 = \frac{E/2}{I_1} \ .$$

Thus,

$$Z_S = Z_1 \frac{\left[1 - G_I\left(\dfrac{1-b}{1+b}\right)\right]}{\left[1 + G_I\left(\dfrac{1-b}{1+b}\right)\right]}. \tag{4-26}$$

Equation (4-26) may be used to illustrate the effect on sending-end impedance of providing an image termination ($Z_2$) at terminals 2-2 of the network. When this is done, the value of $Z_r$ in Equation (4-22) becomes zero. Thus, the value of $b$ becomes unity, the quantity $(1 - b)/(1 + b)$ becomes zero, and Equation (4-26) reduces to $Z_S = Z_1$; i.e., the sending-end impedance equals the image impedance.

An expression similar to Equation (4-26) may be derived for the impedance at terminals 2-2 of Figure 4-14. In this case,

$$Z_R = Z_2 \frac{\left[1 - G_I\left(\dfrac{1-a}{1+a}\right)\right]}{\left[1 + G_I\left(\dfrac{1-a}{1+a}\right)\right]}, \tag{4-27}$$

where $a$ is a measure of the departure of the input terminating impedance from the image impedance. It is written

$$a = \frac{Z_1 + Z_s}{Z_1},$$

where $Z_s$ is the incremental impedance when the terminating impedance is not the image impedance.

All of the quantities in Equations (4-26) and (4-27) are complex, and the labor involved in their evaluation is sometimes considerable. Detailed calculations may be performed on a digital computer and, in some cases, tables are available for the evaluation of expressions like those in the two equations above. For ordinary engineering application, however, it is frequently desirable to make quick calculations that need not be extremely accurate. For these purposes, alignment charts have been prepared.

## Alignment Charts

The laboriousness of computation arises from the repetitive use of terms in the form of $(1 - b)/(1 + b)$ where $b$ is complex; therefore, it is desirable to reduce this to a single complex quantity in the polar form, $Q \angle \phi$. If $b$ is written in polar form as $X \angle \theta$, the values of $X$ and $\theta$ may be written as $X = |\alpha + j\beta|$ or $X = \sqrt{\alpha^2 + \beta^2}$ and $\theta = \tan^{-1} \beta/\alpha$.

Four alignment charts, Figures 4-15, 4-16, 4-17, and 4-18 may be used to solve expressions in the form $\dfrac{1 - X \angle \theta}{1 + X \angle \theta} = Q \angle \phi$. Figures 4-15 and 4-16 give values for $Q$ for various combinations of $X$ and $\theta$, while Figures 4-17 and 4-18 give values for $\phi$ for various combinations of $X$ and $\theta$. The following examples illustrates the use of the charts.

*Example 4-1: Use of Alignment Charts*

$$\text{Given: } b = 4 \angle 70° = X \angle \theta$$

$$\text{To evaluate: } \frac{1 - b}{1 + b}$$

First, refer to Figure 4-15. Mark $X = 4$ on the left-hand vertical scale and $\theta = 70°$ on the right-hand vertical scale. Use a straight edge to connect these points and read $Q = 0.848$ on the left side of the $Q$ scale. Next refer to Figure 4-17. Again mark the points $X = 4$ and $\theta = 70°$. With the straight edge, read $\phi = -153.5°$ on the right-hand side of the $\phi$ scale. Thus, $(1 - b)/(1 + b) = 0.848 \angle -153.5°$.

Note that the charts in Figures 4-15 through 4-18 can be used for a quick evaluation of changes in either the magnitude or phase angle, or both, of a termination on a network. These evaluations, useful in determining the performance of circuits and in judging what may be done to improve performance, may be made by using alignment charts or may be even more conveniently made by use of a digital computer.

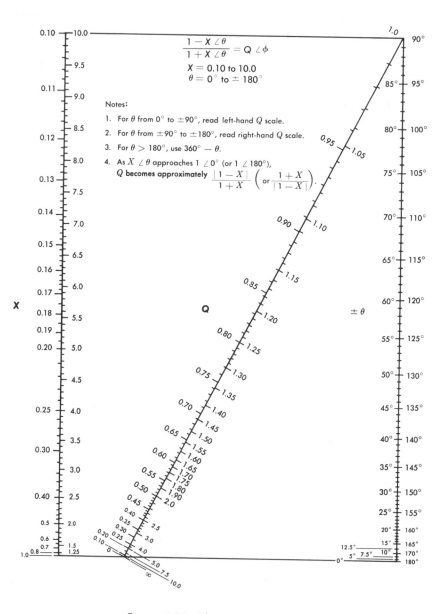

Figure 4-15. Alignment chart Q1.

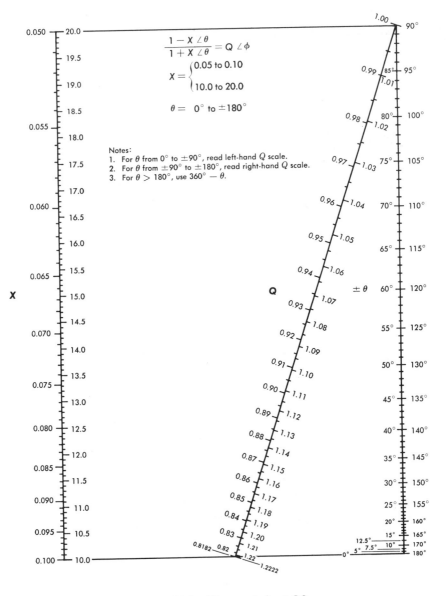

Figure 4-16. Alignment chart Q2.

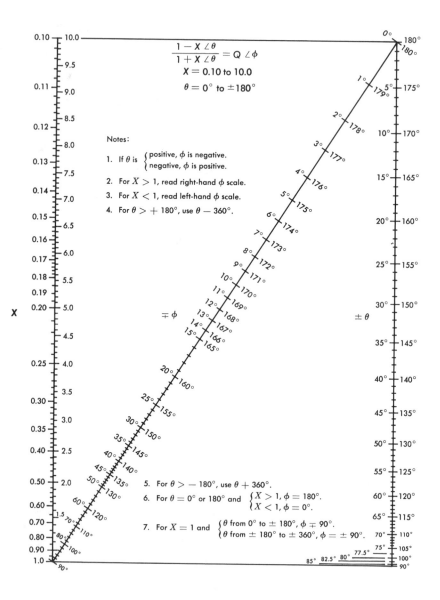

Figure 4-17. Alignment chart $\phi$1.

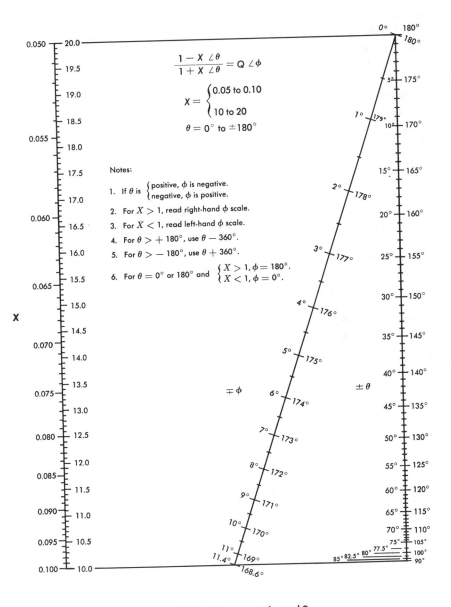

$$\frac{1 - X \angle \theta}{1 + X \angle \theta} = Q \angle \phi$$

$$X = \begin{cases} 0.05 \text{ to } 0.10 \\ 10 \text{ to } 20 \end{cases}$$

$$\theta = 0° \text{ to } \pm 180°$$

Notes:

1. If $\theta$ is $\begin{cases} \text{positive, } \phi \text{ is negative.} \\ \text{negative, } \phi \text{ is positive.} \end{cases}$

2. For $X > 1$, read right-hand $\phi$ scale.

3. For $X < 1$, read left-hand $\phi$ scale.

4. For $\theta > + 180°$, use $\theta - 360°$.

5. For $\theta > - 180°$, use $\theta + 360°$.

6. For $\theta = 0°$ or $180°$ and $\begin{cases} X > 1, \phi = 180°. \\ X < 1, \phi = 0°. \end{cases}$

$\mp \phi$          $\pm \theta$

Figure 4-18. Alignment chart $\phi2$.

## Insertion Loss and Phase Shift

The insertion loss and insertion phase shift of a four-terminal network placed between two impedances may be determined by Equations (3-13), (3-14), and (3-15) and by using relationships similar to those of Equations (4-26) and (4-27) for impedance values. However, the general case, when the terminating impedances and the input and output image impedances are all different, contains many interaction terms which may be difficult to evaluate. Furthermore, this most general situation is usually of only academic interest; since all terminals are accessible, it is sometimes easier to measure the insertion loss than to compute it. Often, the subject network is either symmetrical and has only one value of image impedance, or it is a transmission line of characteristic impedance, $Z_0$.

## Return Loss

The return loss is a measure of the loss in the return path due to an impedance mismatch. In the analysis of speech transmission, it is a convenient measure of the echo caused by a mismatch. The return loss is related to the reciprocal of the absolute value of the *reflection coefficient*, a term which relates impedances at a point of connection in such a way as to give a measure of the voltage or current reflected from the mismatch point towards the transmitting end of a circuit.

From Figure 4-19, the reflection coefficient, $\rho$, at the terminals of $Z_L$ may be written for voltage,

$$\rho_v = \frac{Z_L - Z_G}{Z_L + Z_G},\qquad (4\text{-}28)$$

or for current,

$$\rho_i = \frac{Z_G - Z_L}{Z_G + Z_L}.\qquad (4\text{-}29)$$

The return loss at these terminals is given by 20 log $(1/|\rho|)$ dB; i.e.,

$$\text{Return loss} = 20 \log \frac{1}{|\rho|} = 20 \log \left|\frac{Z_G + Z_L}{Z_G - Z_L}\right| .\qquad (4\text{-}30)$$

The expression for return loss, Equation (4-30), may also be written in the form

$$20 \log \frac{1}{|\rho|} = 20 \log \left| \frac{1}{\left(1 - \frac{Z_L}{Z_G}\right) \Big/ \left(1 + \frac{Z_L}{Z_G}\right)} \right|.$$

The bracketed expression in this equation may be written in polar form as $(1 - X \angle \theta)/(1 + X \angle \theta)$. Thus, the alignment charts of Figures 4-15 to 4-18 may be used to determine the return loss at a junction between two impedances such as that shown in Figure 4-19.

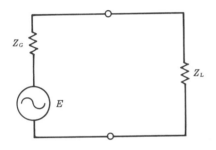

Figure 4-19. Junction between two impedances.

The actual voltage across the load, $Z_L$, is equal to the voltage which would be present across an impedance matched to $Z_G$ plus the reflected voltage. As $Z_L$ approaches zero, the reflection coefficient approaches $-1$, the measured voltage across $Z_L$ approaches zero, the return loss approaches 0 dB, and all the energy is reflected back to $Z_L$. As $Z_L$ approaches infinity, the reflection coefficient approaches $+1$, the voltage across $Z_L$ approaches its open-circuit value (twice the value across $Z_L$ under matched conditions), and the return loss again approaches 0 dB. When $Z_L$ equals $Z_G$, the reflection coefficient is zero in magnitude and angle, the voltage across $Z_L$ is one-half the open-circuit value, and the return loss is infinite.

The effects of impedance mismatch on return loss are illustrated in Figure 4-20 which shows that the return loss increases as the angle, $\theta$, decreases and as the ratio of $|Z_L/Z_G|$ or $|Z_G/Z_L|$ approaches unity. The angle $\theta$ is that between the load and generator impedances. The values of the parameters of Figure 4-20 may be written $Z_L = |Z_L| \angle \theta_L$, $Z_G = |Z_G| \angle \theta_G$, and $\theta = |\theta_L - \theta_G|$.

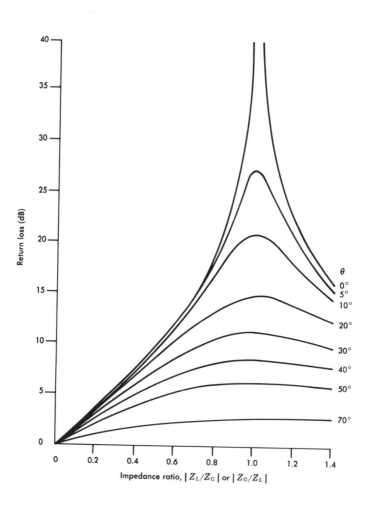

Figure 4-20. Return loss variations.

## Echo — Magnitude and Delay

Return now to Figure 4-14. The condition at terminals 2-2 is one of mismatch; the effect of the mismatch could be evaluated in terms of return loss, as above, simply by using values in Equation (4-30) such that $Z_G = Z_2$, and $Z_L = bZ_2$. However, it is often desirable to evaluate the magnitude of the reflected voltage or current wave and to determine the delay encountered by the reflected wave in transmission through the network.

Consider first the magnitude of the reflected wave. From Equation (4-22), $b = (Z_2 + Z_r)/Z_2$, or $Z_r = Z_2(b - 1)$. The total current at the output is

$$I_R = \frac{E_2}{2Z_2 + Z_r} = \frac{E_2}{Z_2(b + 1)}. \qquad (4\text{-}31)$$

The voltage across $Z_r$ may be regarded as the reflected voltage due to the mismatch. It is written

$$E_r = I_R Z_r = I_R Z_2(b - 1).$$

Substitution of Equation (4-31) yields

$$E_r = E_2 \left(\frac{b - 1}{b + 1}\right) = -E_2 \left(\frac{1 - b}{1 + b}\right).$$

The output current, from Figure (4-14), may be written also as

$$I_R = I_2 + I_r.$$

From Ohm's law,

$$I_2 = \frac{E_2}{2Z_2}$$

and

$$I_r = -\frac{E_r}{2Z_2} = \frac{E_2}{2Z_2}\left(\frac{1 - b}{1 + b}\right).$$

A useful expression for the ratio of reflected to incident current is obtained by dividing $I_r$ by $I_2$:

$$\frac{I_r}{I_2} = \frac{1 - b}{1 + b}. \qquad (4\text{-}32)$$

It can be shown that a similar relationship exists for a mismatch at the input; i.e.,

$$\frac{I_s}{I_1} = \frac{1-a}{1+a}.$$

(4-33)

Equations similar to (4-32) and (4-33) can also be developed to show the ratio of reflected to incident voltages.

Echo evaluations must, of course, take into account the loss encountered in transmission through the network an appropriate number of times. Successive reflections become increasingly attenuated and at some point may be ignored.

The time delay or transit time for a wave to propagate through a four-terminal network may be shown to be

$$T = \frac{\theta}{2 \times 360° \times f}$$

(4-34)

where $\theta$ is the angle of $G_I$ in degrees and $f$ is the frequency in hertz. Then, the round-trip delay for an echo to be transmitted through a network and back again is

$$T_2 = \frac{\theta}{360° \times f}.$$

(4-35)

## Power Transfer

In Figure 4-19, $E$ and $Z_G$ represent a source of power. This source may be a telephone instrument, a repeater amplifier, or the sending side of any point in a telephone connection. The impedance $Z_L$ is the load which receives the power transmitted. It may be another telephone instrument or a radio antenna—the receiving side of any point in a connection. The amount of power transferred from the source to the load may be determined by the relative values of $Z_G$ and $Z_L$. The power transferred can be shown to be a maximum under three different assumptions as follows:

(1) If $Z_G$ is a fixed impedance and there is no restriction on the selection of $Z_L$, the power transferred is a maximum value when $Z_L$ is the conjugate of $Z_G$, that is, when $Z_L$ and $Z_G$ have equal components of resistance and their reactive components are equal and opposite. This may be written $Z_G = R + jX$, and $Z_L = Z_G^* = R - jX$.

(2) If $Z_G$ is a fixed impedance and the magnitude of $Z_L$ can be selected but not its angle, the power transferred is a maximum when the absolute values of $Z_L$ and $Z_G$ are equal ( $|Z_L| = |Z_G|$ ). That is, the impedances are equal disregarding phase.

(3) If both $Z_G$ and $Z_L$ are pure resistances, the power transferred is a maximum when the source and load resistances are equal ($R_G = R_L$).

Figure 4-21 shows power and efficiency relationships for case (3) over a range of load resistance values from 0 to $2R_G$. Curve (A) shows that the power delivered to the load, $R_L$, is zero when $R_L = 0$, increases to 25 percent of the maximum possible when $R_L = R_G$, and then gradually decreases as $R_L$ is further increased. The total power that can be developed, designated as 100 percent, is that delivered to the internal resistance of the generator when $R_L = 0$, i.e., a short

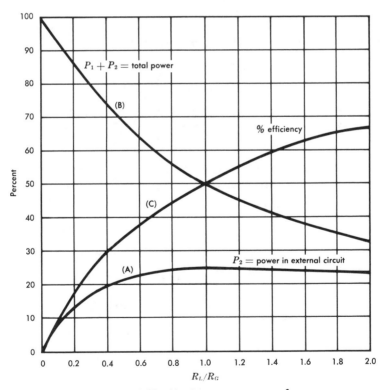

Figure 4-21. Maximum power transfer.

circuit. As the value of $R_L$ is increased the total power developed decreases as shown by curve (B). When $R_L = R_G$, the total power is 50 percent of the maximum, half delivered to the generator resistance and half to the load resistance. The total power decreases to zero when $R_L = \infty$, an open circuit. Curve (C) shows the efficiency of the circuit, i.e., the percentage of the total generated power that is delivered to the load as a function of the ratio of the load resistance to the generator resistance. The condition of maximum power delivered to the load, $R_L = R_G$, approximates the most desirable condition in telephony since, in most applications, the primary interest is in delivering maximum power to the load regardless of the efficiency.

However, in telephony another transmission parameter must be considered, the generation of reflections or echoes. The necessity for compromise between delivering maximum power to a load and maintaining reasonable performance in respect to reflections can best be illustrated by an example.

### Example 4-2: Power Transfer and Return Loss

In Figure 4-19, let $Z_G = 900 - j200$ ohms and let $E = 1$ volt rms at 1000 Hz.

(a) What is the return loss at 1000 Hz at the junction between $Z_G$ and $Z_L$ and what is the power delivered to $Z_L$ when $Z_L = 900 + j200$?

(b) What is the return loss at 1000 Hz at the junction between $Z_G$ and $Z_L$ and what is the power delivered to $Z_L$ when $Z_L = 922 + j0$?

*Case a:*

$$\text{Return loss} = 20 \log \frac{1}{|\rho|} = \left| \frac{1}{\left(1 - \frac{Z_L}{Z_G}\right) \Big/ \left(1 + \frac{Z_L}{Z_G}\right)} \right|$$

$$Z_G = 900 - j200 = 922 \angle -12.5°$$

$$Z_L = 900 + j200 = 922 \angle +12.5°$$

$$\frac{Z_L}{Z_G} = \frac{922 \angle -12.5°}{922 \angle +12.5°} = 1 \angle -25°$$

$$\text{Return loss} = 20 \log \left| \frac{1}{(1 - 1 \angle -25°)/(1 + 1 \angle -25°)} \right|$$

$$= 20 \log \left| \frac{1}{0.22 \angle +90°} \right|$$

$$= 13.2 \text{ dB},$$

where the value of $(1 - 1 \angle -25°)/(1 + 1 \angle -25°)$ is found from Figures 4-15 and 4-17. Alternatively, the return loss may be determined by interpolation in Figure 4-20.

To determine the power delivered to the load, the current may first be determined:

$$I = \frac{E}{Z_G + Z_L} = \frac{1 \angle 0}{1800} = 0.000554 \text{ ampere, rms.}$$

Then,

$$P_L = I^2 R_L = 0.000554^2 \times 900 = 0.000276 \text{ watt.}$$

*Case b:*

$$Z_G = 900 - j200$$

$$Z_L = 922 + j0$$

$$\frac{Z_G}{Z_L} = \frac{922 \angle -12.5°}{922 \angle 0} = 1 \angle -12.5°$$

$$\text{Return loss} = 20 \log \left| \frac{1}{(1 - 1 \angle -12.5°)/(1 + 1 \angle -12.5°)} \right|$$

$$= 20 \log \frac{1}{0.11}$$

$$= 19.2 \text{ dB.}$$

The power delivered to the load is computed as follows:

$$I = \frac{E}{Z_G + Z_L} = \frac{1 \angle 0}{900 - j200 + 922} = \frac{1 \angle 0}{1822 - j200}$$

$$= \frac{1822 + j200}{3,360,000} = 0.000542 + j0.000060$$

$$= 0.000545 \text{ ampere.}$$

$$P_L = I^2 R_L = 0.000545^2 \times 922$$

$$= 0.000274 \text{ watt.}$$

Thus, an increase of $19.2 - 13.2 = 6$ dB in return loss (resulting in a 6-dB reduction in echo amplitude) is achieved by providing a resistive termination of a value equal to the absolute value of the source, 922 ohms. For this improvement, the delivered power of 0.000276 watt is reduced to 0.000274 watt, a negligible penalty of $10 \log 0.000276/0.000274 = 0.03$ dB.

## Stability

When a circuit is unstable, it is said to be *singing;* that is, unwanted signal currents and voltages flow in the circuit without an external source of applied signal energy. In Figure 4-22, such con-

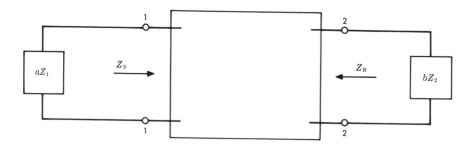

Figure 4-22. Four-terminal network; mismatched at input and output.

ditions of instability would result in signal current flow in $aZ_1$, $bZ_2$, and in the network.

The complete development of mathematical criteria for absolute stability or absolute instability is not undertaken here. However, stability criteria are presented with some background to indicate how they are derived.

If currents are circulating in the input circuit of Figure 4-22 without an external source of energy, there must be zero impedance at the frequency at which current is observed. That is, if such a current is circulating through the input, then, at that frequency

$$aZ_1 + Z_S = 0.$$

It can be shown that when such a condition exists a similar condition exists at the output; that is,

$$bZ_2 + Z_R = 0.$$

For such a condition to exist, the sending-end impedance, $Z_S$, must have a negative resistance component equal to the positive resistance component of $aZ_1$, and the reactive component of impedance $Z_S$ must be equal in magnitude but opposite in sign to the reactive component of $aZ_1$. Thus, stability is guaranteed if neither $aZ_1$ nor $Z_S$ has a negative resistance component and at least one has a positive resistance component.

A *stability index* can be derived in terms used earlier in this chapter. It may be written

$$\text{Stability index} = 1 - G_I \left( \frac{1-a}{1+a} \right) \left( \frac{1-b}{1+b} \right). \qquad (4\text{-}36)$$

The two parenthetical expressions are of a form which can be evaluated by the alignment charts of Figures 4-15 through 4-18.

Note that if the network is terminated at either end in its image impedance, it cannot be made to sing. Under these conditions $a = 1$ or $b = 1$, and the stability index $= 1$, a criterion for absolute stability.

The condition for singing is that the stability index $= 0$, i.e., that

$$G_I = \cfrac{1}{\left(\dfrac{1-a}{1+a}\right)\left(\dfrac{1-b}{1+b}\right)}. \qquad (4\text{-}37)$$

The circuit will not sing provided the magnitude of $G_I$ is slightly less than that given by Equation (4-37). A sample calculation of this type circuit is given in Figure 4-23.

Usually, a network is terminated in impedances such that the stability index falls between the extremes of 0 and 1. The margin against singing may be found by

$$\text{Singing margin} = 20 \log \left| G_I \left(\frac{1-a}{1+a}\right)\left(\frac{1-b}{1+b}\right)\right|. \qquad (4\text{-}38)$$

## 4-4  NETWORK ANALYSIS

The preceding material on the basic network laws and their applications provides the tools for network analysis. Some extensions of these tools and some added sophistication in mathematical manipulations make the analysis job applicable to very complex network configurations.

### Mesh Analysis

A circuit of any complexity may be analyzed by considering each mesh of the circuit independently and writing an equation for the voltage relations in each. To do this, of course, it is first necessary to define a mesh.

In Figure 4-2, for example, *nodes* are defined as those points at which individual series combinations of components are interconnected. The series combinations are called *branches* (each $Z$ in Figure 4-2 may be made up of series-connected elements in any combination). A *mesh* may then be regarded as openings in the network schematic such as those that might be observed in a fish net. The boundary of a mesh, called the mesh contour, is made up of network

| $\theta_1$ | $\theta_{aZ_1}$ | $\theta_a$ | $\dfrac{1-a}{1+a}$ "Q" | $\theta_2$ | $\theta_{bZ_2}$ | $\theta_b$ | $\dfrac{1-b}{1+b}$ "Q" | $\dfrac{1}{\left(\dfrac{1-a}{1+a}\right)} \times \dfrac{1}{\left(\dfrac{1-b}{1+b}\right)} = \|G_I\|$ | MIN TOTAL LOSS (DB) |
|---|---|---|---|---|---|---|---|---|---|
| $+50°$ | $-90°$ | $-140°$ | 2.70 | $+30°$ | $-90°$ | $-120°$ | 1.73 | (0.370) (0.578) = 0.214 | 6.7 |
| $0°$ | $-90°$ | $-90°$ | 1.00 | $0°$ | $-90°$ | $-90°$ | 1.00 | (1.0)　(1.0)　= 1.0 | 0.0 |
| $+10°$ | $-90°$ | $-100°$ | 1.19 | $+10°$ | $-90°$ | $-100°$ | 1.19 | (0.841) (0.841) = 0.708 | 1.5 |
| $+20°$ | $-90°$ | $-110°$ | 1.43 | $+20°$ | $-90°$ | $-110°$ | 1.43 | (0.700) (0.700) = 0.490 | 3.1 |
| $+30°$ | $-90°$ | $-120°$ | 1.73 | $+30°$ | $-90°$ | $-120°$ | 1.73 | (0.578) (0.578) = 0.334 | 4.8 |
| $+40°$ | $-90°$ | $-130°$ | 2.15 | $+40°$ | $-90°$ | $-130°$ | 2.15 | (0.465) (0.465) = 0.216 | 6.7 |
| $+50°$ | $-90°$ | $-140°$ | 2.70 | $+50°$ | $-90°$ | $-140°$ | 2.70 | (0.370) (0.370) = 0.137 | 8.7 |
| $+60°$ | $-90°$ | $-150°$ | 3.70 | $+60°$ | $-90°$ | $-150°$ | 3.70 | (0.270) (0.270) = 0.073 | 11.4 |
| $+70°$ | $-90°$ | $-160°$ | 5.50 | $+70°$ | $-90°$ | $-160°$ | 5.50 | (0.182) (0.182) = 0.033 | 14.8 |
| $+80°$ | $-90°$ | $-170°$ | 12.0 | $+80°$ | $-90°$ | $-170°$ | 12.0 | (0.083) (0.083) = 0.0069 | 21.5 |
| $+90°$ | $-90°$ | $-180°$ | $\infty$ | $+90°$ | $-90°$ | $-180°$ | $\infty$ | (0)　(0)　= 0 | $\infty$ |

Notes:

$\theta_1$ = angle of $Z_1$, the input image impedance.

$\theta_{aZ_1}$ = angle of $aZ_1$, assumed to be $-90°$.

$\theta_a$ = worst angle of $a$ in $\left(\dfrac{1-a}{1+a}\right)$.

$\|a\| = 1.$

$\theta_2$ = angle of $Z_2$, the output image impedance.

$\theta_{bZ_2}$ = angle of $bZ_2$, assumed to be $-90°$.

$\theta_b$ = worst angle of $b$ in $\left(\dfrac{1-b}{1+b}\right)$.

$\|b\| = 1.$

Figure 4-23. Computations for guaranteed stability.

branches. The least number of independent loops, or closed meshes, is one greater than the difference between the number of branches and the number of nodes. The number of independent loops determines the number of independent mesh equations needed to solve the network problem. Examination of Figure 4-2 shows that there are three branches and two nodes. Application of the rule indicates there are two independent meshes.

In mesh analysis, the parameters of the branches are expressed as impedances, the independent variables are the voltages and the voltage drops in each of the branches of a mesh, and the dependent variables are the currents in each branch of a mesh. A simple example of mesh analysis of the circuit of Figure 4-2 may be performed by using the rule above regarding the number of independent meshes in the circuit and by applying Kirchoff's laws.

Thus, the equations

$$E - I_1 Z_A - (I_1 - I_2) Z_C = 0$$

and

$$E - I_1 Z_A - I_2 Z_B = 0$$

provide the two independent equations for the two independent meshes. If the values of $E$, $Z_A$, $Z_B$, and $Z_C$ are known, the two mesh currents can be determined from these equations.

## Nodal Analysis

In nodal analysis, the branch parameters are most conveniently expressed as admittances (recall that admittance is the reciprocal of impedance; i.e., $Y = 1/Z$), the dependent variables are the voltages at the individual nodes, and the independent variables are the currents entering and leaving each node. Simultaneous equations are written for node currents, and their solution is the nodal analysis of the network. The number of independent nodal equations that may be written is one less than the number of nodes.

There is, of course, a direct correspondence between mesh and nodal equations. One approach is often found superior to the other, and the choice, while theoretically a matter of indifference, is often important from the points of view of convenience and flexibility in treating such things as parasitic circuit elements or active device parameters. The more complex circuits are usually more easily analyzed by the nodal approach.

Finding solutions to mesh or nodal circuit equations can become quite complex when all circuit elements are considered. Such equations, except in the simplest cases, are now usually solved by the use of an electronic computer.

## Determinants

In simple networks, brute-force solution of simultaneous equations by successive substitution of one equation in another is generally simple and straightforward. Only modest amounts of network complexity, however, make this approach to finding solutions prohibitive in the amount of time consumed. Further, the processes become so involved that the accuracy of the work must always be carefully checked to guard against error.

The coefficients of the dependent variables of the simultaneous equations may be arranged in rows and columns corresponding to the terms of the equations. If the resulting array is square (i.e., if it has the same number of rows and columns), solutions to the simultaneous equations can be found by the methods of determinants [2].

## Matrix and Linear Vector Space Analyses

While it is often possible to determine significant but not complete characteristics of a network by means of voltage, current, and impedance measurements made at the terminals, such expressions may not completely define the network. These expressions, however, are often useful in relating the network performance to its interaction with other interconnected networks and in defining certain properties of the subject network. The coefficients of terms in the mathematical expressions derived from such measurements and observations may be arranged in matrix form; the matrix may or may not be square. Mathematical manipulation of the matrix expressions provides a convenient method of network analysis. This may be regarded as a "black box" approach to analysis, which ignores the internal structure of the network but permits specification of its external behavior. The application of the concepts of linear vector spaces to matrix analysis adds a significantly greater amount of power to network analysis [3].

## 4-5 TRANSFORMERS

Many types of transformers, sometimes called repeat coils, are used in telecommunications circuits. In most cases, the applications differ significantly from those applying to alternating current power distribution systems where the principal use is to step alternating voltages up or down. In communications circuits, in addition to voltage transformation, transformers are used to match impedances, to split and combine transmission paths, to separate alternating and direct currents, and to provide dc isolation between circuits. Impedance matching and the splitting and combining of transmission paths are discussed in some detail because of their importance in transmission.

### Impedance Matching

Unequal ratio transformers are used to match unequal impedances to permit maximum energy transfer. The currents through any two windings of such a transformer are inversely proportional to the number of turns in the two windings. The voltages across the two windings are directly proportional to the number of turns in the two windings. Thus,

$$\frac{V_S}{V_L} = \frac{N_1}{N_2}, \text{ or } V_L = \left(\frac{N_2}{N_1}\right) V_S \tag{4-39}$$

where $N_1$ and $N_2$ are the number of turns on the primary and secondary windings, respectively.

No power is dissipated in an ideal transformer, illustrated in Figure 4-24, and in addition the phase relation between the voltage and current on the two sides of the transformer is exactly the same. Therefore, the product of voltage $V_S$ across the primary winding and current $I_S$ through the primary winding is equal to the corresponding

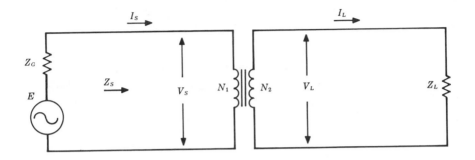

Figure 4-24. Transformer circuit.

product for the secondary winding; that is,

$$V_S I_S = V_L I_L, \text{ or } V_S/V_L = I_L/I_S.$$

Then, substituting this value of $V_S/V_L$ in Equation (4-39),

$$I_L = \left(\frac{N_1}{N_2}\right) I_S. \tag{4-40}$$

From Ohm's law, $V_L = I_L Z_L$. Substituting the values of $V_L$ and $I_L$ from Equations (4-39) and (4-40),

$$V_S\left(\frac{N_2}{N_1}\right) = \left(\frac{I_S N_1}{N_2}\right) Z_L. \tag{4-41}$$

Then,

$$\frac{V_S}{I_S} = Z_S = \left(\frac{N_1}{N_2}\right)^2 Z_L \tag{4-42}$$

and

$$\frac{Z_S}{Z_L} = \left(\frac{N_1}{N_2}\right)^2. \tag{4-43}$$

The relationship shown in Equation (4-43) is used when a transformer is being designed for the purpose of providing an impedance match, i.e., to provide a design in which $Z_S = Z_G$.

Commercial transformers approach the efficiency of ideal transformers very closely. Small losses are occasioned by currents induced in the core (eddy current losses), by flux in the core (hysteresis losses), and by current flowing in the copper windings.

## Separating and Combining

A common means of separating and combining transmission paths is by the use of a transformer called a *hybrid coil*, a complex circuit component. Although the operation of a hybrid coil is somewhat difficult to analyze, a simplified and idealized coil structure may suffice to illustrate how it is used in some common circuit applications.

In the circuit configuration of Figure 4-25(a), the hybrid coil characteristics are such that, if $Z_d$ and/or $Z_c$ are signal sources, the energy is divided equally between the two loads $Z_a$ and $Z_b$ provided certain impedance relationships are satisfied. If $Z_a$ and/or $Z_b$ are signal sources, the energy is similarly divided equally between $Z_d$ and $Z_c$. These hybrid coil characteristics are exploited in combining and separating analog signals (including pilots and test signals), in providing parallel transmission paths in protection switching systems, and in providing the interface between two-wire and four-wire voice-grade facilities (four-wire terminating sets).

In the circuit of Figure 4-25(b), assume that $Z_a = Z_b$, $Z_c = Z_d$, and the number of turns on each of the three windings of the hybrid transformer are the same. With these assumptions, assume a signal source in the branch containing $Z_d$ as shown in Figure 4-25(c). The currents in the right-hand branches of the circuit divide equally between $Z_a$ and $Z_b$ and are cancelled in $Z_c$. Thus, there is no transmission from $Z_d$ to $Z_c$. If the signal source were in series with $Z_c$, the currents would again divide equally between $Z_a$ and $Z_b$. Their

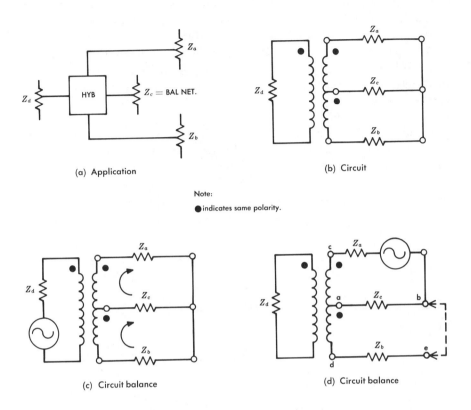

(a) Application

(b) Circuit

Note:

● indicates same polarity.

(c) Circuit balance

(d) Circuit balance

Figure 4-25. Hybrid circuit relationships.

effects would cancel, however, due to the polarity of the magnetic fields in the center-tapped winding of the transformer. Hence, there would be no transmission from $Z_c$ to $Z_d$.

Now assume the signal source to be in series with $Z_a$ as shown in Figure 4-25(d). It is convenient to imagine the circuit to be opened between points b and e. Under these conditions and with $Z_c = Z_d$, the voltage induced between points a and d is exactly equal to the voltage drop in $Z_c$ so that the voltage at b equals that at e. Then, since points b and e are at the same potential, they may be connected without causing current to flow in $Z_b$. Thus, there is no transmission from $Z_a$ to $Z_b$.

In each of the above examples, transmission from one impedance to another involves an equal division of energy to two other impedances; each load impedance dissipates half the power from the source, a loss of 3 dB. In addition, core and copper losses are typically about 0.5 dB. Thus, in designing or analyzing transmission circuits in which equal ratio hybrids* are used, 3.5-dB loss is usually assumed for the hybrid.

A common application of hybrid circuits is at the interface between two-wire and four-wire facilities. Such a circuit, illustrated in Figure 4-26, is known as a four-wire terminating set. Transmission

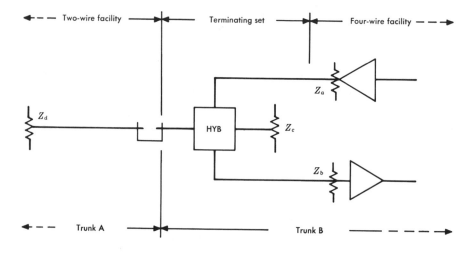

Figure 4-26. Hybrid application—four-wire terminating set.

is from the amplifier with output impedance $Z_a$ to the two-wire trunk with impedance $Z_d$, and from the two-wire trunk $Z_d$ to the amplifier with impedance $Z_b$. When transmitting from $Z_d$ to $Z_b$, half the energy is dissipated in $Z_a$, but the signal is not transmitted through the amplifier because of its one-way transmission characteristics. When transmitting from $Z_a$ to $Z_d$, half the power is lost in $Z_c$. The important thing, however, is that no energy reaches $Z_b$. If this were not so, the signal would circulate through the two sides of the four-wire

---

*In some applications, unequal ratio hybrids are used. The design of such hybrids involves careful selection of impedances and turns ratios, a process too complex to be covered here.

trunk and the hybrid circuits at each end, being amplified by the amplifier circuits each time. This could result in circuit instability, or singing, as previously discussed.

The loss between $Z_a$ and $Z_b$ or between $Z_c$ and $Z_d$ is known as hybrid balance. In carefully controlled laboratory circuits, a balance of 50 dB is easily achievable. However, in the application described, impedance $Z_d$ represents any of a large number of two-wire trunks which may be switched into the connection. The impedances of these trunks vary widely, and so only a compromise value may be used for $Z_c$ to provide control of echoes that are returned to the speaker at four-wire terminating sets.

When supposedly matched impedances are in reality unequal in either their resistive or reactive components, or both, the hybrid balance deteriorates so that the achievable balance may be much less than 50 dB. When this occurs, echo is produced in the transmission circuit. The echo may be evaluated in terms of the return loss at the junction between the two-wire facility and the hybrid which may be calculated as described previously.

## 4-6   RESONANT CIRCUITS

By an appropriate combination of resistors, inductors, and capacitors, circuits may be designed to resonate, i.e., to have extremely high or low loss at a selected frequency. Such circuits, which may be either series or parallel, are often designed as two-terminal networks which then are used as components of a larger, more complicated four-terminal network. Resonance occurs when the inductive and capacitive components of reactance are equal. That is, when

$$ |X_L| = |X_C| = 2\pi f_r L = \frac{1}{2\pi f_r C}. \tag{4-44} $$

The resonant frequency may be found by solving Equation (4-44) for $f_r$:

$$ f_r = \frac{1}{2\pi \sqrt{LC}}. \tag{4-45} $$

Selectivity, i.e., the difference in transmission between the resonant frequency and other frequencies, is determined by the amount of resistance in the circuit. Since the resistance is usually concentrated in the inductor, the objective is to have the ratio of the reactance of the inductor to its resistance as high as possible. This ratio is known as the quality factor, or $Q$, of the inductor and is expressed by

$$Q = \frac{X_L}{R} = \frac{2\pi f L}{R}. \tag{4-46}$$

**Series Resonance.** In a resonant circuit having the inductance and capacitance in series, the circuit reactance is *zero* at the resonant frequency, where the inductive and capacitive reactances are equal as in Equation (4-44); the impedance has a minimum value at this frequency and is equal to the resistance of the circuit. If this resistance is small, the resonant frequency current is large compared to that at other frequencies, as shown in Figure 4-27. One application of series resonance is in the use of a capacitor of proper value in series with a telephone receiver winding, repeating coil winding, or other inductance, where it is desired to increase the current at specific frequencies.

**Parallel Resonance.** In a parallel (often called anti-resonant) circuit, one having the inductance and capacitance in parallel, the impedance of the combination is a *maximum* at the resonant frequency, where the inductance and capacitive reactances are equal. Since the impedance is a maximum, the current is a minimum at the resonant frequency. The selectivity of the circuit is decreased as the resistance is increased, reaching a point where the circuit essentially loses its resonant characteristics. This is shown in Figure 4-28. A parallel resonant circuit is often called a tank circuit since it acts as a storage reservoir for electric energy.

## 4-7 FILTERS

An electrical network of inductors and capacitors designed to permit the flow of current at certain frequencies with little or no attenuation and to present high attenuation at other frequencies is called an electric wave filter. Simple filter configurations and characteristics are illustrated in Figures 4-29 and 4-30. Low-pass and high-pass filters can be combined to give the characteristics of band-

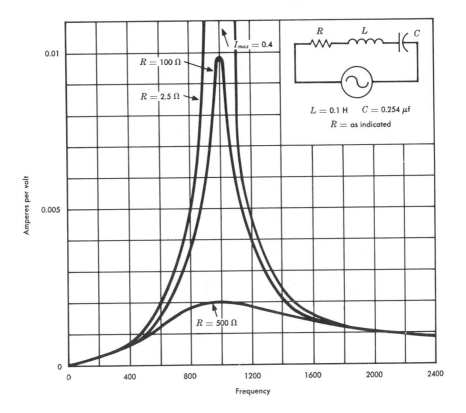

Figure 4-27. Curves of current values in series resonant circuit.

pass or band-elimination filters. These are used to pass or to stop an intermediate band of frequencies.

Inductances and capacitances are opposite in their responses to varying frequencies. An inductance passes low frequencies readily and offers an increasing series impedance with increase in frequency, while the reverse is true of capacitance. Advantage of these characteristics is taken in the design of filters.

The presence of resistance in the inductors used in filter sections introduces additional losses in the transmitting bands and reduces the sharpness of cutoff. One of the most practicable ways to obtain

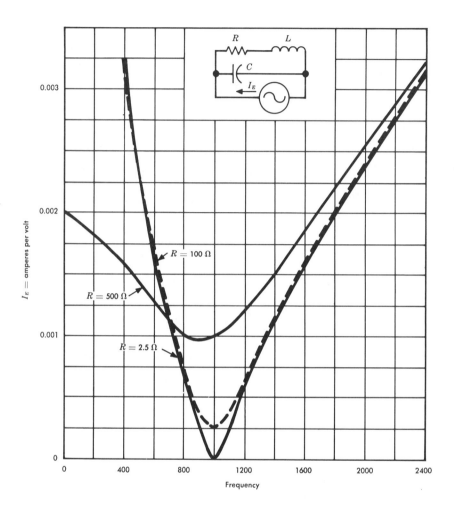

Figure 4-28. Curves of current values in parallel resonant circuit.

a high ratio of reactance to resistance is to use mechanical vibrating systems, such as the piezo-electric crystal. In an electric circuit such as a filter, a crystal acts as an impedance exhibiting both resonant and anti-resonant properties. Crystal filters find wide application in broadband carrier systems.

Older and well known filter structures such as the $m$-derived and constant-$k$ image parameter designs are still used. They can provide

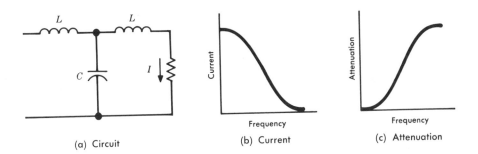

(a) Circuit                (b) Current                (c) Attenuation

Figure 4-29. Low-pass filter.

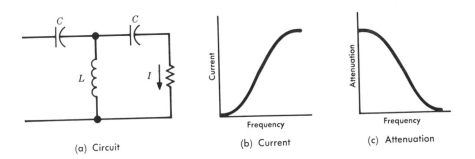

(a) Circuit                (b) Current                (c) Attenuation

Figure 4-30. High-pass filter.

characteristics that satisfy many needs, and they have the further attributes of being relatively easy to design, synthesize, and realize; however, more sophisticated approaches have become necessary as requirements have become more stringent, bandwidths have become wider, and useful frequencies have been pushed higher in the spectrum. Some of the distinguished scientists who have made significant contributions in this field are Bode, Butterworth, Campbell, Darlington, Foster, Guillemin, Johnson, Shea, and Zobel.

REFERENCES

1. Guillemin, E. A. *Communication Networks,* Vol. 1 (New York: John Wiley and Sons, Inc., 1931).

2. Bode, H. W. *Network Analysis and Feedback Amplifier Design* (Princeton, N. J.: D. Van Nostrand Company, Inc., 1945).

3. Huelsman, L. P. *Circuits, Matrices, and Linear Vector Spaces* (New York: McGraw-Hill Book Company, Inc., 1963).

4. Kuo, F. F. *Network Analysis and Synthesis* (New York: John Wiley and Sons, Inc., 1962).

5. Johnson, W. C. *Transmission Lines and Networks* (New York: McGraw-Hill Book Company, Inc., 1950).

6. *Reference Data for Radio Engineers* (Indianapolis: Howard W. Sams and Company, Inc., Oct. 1968).

Chapter 5

# Transmission Line Theory

Electromagnetic wave theory provides the basis for transmission line analysis. Involved are such factors as impedance, impedance matching, loss, velocity of propagation, reflection, and transmission. Knowledge of all these parameters is necessary to an understanding of electrical signal transmission over metallic wire media.

Transmission networks are made up of resistors, capacitors, and inductors. The characteristics of such components are called lumped constants. A transmission line is an electrical circuit whose constants are not lumped but are uniformly distributed over its length. With care, the theory of lumped constant networks can be applied to transmission lines, but lines exhibit additional characteristics which deserve consideration.

The detailed characterization of a given type of cable is dependent on the physical design of the cable. Wire gauge, type of insulation, twisting of the wire pairs, etc., have important effects on attenuation, phase shift, impedance, and other parameters. For the most part, the treatment here pertains to two-wire parallel conductors. However, the analysis is also extended and applied to a coaxial conductor configuration.

## 5-1 DISCRETE COMPONENT LINE SIMULATION

It is convenient to analyze transmission line characteristics in terms of equivalent discrete-component, four-terminal networks. The conductors of an ideal simple transmission line, evenly spaced and extending over a considerable distance, have self-inductance, $L$, and resistance, $R$, which are series-connected elements that must be included in the discrete component equivalent network. Since the line appears electrically the same when viewed from either end, the equivalent circuit must be symmetrical. Thus, the components of the equivalent network are split and connected as shown in Figure 5-1. The insulation between the wires is never perfect; there is some leakage between them. The leakage resistance may be very large, as in a dry cable, or it may be fairly small, as in the case of a wet open-wire pair. Hence, the equivalent circuit must contain a conductance, $G$, in shunt between the line conductors. Also, any two conductors in close proximity to one another have the properties of a capacitor; therefore, the circuit must have shunt capacitance, $C$.

These line parameters (resistance, inductance, conductance, and capacitance) are the *primary constants* and are usually expressed in per-mile values of ohms, millihenries, micromhos, and microfarads. Derived from these are the characteristic impedance and propagation constant; they are the *secondary constants,* both of which are functions of frequency. Although all primary and secondary constants vary with changes in temperature, they are usually expressed as constants at 68°F with correction factors for small changes in temperature. Both primary and secondary constants are often used to characterize transmission lines or equivalent circuits.

In the discussion of networks in Chapter 4, it was suggested that any circuit could be simulated by a T structure. It is not surprising then to find that a useful equivalent circuit for a transmission line is the T network shown in Figure 5-1. For convenience, series constants $R$ and $L$ can be combined as impedance $Z_A$, and shunt constants $C$ and $G$ as impedance $Z_C$ as shown in Figure 5-2. For the present, the relationship to line length is ignored.

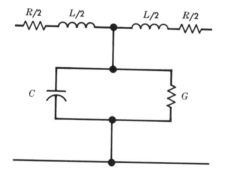

Figure 5-1. Primary constants of a section of uniform line.

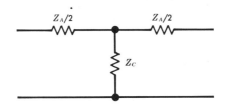

Figure 5-2. Equivalent network of a section of uniform line.

The equivalent circuits in Figure 5-1 or 5-2 are poor approximations of a real transmission line because all of the distributed constants have been concentrated at one point. The approximation is improved by having two T sections in tandem and, in the ultimate, the best representation is an infinite number of tandem-connected T sections, each having the constants of an infinitely short section of the real line.

### Characteristic Impedance

An example of the simulation of a very long uniform transmission line by an infinite number of identical, recurrent, and symmetrical T networks is shown in Figure 5-3. The input impedance at point

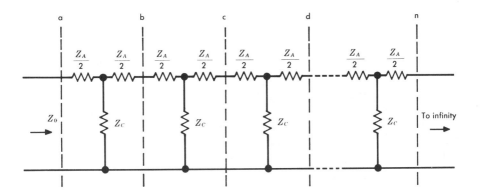

Figure 5-3. Uniform line simulated by an infinite number of identical networks.

a is $Z_0$. Now open the line at b and again measure the impedance of the line towards the right. Removing one section from an infinite number of sections produces no effect, so the impedance still measures $Z_0$.

The impedance at point a was $Z_0$ when the first section was connected to an infinite line presenting impedance $Z_0$ at b. If the first section is terminated at b by a discrete-component network of impedance $Z_0$, the impedance would still measure $Z_0$ at point a. In a transmission line, $Z_0$ is called the *characteristic impedance*. It is related to the equivalent T structure in Figure 5-2 by the expression

$$Z_0 = \sqrt{\frac{Z_A{}^2}{4} + Z_A Z_C} \qquad \text{ohms.} \qquad (5\text{-}1)$$

Impedances $Z_A$ and $Z_C$ contain inductance and capacitance. Since the reactances of inductors and capacitors are functions of frequency, the characteristic impedance of a real or simulated transmission line is also a function of frequency. This property must be recognized when selecting a network which is to terminate a line in its characteristic impedance over a band of frequencies.

It is often more convenient to determine $Z_0$ by test than by computation. This can be done by measuring the impedance presented by the line when the far end is open-circuited ($Z_{oc}$) and when it is short-circuited ($Z_{sc}$). Then, it can be shown that

$$Z_0 = \sqrt{Z_{oc} Z_{sc}} \qquad \text{ohms,} \qquad (5\text{-}2)$$

an equation similar to those given for network image impedances, Equations (4-4) and (4-5).

To summarize, every transmission line has a characteristic impedance, $Z_0$. It is determined by the materials and physical arrangement used in constructing the line. For any given type of line, $Z_0$ is by definition independent of the line length but is a function of frequency. The input impedance to a line is dependent on line length and on the termination at the far end. As the length increases, the value of the input impedance approaches the characteristic impedance irrespective of the far-end termination.

The term *characteristic impedance* is properly applied only to uniform transmission lines. The corresponding property of a discrete component network is called image impedance as previously discussed. However, if the network is symmetrical, it has a single image impedance which is analogous to the characteristic impedance of a uniform line and, as the number of network sections is increased, the impedance approaches the characteristic impedance of the line being simulated.

### Attenuation Factor

If a symmetrical T section is terminated in its image impedance $(Z_0)$ and voltage $E_1$ is applied to the input terminals, current $I_1$ flows at the input. In general, the output voltage and current, $E_2$ and $I_2$ is less than $E_1$ and $I_1$. Let $I_1/I_2$ be designated by $a$; this is the *attenuation factor for the T section*. Also, let $\ln| I_1/I_2 | = \alpha$; this is known as the *attenuation constant for the T network*. Then,

$$\left|\frac{I_1}{I_2}\right| = \left|\frac{E_1}{E_2}\right| = a = e^\alpha. \tag{5-3}$$

If a number of symmetrical T sections of image impedance $Z_0$ are connected in tandem and terminated in $Z_0$ as shown in Figure 5-4, each T section is terminated in $Z_0$, and the ratio of its input current to its output current is $a$. Thus, from the figure

$$\left|\frac{I_1}{I_2}\right| = \left|\frac{I_2}{I_3}\right| = \left|\frac{I_3}{I_4}\right| = \left|\frac{I_4}{I_5}\right| = a. \tag{5-4}$$

To find the ratio of the input current to the output current for the series, the terms of Equation (5-4) may be multiplied to give

$$\left|\frac{I_1}{I_5}\right| = \left|\frac{I_1}{I_2}\right| \cdot \left|\frac{I_2}{I_3}\right| \cdot \left|\frac{I_3}{I_4}\right| \cdot \left|\frac{I_4}{I_5}\right| = a^4$$

or, for $n$ sections of identical series-connected T networks terminated in $Z_0$ at both ends,

$$\frac{I_1}{I_{n+1}} = a^n = e^{n\alpha}, \tag{5-5}$$

where $a^n = e^{n\alpha}$ is the *attenuation factor for the n series-connected T networks*.

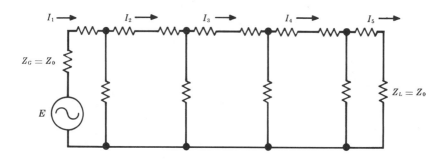

Figure 5-4. Line composed of identical T sections.

## Propagation Constant

The ratio of input to output current, $I_1/I_2$, in a symmetrical T section terminated in its image impedance, $Z_0$, is generally a complex number which may be expressed in a number of forms to indicate a change of both magnitude and phase. For the network of Figure 5-2 when it is terminated in $Z_0$ at both input and output, the current ratio is

$$\frac{I_1}{I_2} = 1 + \frac{Z_A}{2Z_C} + \frac{Z_0}{Z_C} = 1 + \frac{Z_A}{2Z_C} + \sqrt{\frac{Z_A}{Z_C} + \left(\frac{Z_A}{2Z_C}\right)^2} \quad \text{(5-6a)}$$

or, in more general terms,

$$\frac{I_1}{I_2} = e^\gamma = e^{(\alpha + j\beta)} \quad \text{(5-6b)}$$

where $\gamma$ is a complex number called the *propagation constant*, or *complex attenuation constant*. Its real and imaginary parts are defined by

$$\gamma = \alpha + j\beta, \quad \text{(5-7)}$$

where $\alpha$ is the *attenuation constant*, which represents a change in magnitude, and $\beta$ is the *wavelength constant*, or *phase constant*, which represents a change in phase. When applied to actual trans-mission lines with distributed parameters, both constants are usually expressed in terms of units of distance; $\alpha$ is usually expressed as nepers per mile or dB per mile, and $\beta$ is usually expressed as radians per mile or degrees per mile.

For the network of Figure 5-2, the value of $\gamma$ may be computed from Equations (5-6a) and (5-6b) as

$$\gamma = \ln\left[\, 1 + \frac{Z_A}{2Z_C} + \sqrt{\frac{Z_A}{Z_C} + \left(\frac{Z_A}{2Z_C}\right)^2}\,\right]. \qquad (5\text{-}8)$$

If the bracketed expression in Equation (5-8) is written in polar form as $A \angle \beta$, where $A$ is the absolute value and $\beta$ the angle, then

$$\gamma = \ln A e^{j\beta} = \ln A + j\beta. \qquad (5\text{-}9)$$

Since the real and imaginary parts of this equation must equal the corresponding parts of Equation (5-7),

$$\alpha = \ln A = \ln\left|\frac{I_1}{I_2}\right| = \ln\left|\frac{Z_A/2 + Z_C + Z_0}{Z_C}\right| \text{ nepers/T section} \quad (5\text{-}10\text{a})$$

and

$$\beta = \arg \frac{Z_A/2 + Z_C + Z_0}{Z_C} \text{ radians/T section} \qquad (5\text{-}10\text{b})$$

where *arg* stands for *argument* or *angle of.*

Since the attenuation constant can also be expressed in decibels per T section, Equation (5-10a) may be written

$$\alpha = \ln A \text{ nepers/section} = 8.686 \ln A \text{ dB/section} = 20 \log A \text{ dB/section.}$$

This relationship must not be used indiscriminately; it applies only to an infinite line and a line or discrete-component simulation terminated in $Z_0$.

## 5-2   LINE WITH DISTRIBUTED PARAMETERS

Characterization of transmission lines involves the same electrical parameters (primary and secondary constants) as those used in characterizing discrete component networks. In lines, however, these parameters are not discrete and concentrated. They are distributed uniformly along the line. Relationships between primary and secondary constants and between these parameters and other transmission line characteristics (including velocity of propagation, reflections and reflection loss, standing wave ratios, impedance matching, insertion loss, and return loss) must therefore be established in terms of distributed parameters.

## Characteristic Impedance

A single T section may represent a line having distributed elements but at one frequency only. In order to construct an artificial line in simple T-section configurations of discrete elements to simulate a real line, it is necessary to construct many tandem-connected T-sections to simulate even very short sections of line. As the number of sections is increased and the elemental length of line is reduced, the artificial line approaches the actual line in its characteristics over a wide band of frequencies.

Consider an elemental length of line, $\Delta l$. Let $Z$ be the impedance per unit length along the line and $Y$ be the admittance per unit length across the line. The T section of Figure 5-5 represents the

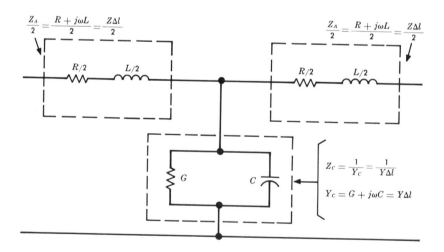

Figure 5-5. T-section equivalent of a short length of line.

equivalent network of a line having length $\Delta l$. The value of $Z_A$ and $Z_C$ of Figure 5-5 may be written

$$Z_A = Z\Delta l$$

and

$$Z_C = \frac{1}{Y\Delta l}.$$

Substituting these values in Equation (5-1) gives the characteristic impedance for a lumped constant network as

$$Z_0 = \sqrt{\frac{(Z\Delta l)^2}{4} + \frac{Z}{Y}} \qquad \text{ohms.}$$

For a line with distributed parameters, let $\Delta l$ approach zero; then, the characteristic impedance is

$$\lim_{\Delta l \to 0} Z_0 = \sqrt{\frac{Z}{Y}} \qquad \text{ohms} \qquad (5\text{-}11a)$$

or, in terms of primary constants,

$$Z_0 = \sqrt{\frac{R + j\omega L}{G + j\omega C}} \qquad \text{ohms.} \qquad (5\text{-}11b)$$

## Propagation Constant

From the previous derivation of the expression for the propagation constant of an equivalent network, Equation (5-8) may be rewritten

$$e^\gamma = 1 + \frac{Z_A}{2Z_C} + \sqrt{\frac{Z_A}{Z_C} + \left(\frac{Z_A}{2Z_C}\right)^2}.$$

By expanding the terms under the radical sign by the binominal theorem and rearranging terms, this expression may be written

$$e^\gamma = 1 + \left(\frac{Z_A}{Z_C}\right)^{\frac{1}{2}} + \frac{Z_A}{2Z_C} + \cdots \qquad (5\text{-}12)$$

Also, expanding $e^\gamma$ as a power series yields

$$e^\gamma = 1 + \gamma + \frac{\gamma^2}{2!} + \cdots \qquad (5\text{-}13)$$

As shown in Figure 5-5, the terms $Z_A$, $Z_C$, and $\gamma$ all place $\Delta l$ in the numerators of Equations (5-12) and (5-13). As $\Delta l$ approaches zero, the higher terms of these expansions become insignificant. Combining the two equations and truncating the series yields

$$1 + \gamma + \frac{\gamma^2}{2!} = 1 + \left(\frac{Z_A}{Z_C}\right)^{\frac{1}{2}} + \frac{1}{2}\left(\frac{Z_A}{Z_C}\right).$$

This equation may be solved algebraically to give

$$\gamma = \sqrt{\frac{Z_A}{Z_C}} = \sqrt{ZY}\,\Delta l$$

for the conditions of Figure 5-5 and for length $\Delta l$. For any length, $l$, made up of $l/\Delta l$ sections,

$$\gamma = \sqrt{ZY}\,l,$$

and for a unit length, $l = 1$, the propagation constant is

$$\gamma = \alpha + j\beta = \sqrt{ZY} = \sqrt{(R+j\omega L)\,(G+j\omega C)}, \qquad (5\text{-}14)$$

where $\alpha$ is in nepers or dB per unit length and $\beta$ is in radians or degrees per unit length. The values of $Z$ and $Y$ are found from the primary constants of the line.

### Attenuation Factor

Previously, the attenuation factor for a number of identical, symmetrical T sections having lumped constants was defined for tandem connections of such sections. Here, where the transmission line is made up of distributed parameters, as the length of an elemental section $\Delta l$ approaches zero, the attenuation constant becomes $\alpha$ nepers per unit length. Then the expression for the attenuation factor analogous to that of Equation (5-5) is given as

$$\text{Attenuation factor} = e^{\alpha l}. \qquad (5\text{-}15)$$

## Velocity of Propagation

The phase shifts represented by the $\beta$ term in Equation (5-6b) express the angular difference in radians between the input and output signals; they imply a time delay between the input signal current (or voltage) and the output signal current (or voltage). This can be used to compute the velocity of propagation through the network or transmission line. The velocity is given by

$$v = \frac{\omega}{\beta},\qquad(5\text{-}16)$$

typically expressed as miles per second.

Equation (5-16) shows that the velocity of propagation is a function of frequency, since $\omega = 2\pi f$. However, $\beta$ is also frequency-dependent since it is made up of reactances derived from $Z_A$ and $Z_C$ of Equation (5-8). Thus, while a transmission line having either discrete or distributed elements tends to introduce delay distortion (a non-linear phase/frequency characteristic) because the velocity of propagation is different at different frequencies, it is theoretically possible to design a line having no delay distortion by designing $\beta$ to be directly proportional to $\omega$.

## Reflections

Only lines which are uniform and which are terminated in their characteristic impedance have so far been considered. As long as the signal is presented with the same impedance at all points in a connection, the only loss is attenuation.

However, if one line with characteristic impedance $Z_G$ is joined to a second line with characteristic impedance $Z_L$, an additional transmission loss is observed. While the signal is traveling in the first line, it has a voltage-to-current ratio $E_G/I_G = Z_G$. Before the signal can enter the second line, it must adjust to a new voltage-to-current ratio $E_L/I_L = Z_L$. In making this adjustment, a portion of the signal is reflected back towards the sending end of the connection.

It is not surprising that there should be a reflection at an abrupt change in the electrical characteristics of a line. There is a disturbance in any form of wave energy at a discontinuity in the transmission medium. For example, sound is reflected from a cliff; light is reflected by a mirror. These conditions are equivalent to a line terminated in either an open circuit or a short circuit; all of the energy in the incident wave is reflected. A less abrupt change in impedance would cause a partial reflection. For example, a landscape is mirrored in the surface of a pool of water because part of the light falling on the water is reflected. The bottom of the pool is also visible if the pool is not too deep, since part of the light falling on the water passes through the discontinuity of the air-water junction and illuminates the bottom. Such a partial reflection occurs when two circuits with different impedances are joined, as in Figure 5-6. The power,

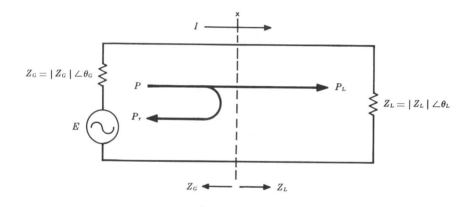

Figure 5-6. Reflection at an impedance discontinuity.

$P$, in the signal arriving at junction x is divided. A portion of the signal, $P_L$, is transmitted through the junction to the load $Z_L$. The remainder of the signal, $P_r$, is reflected and travels back towards the source. If $Z_G$ and $Z_L$ were equal, the power, $P$, would all be delivered to $Z_L$; there would be no reflection.

**Reflection Loss.** A concept frequently used to describe the effect of reflections is that of *reflection loss*, which is defined as the difference in dB between the power that is actually transferred from one circuit to the next and the power that would be transferred if the impedance of the second circuit were identical to that of the first.

Consider the circuit of Figure 5-6 where the impedances do not match. The current in the circuit is $I = E/(Z_G + Z_L)$. The power delivered to $Z_L$, the load, is

$$P_L = I^2 R_L = \frac{E^2 R_L}{(Z_G + Z_L)^2},$$

where $R_L$ is the resistive component of the load impedance. If impedance $Z_L$ matched $Z_G$, that is, $Z_L = Z_G$, the current in the circuit would be $I_m = E/2Z_G$, and the power delivered to the load would be

$$P_m = I_m^2 R_G = \frac{E^2 R_G}{4Z_G^2}.$$

The ratio of the two values of power is

$$\frac{P_m}{P_L} = \frac{(Z_G + Z_L)^2}{4Z_G^2} \times \frac{R_G}{R_L}.$$

Thus, the reflection loss may be written

$$\text{Reflection loss} = 10 \log \frac{P_m}{P_L} = 20 \log \sqrt{\frac{P_m}{P_L}}$$

$$= 20 \log \left| \frac{Z_G + Z_L}{2Z_G} \sqrt{\frac{R_G}{R_L}} \right|$$

$$= 20 \log \left| \frac{Z_G + Z_L}{2\sqrt{Z_G Z_L}} \sqrt{\left| \frac{R_G}{Z_G} \right| \cdot \left| \frac{Z_L}{R_L} \right|} \right|$$

$$= 20 \log \left| \frac{Z_G + Z_L}{2\sqrt{Z_G Z_L}} \right| + 20 \log \sqrt{\frac{\cos \theta_G}{\cos \theta_L}} \qquad \text{dB.}$$

$$(5\text{-}17)$$

Thus, the reflection loss has two components. The first is related to the inverse of the *reflection factor*, defined as $K$, where

$$K = \left| \frac{2\sqrt{Z_G Z_L}}{Z_G + Z_L} \right|. \qquad (5\text{-}18)$$

The second is dependent on the angular relationships between the two mismatched circuits.

It is possible to have negative reflection loss, or reflection gain. This does not mean that power can be generated at an impedance discontinuity. It results from the choice of identical impedances as the reference condition for zero reflection loss which is not the condition for maximum power transfer, as pointed out in Chapter 4.

**Standing Wave Ratio.** A second concept that is useful in describing the effect of reflections is that of a *standing wave ratio*. This concept may be approached from the point of view of a theoretically lossless transmission line.

Equation (5-11) gives the expression for the characteristic impedance of a line as

$$Z_0 = \sqrt{\frac{R + j\omega L}{G + j\omega C}} \qquad \text{ohms.}$$

In this equation, the components that produce loss are the resistance and conductance, $R$ and $G$. When these are negligible relative to $j\omega L$ and $j\omega C$, they may be ignored. This is often true at very high frequencies because of the $\omega$ terms in the expression for $Z_0$. Under these conditions, the characteristic impedance reduces to

$$Z_0 = \sqrt{L/C} \qquad \text{ohms.} \tag{5-19}$$

When the $R$ and $G$ terms are negligible, the propagation constant for the lossless line is determined from Equation (5-14) to be

$$\gamma = j\omega \sqrt{LC}. \tag{5-20}$$

Thus, Equations (5-19) and (5-20) show that for a lossless line the characteristic impedance is a pure real number and the propagation constant is a pure imaginary number. When the propagation constant is expressed as $\gamma = \alpha + j\beta$, the value of attenuation then becomes $\alpha = 0$, and the value of the phase shift constant becomes $\beta = \omega \sqrt{LC}$.

The voltage at any point, $x$, on a lossless transmission line may be shown to be

$$V_x = V_1 \, e^{j\beta x} + V_2 \, e^{-j\beta x} \text{ volts,} \tag{5-21}$$

where $V_1$ and $V_2$ represent the incident and reflected voltages, respectively, when the line is *not* terminated in its characteristic impedance, $Z_0$, and where $x$ is the distance from the *load* to the point of measurement [5].

The velocity of propagation may be found for the lossless line by substituting the value $\beta = \omega \sqrt{LC}$ in Equation (5-16). Thus,

$$v = \frac{\omega}{\beta} = \frac{1}{\sqrt{LC}} \ .$$

For some structures (such as a coaxial transmission line), the velocity of propagation approaches 186,300 miles per second, the velocity of light propagation.

It can be seen from Equation (5-21) that voltage $V_x$ is represented as the sum of two traveling waves, the incident and reflected voltage waves. The voltage may also be expressed in terms of trigonometric functions [5]:

$$V_x = V_L \left( \cos \beta x + j \frac{Z_0}{Z_L} \sin \beta x \right)$$

$$= V_L \cos \beta x + j I_L Z_0 \sin \beta x, \qquad (5\text{-}22)$$

where $I_L$, $V_L$, and $Z_L$ are, respectively, the current and voltage at the load and the impedance of the load.

For a given frequency and value of $x$ (distance from the load), Equation (5-22) can show $V_x = 0$, e.g., when $V_L = 0$ or when $\cos \beta x + j (Z_0/Z_L) \sin \beta x = 0$. This can occur when $\beta x = n\pi$ ($n$, an odd integer) and when $Z_L$ is simultaneously infinite, an open-circuit termination. It can also occur when $Z_L = 0$ (short circuit) and $\beta x = n\pi$ ($n$, an even integer) simultaneously. These are two cases of special interest which produce *standing waves*, i.e., waves which do not propagate along the line but which pulsate between minimum and maximum values at all points except those at which $V_x = 0$.

In the more general case of a line having loss and not terminating in $Z_0$, standing waves are also produced but not necessarily with null points at which $V_x = 0$. The ratio of maximum to minimum voltages in the envelope along the line is known as the *voltage standing wave ratio* (VSWR). The VSWR may vary from one to infinity. When a line is terminated in its characteristic impedance, there is no reflected wave and, as a result, the maximum and minimum are the same value to give the ratio of unity. For the boundary conditions of $Z_L = 0$ or $\infty$ or for $\alpha = 0$, the minima of the envelope are nulls, the value of $V_x = 0$, and the VSWR is infinite.

For cases not involving the boundary conditions of $Z_L = 0$, $Z_L = \infty$, and $\alpha = 0$, the transmission phenomenon can often be analyzed to advantage in terms of a combination of traveling and standing waves; then, the VSWR is used as a measure of the required degree of impedance match. If it is sometimes convenient to perform analyses in terms of currents instead of voltages, analogous expressions are used.

**Impedance Matching.** An impedance discontinuity can sometimes be eliminated by introducing a transformer as an impedance matching device at the junction. In Figure 5-7, the impedance to the left of x, $Z_1$, does not match the impedance to the right of y, $Z_2$. By connecting a transformer of turns ratio, $N_x/N_y = \sqrt{Z_1/Z_2}$, into the circuit between x and y, the line to the left of x is made to look into an impedance $Z_1$, while the line to the right of y looks back into $Z_2$. In

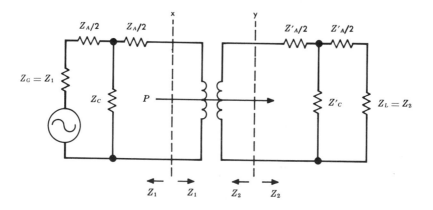

Figure 5-7. Unequal impedances matched by a transformer.

practice, such a transformer would have a loss of a fraction of a dB, but the reflection loss is reduced to near zero. Typical examples of this technique are the transformer in a telephone station set and the input and output transformers in a repeater amplifier.

A network (or pad) with image impedances of $Z_1$ and $Z_2$ would give the same result as the transformer in Figure 5-7. Impedance matching pads find limited application because they have a minimum loss determined by their image impedance ratio. For example, a pad with image impedances of 600 and 500 ohms (a ratio of 1.2) would have a loss of at least 3.75 dB; one with a ratio of 2 would have a minimum loss of about 8 dB.

Advantage is often taken of the standing wave phenomenon in high-frequency transmission lines. A proper connection of a short-circuited stub of line onto a transmission line at the proper point, one-quarter or one-half wavelength from the load, often provides a means for good impedance matching.*

At very high frequencies, in addition to the short-circuited stub technique it is sometimes practical to match impedances by introducing a section of transmission line with gradually changing dimensions. The most common application of this technique is the tapered open-wire line between a TV antenna and the twin lead running to the receiver.

**Insertion Loss.** Insertion loss, discussed in Chapters 3 and 4, may be defined as the loss resulting from the insertion of a network between a source and a load. Further consideration of this factor is desirable because important contributions to insertion loss occur as a result of reflections due to impedance mismatches.

Consider the circuits of Figure 5-8. If the four-terminal network or transmission line is not present, as in Figure 5-8(a), the current supplied to $Z_L$ is

$$I_1 = E/(Z_G + Z_L).$$

---

*Standing wave ratios, reflection coefficients, wavelengths, complex impedances, and other transmission relationships are conveniently computed by Smith charts, nomographic diagrams that display these relationships in a convenient form [4, 5, and 7]. Accurate and extensive calculations of these values and relationships are today carried out on an electronic computer.

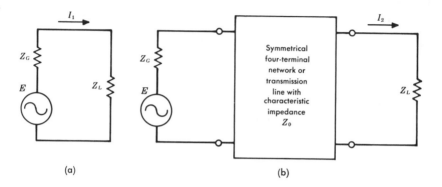

Figure 5-8. Circuits for insertion loss analysis.

When the network is inserted in the circuit, Figure 5-8(b), the current in $Z_L$ may be shown to be

$$I_2 = \frac{2E\,Z_0\,e^{-\gamma}}{(Z_0 + Z_G)\,(Z_0 + Z_L) - (Z_0 - Z_G)\,(Z_0 - Z_L)\,e^{-2\gamma}}.$$

Then,

$$\frac{I_1}{I_2} = \frac{(Z_0 + Z_G)\,(Z_0 + Z_L) - (Z_0 - Z_G)\,(Z_0 - Z_L)\,e^{-2\gamma}}{2\,(Z_G + Z_L)\,Z_0\,e^{-\gamma}}$$

which, when factored, gives

$$\frac{I_1}{I_2} = \left[\frac{(Z_0 + Z_G)\,(Z_0 + Z_L)}{2\,(Z_G + Z_L)\,Z_0\,e^{-\gamma}}\right]\left[1 - \frac{(Z_0 - Z_G)\,(Z_0 - Z_L)\,e^{-2\gamma}}{(Z_0 + Z_G)\,(Z_0 + Z_L)}\right]. \quad (5\text{-}23)$$

The second bracketed term is usually neglected. It results from interactions due to network termination impedances not being the proper image impedances for the network. If the network is a transmission line of any significant length, the term $e^{-2\gamma}$ makes the interaction effect negligible. If either $Z_G$ or $Z_L$ is equal to $Z_0$ or, in the case of a network, is the true image impedance, the interaction effect vanishes completely.

Thus, the insertion loss is usually written,

$$\text{Insertion loss} \approx 20 \log \left| \frac{(Z_0 + Z_G)(Z_0 + Z_L)}{2(Z_G + Z_L) Z_0 e^{-\gamma}} \right| \qquad \text{dB.}$$

The right side of this equation may be rearranged to give

$$\text{Insertion loss} \approx 20 \log \left| \frac{(Z_0 + Z_G)}{2\sqrt{Z_0 Z_G}} \right| + 20 \log \left| \frac{(Z_0 + Z_L)}{2\sqrt{Z_0 Z_L}} \right|$$

$$- 20 \log \left| \frac{Z_G + Z_L}{2\sqrt{Z_G Z_L}} \right| + 8.686\,\alpha \qquad \text{dB.}$$

$$(5\text{-}24)$$

where $\alpha$ is the real component of the propagation constant $\gamma$.

The approximation to the insertion loss given in Equation (5-24) contains three reflection terms and the attenuation constant, $\alpha$. The three reflection terms are seen to be related to the reflection factor defined in Equation (5-18). The first two of these increase the insertion loss as a result of the mismatch between either $Z_0$ and $Z_G$ or $Z_0$ and $Z_L$. The third is a negative term representing the reflection loss due to the mismatch between $Z_G$ and $Z_L$ when the intermediate network is not present.

Terms corresponding to the term $20 \log \sqrt{\cos \theta_G / \cos \theta_L}$ of Equation (5-17) do not appear in the equation for insertion loss because, in the insertion loss definition, the ratio of currents (or, more precisely, powers) is taken with respect to the same impedance, $Z_L$. Thus, the power factor term reduces to unity.

If the transmission line is short or for any other reason has very low loss, the insertion loss expression of Equation (5-24) must be modified by the interaction (second bracketed) term in Equation (5-23). The evaluation of this equation is usually quite laborious but computer programs, tables, and nomographs are available to simplify computations.

**Return Loss.** In the design of two-wire circuits, the amount of signal reflected at a junction is of interest because the reflected energy represents an echo of which the amplitude must be controlled. As discussed in Chapter 4, the difference in dB between the incident current or voltage and the reflected current or voltage at an impedance discontinuity is called *return loss*. If the two impedances at a junction are matched, the return loss is infinite, since there is no energy reflected. The greater the difference between impedances on each side of a junction, the lower the return loss. At the junction of two lines of impedance $Z_1$ and $Z_2$,

$$\text{Return loss} = 20 \log \frac{1}{|\rho|} = 20 \log \left| \frac{Z_1 + Z_2}{Z_1 - Z_2} \right| \qquad \text{dB}, \qquad (5\text{-}25)$$

where $\rho$ is the reflection coefficient.

In the telephone plant, a serious source of low return loss is the interface between two-wire and four-wire facilities; four-wire terminating sets contain a compromise balancing network that may not match the impedance of the office wiring or other connected two-wire circuits. The most serious problem occurs at class 5 switching offices where toll connecting trunks are switched to a variety of loops with a wide range of impedances.

In the Bell System, the term *structural return loss* is commonly used as a measure of the departure of the characteristic impedance of a transmission medium from its design or nominal value. The term is descriptive of the fact that such impedance departures are primarily due to systematic and repetitive deformations of the physical structure of the medium. The values of $Z$ used in Equation (5-25) may be regarded as the nominal and measured values of the characteristic impedance of the medium to determine the structural return loss.

## 5-3  LOADED LINES

In the previous discussion of velocity of propagation, it was pointed out that transmission lines normally introduce delay distortion. It was further suggested that delay distortion can be theoretically eliminated by design. An understanding of the theoretical basis for such a design is of some interest. Of far great practical importance is the application of the relationships involved to the reduction of attenuation over most of the voice-frequency band.

### Analysis

To achieve the theoretical design of a distortionless line, it is necessary that the series impedance, $Z$, and the shunt admittance, $Y$, have the same angle. This requirement may be expressed

$$\frac{\omega L}{R} = \frac{\omega C}{G},$$

or

$$LG = RC. \tag{5-26}$$

The impedance and admittance may be written $Z = R + j\omega L$ and $Y = G + j\omega C$, respectively. Thus,

$$Y = \frac{RC}{L} + \frac{j\omega LG}{R} = \frac{G}{R} \, Z. \tag{5-27}$$

When this expression is substituted in Equation (5-14),

$$\gamma = \sqrt{ZY} = Z\sqrt{G/R} = Z\sqrt{C/L} \quad . \tag{5-28}$$

Then,

$$\gamma = \alpha + j\beta = \sqrt{G/R} \, (R + j\omega L)$$

and

$$\alpha = \sqrt{RG} = R\sqrt{C/L} \quad , \tag{5-29}$$

$$\beta = \omega L\sqrt{G/R} = \omega\sqrt{LC} \quad , \tag{5-30}$$

$$v = \omega/\beta = 1/\sqrt{LC} \quad . \tag{5-31}$$

With Equations (5-26) and (5-27) substituted in Equation (5-11a),

$$Z_0 = \sqrt{Z/Y} = \sqrt{R/G} = \sqrt{L/C} \quad \text{ohms.} \tag{5-32}$$

Thus, when $\omega L/R = \omega C/G$, the attenuation $(\alpha)$, the velocity $(v)$, and the characteristic impedance $(Z_0)$ are independent of frequency, and $Z_0$ is a pure resistance. Such a line, terminated in its characteristic impedance, has no loss distortion or delay distortion. Note, however, that the attenuation and velocity both decrease, and the characteristic impedance increases.

Unfortunately, a transmission line having such optimum characteristics is not readily attainable in practice. In transmission lines made up of pairs in well maintained cables, the value of the conductance, $G$, is very small. It is not desirable to increase it artificially because that would increase the attenuation correspondingly as indicated in Equation (5-29). The value of capacitance cannot be changed appreciably because of practical considerations of spacing between conductors. To approach the optimum condition where $LG = RC$ as indicated by Equation (5-26), it would be necessary either to increase the inductance, $L$, or to reduce the resistance $R$. The latter is not economical (it may be accomplished by increasing the size of the wire), and so the only remaining alternative is to increase the inductance. This practice, known as *inductive loading,* is used to approach the conditions sought, especially to reduce the line attenuation.

### Inductive Loading

In considering the effect of inductive loading, note that the configuration of the T section of Figure 5-5 is basically that of a low-pass filter as illustrated in Figure 4-29. The critical frequency of such a structure is the frequency below which there is very little attenuation (ideally none) and above which the attenuation increases very rapidly. For the structure of Figure 5-5 in which $G = 0$, this frequency is

$$f_c = \frac{1}{\pi \sqrt{LC}} \text{ Hz.} \tag{5-33}$$

When applied to lines loaded with discrete elements, the value of $L$ is the load coil inductance. Although the inductive component of the line impedance of the load section should be added, it is usually negligible. Similarly, the value of $C$ to be used is the primary constant value of capacitance for the medium multiplied by the length of the load section. In theory, this value should be increased by the capacitance of the load coil, but this also is usually negligible.

As previously mentioned, series inductance may be added to reduce the attenuation in a cable pair. Below the critical frequency, $f_c$, the attenuation is reduced as indicated by Equation (5-29). Unfortunately, line inductance cannot be increased by loading without increasing resistance by virtue of the wire used in the load coils. Because of the resistance increase and other frequency-dependent limitations in the application of inductive loading, the attenuation is, in practice, more nearly $\alpha = (R\sqrt{C/L})/2$ than the value of Equation (5-29). Above the critical frequency, the attenuation increases rapidly. The effect of increasing $L$ by installing load coils in a line is illustrated by Figure 5-9.

**Loading Methods.** Loading may be accomplished by either of two practical methods. The first, called continuous loading, involves placing magnetic material (e.g., permalloy tape) around the copper conductors. This method is expensive and has been employed in only a few cases on submarine cable installations. The second method uses discrete inductances introduced along the line at regular intervals.

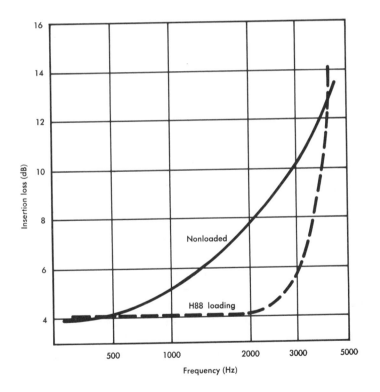

Figure 5-9. Insertion loss of 12,000 feet of 26-gauge cable measured between terminations of 900 ohms in series with 2 $\mu$f.

Below the critical frequency, the transmission line analysis for such an arrangement is reasonably accurate on the assumption that such components are uniformly distributed along the line but this assumption does not hold above the critical frequency.

Load Coil Spacing. Load coils introduce impedance discontinuities into an otherwise smooth or uniform line. This effect is minimized by making the spacing between load coils short compared to the wavelength of the transmitted signal and by spacing the coils precisely along the line. Imprecise coil spacing and coil resistance both introduce transmission irregularities in the passband primarily as a result of the deterioration of the structural return loss of the medium. General rules for the precision of spacing of load coils have been worked out, and manufacturing limits on the allowable variation in coil parameters are imposed. Corrective measures to overcome situations that, due to geographical or other considerations, cannot be adjusted to meet requirements on the uniformity of coil spacing are given in the form of building-out network specifications. General rules on allowable spacing deviations as applied to interlocal trunks are as follows:

(1) The deviation of the average spacing from the nominal or standard value should not exceed ±2 percent.

(2) The deviations of the length of individual sections from the average section length should not exceed ±3 percent.

(3) The percentage of deviation of each section length from the average section length should be determined and the numerical average of these percentages, disregarding signs, should be 1.2 percent or less, where practicable.

End sections, nominally designed as one-half the standard length, are frequently built out by the use of additional discrete components to correct the electrical length of the section (overcoming natural length discrepancies) or to correct the impedance of the structure to conform with impedance matching requirements at the central office.

A number of loading arrangements have been standardized in the Bell System. The spacing between coils, or loading-section length, is designated by a series of code letters. The codes for the more common spacings are as follows:

| | | | |
|---|---|---|---|
| B | 3000 feet | M | 9000 feet |
| C | 929 feet | X | 680 feet |
| D | 4500 feet | Y | 2130 feet |
| E | 5575 feet | Z | 5280 feet |
| H | 6000 feet | | |

This spacing code is combined with numerals to designate wire gauge and load inductance values. For example, the cable of Figure 5-9 is designated as 26H88 loading, indicating 26-gauge wire, 6000-foot spacing, and load coils of 88 millihenries inductance. This inductance is equally divided between two coils wound on a toroidal core. Each coil is connected in one side of the line in such a manner that the two inductances add (series aiding) to give the required value. These loading arrangements have been designed to achieve the greatest practicable reduction in attenuation and not to achieve a minimum of delay distortion. Thus, the criterion for minimum delay distortion given in Equation (5-26) is not met.

## 5-4   COAXIAL CABLE

Consideration of transmission lines thus far has been confined to lines made up of two parallel wire conductors. However, a coaxial configuration of conductors may be used to advantage where high and very high frequencies are involved. The conducting pair consists of a cylindrical tube in which is centered a wire as shown in

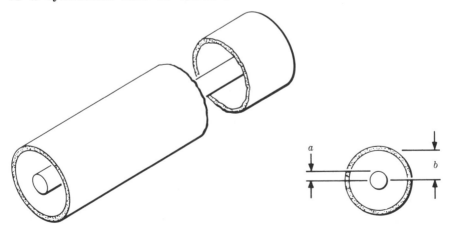

Figure 5-10. Coaxial cable.

Figure 5-10. In practice, the central wire is held in place quite accurately by insulating material which may take the form of a solid or plastic foam core, discs or beads strung along the center conductor, or a spirally wrapped plastic string. In such a conducting pair, equal and opposite currents flow in the central wire and the outer tube, just as equal and opposite currents flow in the more ordinary parallel wires.

At high frequencies, a unit length of coaxial in which the dielectric loss in the insulation is negligible (effectively gaseous) would have an inductance which is about one-half the inductance of two parallel wires separated by a distance equal to the radius of the coaxial tube. The capacitance of the same coaxial is approximately twice that of two parallel wires separated by the same distance and having the same diameter as that of the central coaxial conductor. If the outside radius of the central conductor is designated $a$ and the internal radius of the tube is $b$, the characteristic impedance at high frequencies neglecting leakage is approximately

$$Z_0 = \sqrt{\frac{L}{C}} = 138 \log \frac{b}{a} \qquad \text{ohms.} \qquad (5\text{-}34)$$

The attenuation constant per mile, where both conductors are of the same material, varies as the square root of frequency and is approximately

$$\alpha = \frac{R}{2Z_0} = 24.4 \times 10^{-6} \left[ \frac{\sqrt{f}\left(\dfrac{1}{a} + \dfrac{1}{b}\right)}{\log \dfrac{b}{a}} \right] \text{nepers/mile} \qquad (5\text{-}35)$$

where $a$ and $b$ are in centimeters. From Equation (5-35) it may be determined that minimum attenuation is obtained when the coaxial is so designed that $b/a = 3.6$. With this configuration, $Z_0$ is about 77 ohms. The attenuation varies with temperature by approximately 0.11 percent per degree Fahrenheit.

The present standard coaxial used for transmission in the Bell System employs a copper tube 0.369 inches in inside diameter and a copper center wire 0.1003 inches in diameter. This, it will be noted, approximates the optimum ratio specified above for minimum attenuation. The nominal impedance is 75 ohms, somewhat lower than would be computed by use of Equation (5-34). This is because the insulation, which is a composite of air and polyethylene discs, has a dielectric constant of about 1.1. Velocity of propagation in the coaxial approaches closely the speed of light. A study of the basic characteristics of the coaxial shows that at the high frequencies assumed, the attenuation is substantially less than that of a parallel wire line of comparable dimensions. The attenuation is approximately 3.95 dB/mile at 1 MHz and at 20°C. More important is the fact that at frequencies of interest the shielding effect of the outer cylindrical conductor prevents interference from external sources of electric energy. The shielding effect also prevents radiation losses of the energy being transmitted over the coaxial. Thus, crosstalk between coaxials is minimum.

#### REFERENCES

1. Anner, G. E. and W. L. Everitt. *Communication Engineering* (New York: McGraw-Hill Book Company, Inc., 1956).

2. Schelkunoff, S. A. *Electromagnetic Waves* (Princeton, N. J.: D. Van Nostrand Company, Inc., 1943).

3. Johnson, W. C. *Transmission Lines and Networks* (New York: McGraw-Hill Book Company, Inc., 1950).

4. Ware, L. A. and H. R. Reed. *Communication Circuits* (New York: John Wiley and Sons, Inc., 1955).

5. Skilling, H. H. *Electric Transmission Lines* (New York: McGraw-Hill Book Company, Inc., 1951).

6. Creamer, W. J. *Communication Networks and Lines* (New York: Harper and Brothers, 1951).

7. Smith, P. H. "Transmission Line Calculator," *Electronics* (Jan. 1939 and Jan. 1944).

Chapter 6

# Wave Analysis

A signal may often be represented by a function whose value is specified at every instant of time. In transmission work, however, this type of characterization is not always the most convenient to use because information about transmission lines and networks is usually presented as functions of frequency rather than of time. Therefore, a method is needed for translating between time-domain and frequency-domain expressions of signal and network characteristics. It is possible to pass from one domain to the other by using mathematical transformations. This ability is useful in providing answers to questions which arise when a pulse signal expressed as a time-domain function is applied to a transmission line whose characteristics are known in terms of frequency-domain functions. These questions might be: How does the pulse look at the output of the line? What are the frequency spectra of pulse signals as functions of pulse repetition rates and duty cycles? What is the resultant energy distribution? What are the bandwidth considerations?

The duality between frequency and time domains in describing signals and linear networks is a concept that can become so familiar that a person may often unconsciously transfer from one domain to the other without effort. For instance, a sine wave of frequency $f_0$ might be pictured in the time domain as a curve which crosses the time axis $2f_0$ times per second or in the frequency domain as a narrow spike located at a point $f = f_0$ on the frequency axis and characterized by two numbers giving its amplitude and phase. In this simple case, a method of passing from one of these representations to the other is simple to visualize and to formulate. The frequency, amplitude, and phase can be obtained from a time-domain representation of a sinusoidal wave by merely counting and measuring appropriate dimensions; on the other hand, if the frequency, amplitude, and phase are given, the time-domain waveform can be constructed. However, for more complicated waveforms, this transformation is not so simple, and a well defined mathematical procedure must be employed to pass from one domain to the other.

The *Fourier transform pair* is the mathematical formulation of this useful concept; as such, it is indispensable for dealing with signals and linear networks. A review of some of the important properties of the Fourier transform and illustrations of its uses can give a qualitative understanding of its meaning and application. The reader is referred to standard mathematics texts for more rigorous treatments of the subject.

## 6-1  INSTRUMENTATION

To complement the mathematical procedures of defining signals, interferences, or networks by means of the Fourier transform pair, field or laboratory observations are often needed to study engineering problems of maintenance, design, or performance evaluations. Two types of instrumentation are commonly used to display signals and interferences in the time or frequency domain for visual study. These are the cathode ray oscilloscope, which displays a signal or an interference in the time domain, and the spectrum analyzer, which displays signal or interference components in a frequency band.

Figure 6-1 illustrates, in a simple block diagram, the operation of a cathode ray oscilloscope. The position of the cathode ray spot on the tube face is a function of the voltages impressed on the horizontal and vertical deflection plates. The output signal of the sawtooth generator is impressed on the horizontal deflection plates and causes the spot to move repetitively from left to right. The signal under study, usually a periodic time function, is applied to the vertical deflection plates. This causes the spot to move vertically in accordance with the voltage applied and, when the two signals are properly synchronized, a time-domain representation of the test signal waveform is traced on the tube face.

The operation of a spectrum analyzer is somewhat more complicated. A sawtooth signal is used to deflect the spot from left to right

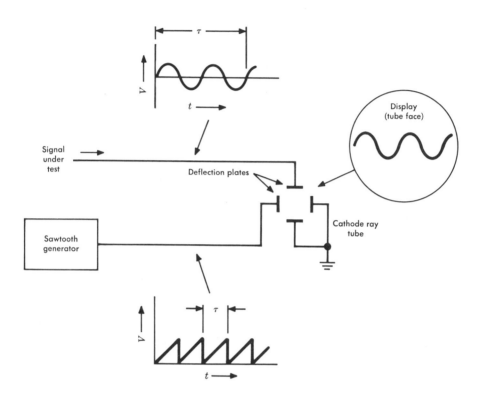

Figure 6-1. Cathode ray oscilloscope operation (time domain).

on the tube face as in the cathode ray oscilloscope. However, the signal applied to the vertical plates must be subjected to a number of transformations before it can be so used. The signal consists of a broad band of frequencies illustrated by the band from $f_B$ to $f_T$ in Figure 6-2. This signal is impressed at the input to the analyzer and mixed with the output of the tunable oscillator. This oscillator is driven by the sawtooth generator. Its output signal varies in frequency to sweep across the band from $f_B$ to $f_T$ in a repetitive fashion as the output voltage of the sawtooth generator increases from its minimum to its maximum value. This process converts the signal, in effect, from a voltage-frequency function to a voltage-time function. The output of the mixer, however, has many unwanted components. A bandpass filter is used to select the desired component, which is then impressed on the vertical deflection plates of the cathode ray

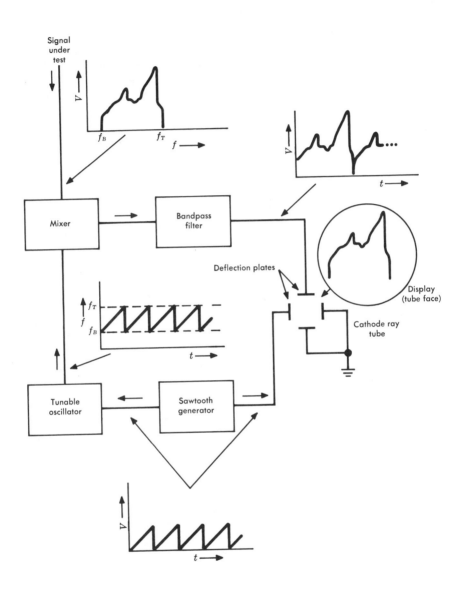

Figure 6-2. Spectrum analyzer operation (frequency domain).

tube. The filtering is usually accomplished through several stages of intermediate frequencies where the bandlimiting is more easily effected.

## 6-2 PERIODIC SIGNALS

The alternative descriptions of periodic signals in the time and frequency domains are based upon the fact that when sine waves of appropriate frequencies, phases, and amplitudes are combined, their sum can be made to approximate any periodic signal. Similarly, any one of these signals can be decomposed into its component sine waves.

### Fourier Series Representation

A good starting point for discussing wave analysis is the familiar Fourier series representation of periodic functions given by

$$f(t) = \frac{A_0}{2} + \sum_{n=1}^{\infty} (A_n \cos n\omega_0 t + B_n \sin n\omega_0 t), \qquad (6\text{-}1)$$

where $A_0$, $A_n$, and $B_n$ are constants that may be computed by the following equations [1]:

$$A_0 = \frac{1}{\pi} \int_0^{2\pi} f(t)\, dt, \qquad (6\text{-}2)$$

$$A_n = \frac{1}{\pi} \int_0^{2\pi} f(t) \cos nt\, dt, \qquad (6\text{-}3)$$

and

$$B_n = \frac{1}{\pi} \int_0^{2\pi} f(t) \sin nt\, dt. \qquad (6\text{-}4)$$

The interval over which the integration is performed, 0 to $2\pi$, is the fundamental period of the function $f(t)$.

The validity of the Fourier series may be demonstrated by displaying a square wave simultaneously on an oscilloscope and on a spectrum analyzer. The spectral components of such a signal may be filtered and displayed on the two instruments in various combinations. To illustrate such a demonstration, Equation (6-1) may be rewritten

$$f(t) = \frac{C_0}{2} + \sum_{n=1}^{\infty} C_n \cos (n\omega_0 t + \phi_n), \tag{6-5}$$

where

$$C_0 = A_0,$$

$$C_n = (A_n^2 + B_n^2)^{1/2},$$

$$\cos \phi_n = A_n/C_n,$$

$$\sin \phi_n = -B_n/C_n.$$

Figure 6-3 illustrates various displays that might be observed; sketches of oscilloscope patterns are at the left and spectrum analyzer displays at the right. In Figure 6-3(a), the output of a square-wave generator is shown at the left. The period of the wave is $1/\omega_0 = T$ seconds. The wave is shown as having an amplitude of unity ($A = 1$) and a pulse width of $T/2$. The corresponding spectrum analyzer display shows a component at as many odd harmonics of the fundamental as are impressed at the input to the analyzer. In the illustration, harmonics are shown up to the eleventh.

Figures 6-3(b), 6-3(c), and 6-3(d) illustrate the displays when the inputs to the measuring sets are limited to the fundamental, third, and fifth harmonics, respectively. It is interesting to note how quickly the oscilloscope display approaches the original square wave.

Consideration of Figure 6-3 and Equation (6-5) shows how the Fourier expansion for the square wave, $f(t)$, may be used to determine certain requirements on a channel that is to be used to transmit the square wave. The extent to which pulse distortion can be tolerated determines the number of signal components that must be transmitted and, therefore, the bandwidth that must be provided. Further detailed study would also show how much distortion (gain and/or phase) can be tolerated. The idealized sketches of Figure 6-3

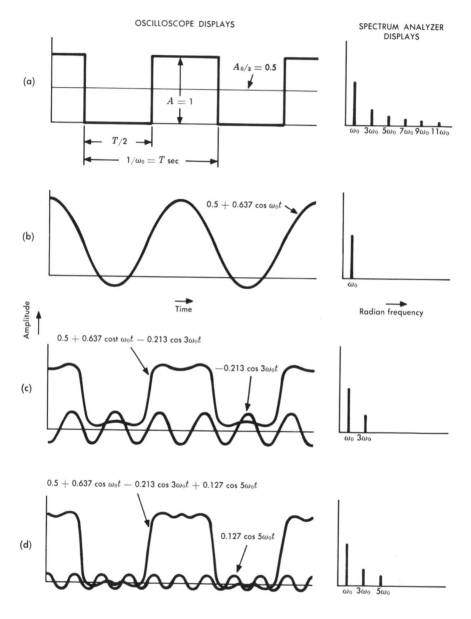

Figure 6-3. Fourier components of a square wave.

indicate no distortion. Gain distortion would change desired relationships among the amplitudes of the signal components, and phase distortion would cause relative shifts of the components along the time axis. Such shifts would also cause distortion of the pulse.

**Symmetry.** The Fourier analysis of certain periodic waveforms can frequently be simplified by observing properties of symmetry in the waveform and by selecting the coordinates about which the waveform varies to take maximum advantage of the observed symmetry properties. It can be shown that, by proper choice of axes, one or more of the coefficients can always be made zero; however, if more than one is to be made zero, the waveform *must* exhibit odd or even symmetry. It is desirable to define these properties of symmetry and to illustrate them mathematically and graphically because by taking advantage of such properties it is possible to reduce greatly the cumbersome mathematics sometimes necessary to evaluate the coefficients $A_n$ and $B_n$ of Equation (6-1).

Periodic functions exhibiting *odd symmetry* have the mathematical property that

$$f(-t) = -f(t). \tag{6-6}$$

That is, the shape of the function, when plotted, is identical for positive and negative values of time, but there is a reversal of sign for corresponding values of positive or negative time. A familiar function exhibiting this property is the sine function. Figure 6-4(a) also illustrates a function having odd symmetry.

A function having odd symmetry contains no cosine terms and, in addition, contains no dc component. Thus, in Equation (6-1), since $A_n = 0$ and $A_0 = 0$, the Fourier series is written

$$f(t)_{\text{odd}} = \sum_{n=1}^{\infty} B_n \sin n\,\omega_0 t. \tag{6-7}$$

Graphically, the function can be seen to have odd symmetry by folding the right side of the time axis over upon the left side and then rotating the folded half 180 degrees about the abscissa, which must be selected as the dc component of the waveform. When the function

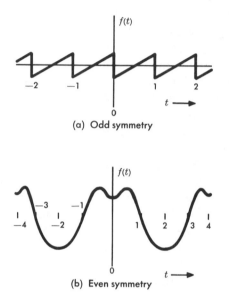

(a) Odd symmetry

(b) Even symmetry

Figure 6-4. Symmetrical functions.

is folded and rotated as indicated, the folded portion is superimposed directly on the unfolded function for negative time.

A function having *even symmetry* contains no sine terms. That is

$$B_n = 0$$

in Equation (6-1). In this case, the Fourier series is written

$$f(t)_{\text{even}} = \frac{A_0}{2} + \sum_{n=1}^{\infty} A_n \cos n\, \omega_0 t. \qquad (6\text{-}8)$$

The mathematical property that such a function exhibits is that

$$f(t) = f(-t). \qquad (6\text{-}9)$$

Graphically, this function may be
seen to be even if the portion to the
right of the vertical axis (positive
time) is folded about the axis to
fall upon the left portion (negative
time). If the function is even, the
**folded portion would fall directly
upon the unfolded left portion.**
Such a function is illustrated in
Figure 6-4(b).

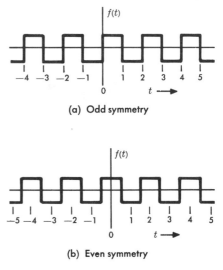

(a) Odd symmetry

(b) Even symmetry

Some functions can be adapted to
have either odd or even symmetry
by the appropriate selection of
axes. One such example is given in
Figure 6-5. In Figure 6-5(a), the
function exhibits odd symmetry.
By shifting the vertical axis to the
right by one-half a unit time in-

Figure 6-5. Symmetry by axis choice.

terval, the function is translated into one having even symmetry.
This is shown in Figure 6-5(b) and is also illustrated in Figure 6-3.

### *Example 6-1: A Fourier Series Application*

It has been shown how the Fourier analysis of a square wave
can be used to illustrate the manner in which such a wave can
be decomposed into its constituent harmonically related components (Figure 6-3). Similarly, a square wave was used to
illustrate how a proper choice of coordinates can simplify a
problem by taking advantage of symmetry properties in the
wave to be analyzed (Figure 6-5).

This example of Fourier analysis demonstrates the effect on
frequency content, harmonic amplitudes, and required relative
bandwidth of changing the period of a periodic rectangular wave.
The waveforms are illustrated in Figure 6-6. In each of the
waveforms illustrated, the pulse amplitude is unity and the pulse
duration is $\tau$ seconds. In Figure 6-6(a) the repetition period
is $T_a = 2\tau$ seconds; in Figure 6-6(b) the period is $T_b = 2T_a$
seconds; in Figure 6-6(c) the period is $T_c = 2T_b$ seconds. In

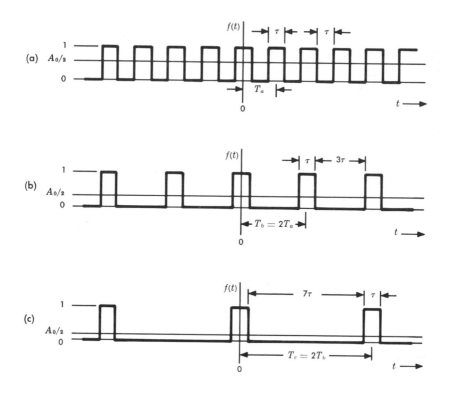

Figure 6-6. Periodic rectangular pulses with different periods.

each, the vertical axis is chosen so that the function exhibits even symmetry. Thus, there are no sine terms in the Fourier series. For each, then, the Fourier series may be written as in Equation (6-8).

$$f(t)_{even} = \frac{A_0}{2} + \sum_{n=1}^{\infty} A_n \cos n\,\omega_0 t.$$

For the three cases in Figure 6-6, the periodic functions may be written, respectively, as follows:

**Figure 6-6(a)**

$$\left.\begin{array}{l} f(t) = 1 \\[2mm] f(t) = 0 \end{array}\right\} \left(m - \frac{1}{2}\right)\tau \leq t \leq \left(m + \frac{1}{2}\right)\tau \quad \left\{\begin{array}{l} \dfrac{m}{2} = 0 \text{ or integer} \\[3mm] \dfrac{m}{2} = \text{fraction} \end{array}\right.$$

In this case, $\tau = \dfrac{T_a}{2} = \dfrac{2\pi}{2} = \pi$.

Therefore,

$$\left.\begin{array}{l} f(t) = 1 \\[2mm] f(t) = 0 \end{array}\right\} \left(m - \frac{1}{2}\right)\pi \leq t \leq \left(m + \frac{1}{2}\right)\pi \quad \left\{\begin{array}{l} \dfrac{m}{2} = 0 \text{ or integer} \\[3mm] \dfrac{m}{2} = \text{fraction} \end{array}\right.$$

**Figure 6-6(b)**

$$\left.\begin{array}{l} f(t) = 1 \\[2mm] f(t) = 0 \end{array}\right\} \left(m - \frac{1}{2}\right)\tau \leq t \leq \left(m + \frac{1}{2}\right)\tau \quad \left\{\begin{array}{l} \dfrac{m}{4} = 0 \text{ or integer} \\[3mm] \dfrac{m}{4} = \text{fraction} \end{array}\right.$$

Now, $\tau = \dfrac{T_b}{4} = \dfrac{2\pi}{4} = \dfrac{\pi}{2}$,

and

$$\left.\begin{array}{l} f(t) = 1 \\[2mm] f(t) = 0 \end{array}\right\} \left(m - \frac{1}{2}\right)\frac{\pi}{2} \leq t \leq \left(m + \frac{1}{2}\right)\frac{\pi}{2} \quad \left\{\begin{array}{l} \dfrac{m}{4} = 0 \text{ or integer} \\[3mm] \dfrac{m}{4} = \text{fraction} \end{array}\right.$$

**Figure 6-6(c)**

$$\left.\begin{array}{l} f(t) = 1 \\[2mm] f(t) = 0 \end{array}\right\} \left(m - \frac{1}{2}\right)\tau \leq t \leq \left(m + \frac{1}{2}\right)\tau \quad \left\{\begin{array}{l} \dfrac{m}{8} = 0 \text{ or integer} \\[3mm] \dfrac{m}{8} = \text{fraction} \end{array}\right.$$

In this instance, $\tau = \dfrac{T_c}{8} = \dfrac{2\pi}{8} = \dfrac{\pi}{4}$.

Thus,

$$\left.\begin{array}{l} f(t) = 1 \\[2mm] f(t) = 0 \end{array}\right\} \left(m - \frac{1}{2}\right)\frac{\pi}{4} \leq t \leq \left(m + \frac{1}{2}\right)\frac{\pi}{4} \quad \left\{\begin{array}{l} \dfrac{m}{8} = 0 \text{ or integer} \\[3mm] \dfrac{m}{8} = \text{fraction} \end{array}\right.$$

In the above equations $m$ is an integer from $-\infty$ to $+\infty$. The value of the dc component, $A_0/2$, may be determined for each case by means of Equation (6-2). Thus,

(a) $A_0/2 = \dfrac{1}{2}$;     (b) $A_0/2 = \dfrac{1}{4}$;     (c) $A_0/2 = \dfrac{1}{8}$.

Note that the value of $A_0/2$, for a periodic function may be determined as the value of $f(t)$ averaged over one period. Where the function represents rectangular pulses, the value of $A_0/2$ is $A\tau/T$ where $A$ is the amplitude of the pulse.

Now, to further illustrate, consider the frequencies of the fundamentals and third and fifth harmonics of the three waveforms of Figure 6-6. The frequency of the fundamental, $f_1$, is the reciprocal of the fundamental period, $T_a$, $T_b$, or $T_c$. For the three cases of interest, the frequencies are

(a) $f_1 = \dfrac{1}{T_a} = \dfrac{1}{2\tau}$;   $f_3 = 3f_1 = \dfrac{3}{2\tau}$;   $f_5 = 5f_1 = \dfrac{5}{2\tau}$.

(b) $f_1 = \dfrac{1}{T_b} = \dfrac{1}{4\tau}$;   $f_3 = 3f_1 = \dfrac{3}{4\tau}$;   $f_5 = 5f_1 = \dfrac{5}{4\tau}$.

(c) $f_1 = \dfrac{1}{T_c} = \dfrac{1}{8\tau}$;   $f_3 = 3f_1 = \dfrac{3}{8\tau}$;   $f_5 = 5f_1 = \dfrac{5}{8\tau}$.

Thus, the frequencies of the fundamentals and their harmonics are seen to decrease as the period, $T$, of the fundamental increases.

Finally, the amplitudes of these signal components may be determined from Equation (6-3):

(a) $A_1 = \dfrac{1}{\pi} \displaystyle\int_0^{2\pi} f(t) \cos t \, dt$

$$= \frac{1}{\pi} \int_0^{\pi/2} f(t) \cos t \, dt + \frac{1}{\pi} \int_{\pi/2}^{3\pi/2} f(t) \cos t \, dt$$

$$+ \frac{1}{\pi} \int_{3\pi/2}^{2\pi} f(t) \cos t \, dt$$

$$= \frac{1}{\pi} \sin t \Big]_0^{\pi/2} + 0 + \frac{1}{\pi} \sin t \Big]_{3\pi/2}^{2\pi}$$

$$= \frac{1}{\pi}\,(1+1)\; = \frac{2}{\pi} = +0.637,$$

$$A_3 = \frac{1}{\pi} \int_0^{2\pi} f(t)\,\cos 3t\, dt$$

$$= \frac{1}{3\pi}\,\sin 3t\,\Big]_0^{\pi/2} \;+\,0\,+\, \frac{1}{3\pi}\,\sin 3t\,\Big]_{3\pi/2}^{2\pi}$$

$$= \frac{1}{3\pi}\,(-1-1) = -\,\frac{2}{3\pi} = -0.213,$$

and

$$A_5 = +\,0.127.$$

(b) $A_1 = \dfrac{1}{\pi} \displaystyle\int_0^{2\pi} f(t)\,\cos t\, dt$

$$= \frac{1}{\pi} \int_0^{\pi/4} f(t)\,\cos t\, dt + \frac{1}{\pi} \int_{\pi/4}^{7\pi/4} f(t)\,\cos t\, dt$$

$$+ \frac{1}{\pi} \int_{7\pi/4}^{2\pi} f(t)\,\cos t\, dt$$

$$= \frac{1}{\pi}\,\sin t\,\Big]_0^{\pi/4} \;+\,0\,+\, \frac{1}{\pi}\,\sin t\,\Big]_{7\pi/4}^{2\pi}$$

$$= \frac{1}{\pi}\,(0.707 + 0.707) = 0.450,$$

$$A_3 = \frac{1}{\pi} \int_0^{2\pi} f(t)\,\cos 3t\, dt$$

$$= \frac{1}{\pi}\,\sin 3t\,\Big]_0^{\pi/4} \;+\,0\,+\, \frac{1}{3\pi}\,\sin 3t\,\Big]_{7\pi/4}^{2\pi}$$

$$= \frac{1}{3\pi}\,(+0.707 + 0.707) = +\,0.151,$$

and

$A_5 = -0.0899.$

(c)    $A_1 = +0.244,$

$A_3 = +0.196,$

and

$A_5 = +0.118.$

While the amplitudes of $A_n$ can be seen to decrease with increasing $n$ for each of the three cases, observe that there is no obvious, simple relationship among the values of $A_n$ from case to case in the example. The value of $A_1$ appears to behave logically, decreasing as $T/\tau$ increases, but $A_3$ and $A_5$ appear to behave erratically in regard to amplitude and sign.

## The $(\sin x)/x$ Function

The lengthy and laborious calculations of Example 6-1 are given to illustrate in detail how the coefficients of a periodic function expressed as a Fourier series can be determined; however, for a number of commonly found waveforms, these coefficients have already been calculated [2]. Many of the expressions for such coefficients contain a term in the form of $(\sin x)/x$. This function is so commonly found that a plot of the function on a normalized scale is given in Figure 6-7.

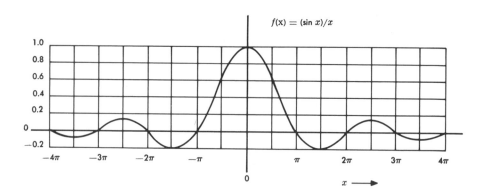

Figure 6-7. The $(\sin x)/x$ function.

In Example 6-1, the coefficient amplitude may be computed for each harmonic component by

$$A_n = A_0 \left( \frac{\sin \dfrac{n\pi\tau}{T}}{\dfrac{n\pi\tau}{T}} \right) \tag{6-10}$$

Values for $n\pi\tau/T = x$ may be found from Figure 6-7 for values of $n$, $\tau$, and $T$ defined as in Example 6-1. Recall, also, that for rectangular pulses, $A_0 = A\tau/T$.

## 6-3  NONPERIODIC SIGNALS

Although the Fourier series is a satisfactory and accurate method of representing a periodic function as a sum of sine and cosine waves as illustrated by Equation (6-1), somewhat broader mathematical expressions, known as the *Fourier transform pair*, must be used to represent nonperiodic signals as functions of time or as functions of frequency. Although these are most useful in characterizing nonperiodic signals, they may also be applied to the analysis or synthesis of periodic signals. Similar mathematical representations may be used to describe the transmission response of a network or transmission line by combining expressions representing signals with those representing network characteristics.

### The Fourier Transform Pair

The determination of the components of a signal can be accomplished by the methods of Fourier analysis. If the signal is periodic, the analysis is relatively simple and can be carried out, as previously described, by a Fourier series representation. If the signal is nonperiodic, the *Fourier transform* may sometimes be used.* It is written

$$g(\omega) = \int_{-\infty}^{\infty} f(t) e^{-j\omega t} dt. \tag{6-11}$$

*Many signals cannot be expressed in terms of Fourier components because the function $f(t)$ is not deterministic. Methods of analyzing these functions depend on expressing them in probabilistic forms, usually in terms of the spectral density function [3].

This equation may be used to determine the function of frequency, $g(\omega)$, given a function of time, $f(t)$, that is single valued, has only a finite number of discontinuities, a finite number of maxima and minima in any finite interval, and whose integral converges.

The inverse function, written

$$f(t) = \frac{1}{2\pi} \int_{-\infty}^{\infty} g(\omega) e^{j\omega t} d\omega, \qquad (6\text{-}12)$$

is known as the *Fourier integral*, or the inverse Fourier transform; this expression is used for Fourier synthesis. Given the function of frequency, $g(\omega)$, of a signal, the signal may be synthesized as a function of time by Equation (6-12). Together, Equations (6-11) and (6-12) are the Fourier transform pair.

Most signals transmitted over the telephone network are random in many parameters such as probability of occurrence, amplitude, or phase. Such signals usually cannot be expressed in terms of Fourier components because the function $f(t)$ is not deterministic.* Much can be learned, however, by examining some random signals, such as the random data signals depicted in Figure 6-8, in terms of the characteristics of one pulse [for which $f(t)$ is deterministic], provided the interaction among pulses is not neglected.

**The Single Rectangular Pulse.** Consider the single rectangular pulse of Figure 6-9. From Equation (6-11) and from examination of the pulse $[f(t) = A$ from $-\tau/2$ to $+\tau/2$ and zero elsewhere], the Fourier transform may be written

$$g(\omega) = A \int_{-\infty}^{\infty} f(t) e^{-j\omega t} dt = A \int_{-\tau/2}^{\tau/2} e^{-j\omega t} dt.$$

Observation of the nature of the function $f(t)$ and subsequent substitution of the limits of integration make this equation tractable. Integrated, the equation becomes

$$g(\omega) = \frac{2A \sin \dfrac{\omega\tau}{2}}{\omega} = \frac{A\tau \sin \dfrac{\omega\tau}{2}}{\dfrac{\omega\tau}{2}}, \qquad (6\text{-}13)$$

*Much work has been done to analyze such signals with digital computers. This procedure has been made more efficient by use of the *fast Fourier transform* [4].

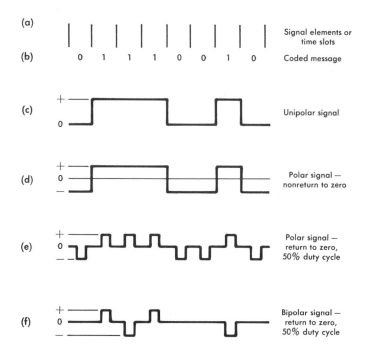

Figure 6-8. Some signal formats for a random signal.

the familiar $(\sin x)/x$ form. Note that the expression has a continuous distribution of energy at all frequencies, rather than at discrete frequencies as indicated for the components of the Fourier series for the periodic function represented by Equation (6-10). The function of Equation (6-13) is a pure real and, therefore, the com-

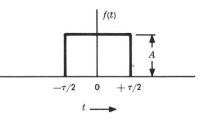

Figure 6-9. A single pulse.

ponents of the signal in the time domain are all cosine functions, in phase at $t = 0$. Values for $g(\omega)$ in Equation (6-13) may be found by using appropriate values for $A$ and $\tau$ and, substituting $x = \omega\tau/2$, by use of the plotted values of $(\sin x)/x$ in Figure 6-7.

**The Impulse.** An impulse is approximated when a rectangular pulse is narrowed without limit while keeping its area ($A\tau$ in Figure 6-9) unchanged. To simplify the treatment, the area may be assumed to be equal to unity. Thus, in the time domain an impulse is a signal having energy but infinitesimal duration.

The corresponding frequency spectrum may be found from Equation (6-13) by noting the assumption that $A\tau = 1$ and that $(\sin \omega\tau/2)/(\omega\tau/2) = 1$ when $\omega\tau/2 = 0$ (see Figure 6-7). Thus, the resulting spectrum contains all frequencies from $-\infty$ to $+\infty$ of equal phase at $t = 0$ and each having an amplitude of unity. This description of an impuse is useful in discussing the impulse response of a network.

## Transmission Response

Transmission of nonperiodic signals through a network or transmission line may be studied by Fourier transform methods in either the frequency or time domain.

**Frequency Response.** The complex frequency spectrum can often be utilized to simplify rather complicated problems. The advantages to be had by operating in the frequency domain arise from the relatively simple relationship between input and output signals transmitted through linear networks or transmission lines when the relationship is specified in that dimension. In a typical problem, the input signal has a spectrum $g_i(\omega)$ and the output $g_o(\omega)$. The transmission path can be described by a frequency function which is its transfer impedance (transfer voltage or current ratio), or what is commonly called its frequency response. This function, $H(\omega)$, can be established by computation from the known circuit constants of the system or network. It can also be found experimentally by applying a sine-wave test signal of known characteristics at the input and measuring amplitude and relative phase at the output.

The relationship between the input and output spectra of a signal applied to a network is particularly simple;

$$g_o(\omega) = H(\omega)\, g_i(\omega), \qquad (6\text{-}14)$$

where $g_o$, $g_i$, and $H$ are, in general, complex functions of the radian frequency, $\omega$. In polar form, the amplitude and phase relationships are, respectively,

$$|\, g_o(\omega)\, | = |\, H(\omega)\, |\, |\, g_i(\omega)\, | \qquad (6\text{-}15)$$

$$\theta_o(\omega) = \theta_h(\omega) + \theta_i(\omega) \quad . \qquad (6\text{-}16)$$

The validity of these relations rests upon the superposition principle since $g_o(\omega)$ is computed by assuming that it is a linear combination of the responses of the network to each frequency component (taken individually) in the input wave. This observation implies that if the response of a linear system to the gamut of sine-wave excitations is known, then its response to any other waveform can be found uniquely by decomposing that wave into its Fourier components and computing the response to each individual component. The output waveform, $f_o(t)$, can be found by evaluating the Fourier integral of $g_o(\omega)$. The principle outlined here is the basis for all sine-wave testing techniques used in practice. It should be noted, however, that it is useful only for *linear systems* since it is only in such systems that superposition is generally valid. In the case of a nonlinear device, such as a rectifier, the response to each input waveform must be computed separately; the complex frequency response of the network does not allow generalization to include other functions.

**Impulse Response.** Transmission through a network can also be completely described in terms of its impulse response, which is defined as the function $h(t)$ that would be found at the output as a result of applying an impulse (previously defined) to the input terminals. Since the time function applied to the input has a flat frequency spectrum, it would be expected that $h(t)$ will have a spectrum which differs from flatness by the frequency characteristic of the network. In other words, $H(\omega)$ gives the frequency and phase spectra of $h(t)$. Expressed analytically, a unit impulse input to a network $H(\omega)$ produces an output $h(t)$ given by

$$\mathrm{F}\,[h(t)] = H(\omega) \tag{6-17}$$

from which it follows that

$$H(\omega) = \int_{-\infty}^{\infty} h(t)\, e^{-j\omega t} dt \tag{6-18}$$

and also

$$h(t) = \frac{1}{2\pi} \int_{-\infty}^{\infty} H(\omega)\, e^{j\omega t}\, d\omega. \tag{6-19}$$

The impulse response is, of course, a real function of time. Certain relationships between $H(\omega)$, $H(-\omega)$, and the conjugate of $H(\omega)$, written $H^*(\omega)$, can be shown [4]. These lead to the following:

$$H(-\omega) = H^*(\omega)$$

$$H_R(\omega) = H_R(-\omega)$$

$$(6\text{-}20)$$

$$-H_I(\omega) = H_I(-\omega)$$

$$|H(\omega)| = |H(-\omega)|$$

where $H_R$ and $H_I$ are the real and imaginary parts, respectively, of $H(\omega)$.

These are extremely important mathematical properties of any physical transmission path — network or transmission line. The first expression in the series of equations numbered (6-20) shows that the transfer impedance of the network, $H(\omega)$, expressed for negative frequencies, $H(-\omega)$, is equal to its conjugate expressed for positive frequencies, $H^*(\omega)$. From this fact, the second expression is derived directly to show that the real part of the impedance function, $H_R(\omega)$, has even symmetry about zero frequency. The third expression shows that the imaginary (phase) component of $H(\omega)$, $H_I$, has odd symmetry about zero frequency. The last expression, showing the relation between absolute values of $H$, follows from the first.

**Bandwidth.** It was previously shown, in the discussion of the single rectangular pulse, that the ability to establish limits of integration led to a useful expression for a frequency domain description of the pulse. In a similar manner, the recognition of the finite bandwidth of a channel makes practical the impulse response analysis of transmission through a network.

An examination of the Fourier integral of Equation (6-19) indicates that in order to determine the function of time corresponding to a particular frequency spectrum, it is necessary to know that spectrum from $-\infty$ to $+\infty$. However, in the application of Fourier synthesis to any real situation, the signal under study is always generated by a source capable of producing only a finite range of frequencies.

Similarly, the signal is carried on a channel capable of transmitting only a finite bandwidth. Hence, it is necessary to examine the spectrum only in this region, and the signal can be assumed to be zero outside this region. Such a finite bandwidth would restrict the number of time functions which can be synthesized to those whose fastest time rate of change is of the same order as the rate of the highest frequency component that may be present.

In practice, limits are used which depend upon the characteristics of the physical system or circuit being dealt with, rather than using the infinite limits given in Equation (6-19). This equation may be modified to account for the finite bandwidth of any real system, and the Fourier integral can be written

$$h(t) = \frac{1}{2\pi}\int_{-\omega_2}^{-\omega_1} H(\omega)\, e^{j\omega t}\, d\omega + \frac{1}{2\pi}\int_{\omega_1}^{\omega_2} H(\omega)\, e^{j\omega t}\, d\omega. \qquad (6\text{-}21)$$

### *Example 6-2: Impulse Response of an Ideal Low-Pass Filter*

As an example of the usefulness of the Fourier transform pair, consider a problem in pulse transmission, where information is being transmitted in digital form. At the transmitting terminal, a pulse is either sent or not sent at times $t_1$, $t_2$, etc. The problem is to tell, after the signal has been transmitted through the transmission medium (represented here by a low-pass filter), whether or not a pulse is present for each signal element or time slot at the receiver.

For this example, assume that the difference between two successive coded signals, illustrated by $S_1$ and $S_2$ in Figure 6-10,

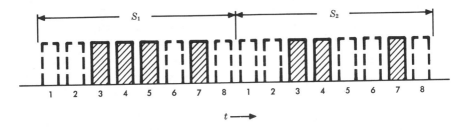

Figure 6-10. Successive code signals.

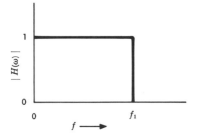

Figure 6-11. Idealized low-pass transmission characteristic.

lies in the fact that $S_1$ has a pulse in position 5, whereas $S_2$ does not. Further assume that these signals are passed through a low-pass filter which has an idealized transmission characteristic shown in Figure 6-11. This idealized transmission characteristic has a constant finite value of attenuation (assumed to be 0 dB for this problem) from zero frequency to $f_1$ and has infinite attenuation above $f_1$. It has no delay distortion for frequencies from zero to $f_1$; delay distortion above $f_1$ is of no consequence since there is no signal transmission above $f_1$. (This is an easy case to analyze; such characteristics are impossible to achieve but can be approximated. More achievable characteristics are more complicated to analyze.) The example, then, illustrates how bandwidth limitation alone can cause energy in the fourth position of $S_2$ to spill over into pulse position 5.

If the transmission characteristic of the network is known, it is next necessary to assume a spectrum for the input pulse at position 4 and, in turn, determine its effect on the pulse or lack of pulse in position 5. Although the first inclination would probably be to assume a rectangular pulse like that of Figure 6-9 (even though real pulses are never exactly rectangular), the problem can be simplified by assuming an impulse. Compare the spectrum of an impulse (flat versus frequency, with no phase reversals) with the spectrum of a rectangular pulse in the region of $\omega = 0$ (almost flat for very low frequencies). It is seen that, if the transmitted bandwidth is small enough compared to the first frequency at which $(\sin x)/x$ becomes zero, the output will be the same whether the input is taken to be a narrow rectangular pulse or an impulse. Since the spectrum of an impulse is easier to handle analytically, the input is assumed to be an impulse. If it is desired to refine the results later, the input spectrum may be modified to have the $(\sin x)/x$ shape, or the frequency response, $H(\omega)$, may be modified.

Moreover, if the input signal is assumed to be an impulse, the task is to determine the signal (as a function of time) at the output of a path having the transmission characteristic shown in Figure 6-11. First notice that $|H(\omega)|$ can be plotted for negative as well as positive frequencies. By the relations of Equations (6-20), the plot would look like Figure 6-12, where $\omega_1 = 2\pi f_1$ has been substituted for $f_1$.

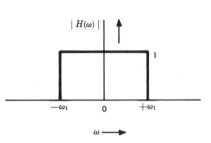

Figure 6-12. Idealized low-pass characteristic (positive and negative frequencies).

If Equation (6-19) is applied to Figure 6-12 and constant delay is ignored, the output pulse may be represented as

$$h(t) = \frac{1}{2\pi} \int_{-\omega_1}^{\omega_1} e^{j\omega t} \, d\omega.$$

The term $H(\omega)$ in Equation (6-19) is shown in Figure 6-12 to be equal to unity in the interval from $-\omega_1$ to $+\omega_1$ and so does not appear in the above expression for $h(t)$.

This equation may be integrated to yield

$$h(t) = \frac{\omega_1}{\pi} \times \frac{\sin \omega_1 t}{\omega_1 t}.$$

This is a $(\sin x)/x$ function of time plotted in Figure 6-13. On this plot, $t = 0$ is arbitrary; for a physical network which approximates the characteristic of Figures 6-11 and 6-12, the zero time point represents the absolute delay of the transmission path. The optimum time for the next

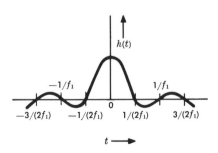

Figure 6-13. $h(t)$ at output of low-pass transmission path.

pulse is at $t = 1/(2f_1)$ because $h(t)$ goes through zero at that point, and interpulse (or intersymbol) interference is minimized.

If the cutoff of the transmission path is at 500 kHz, then the interval between impulses should be 1 microsecond (repetition rate, 1 MHz). A shorter interval would tend to make the receiver think a pulse is present when in fact it is not; a longer interval would result in some cancellation when the following pulse is present. The spacing of pulses to avoid intersymbol interference is one of the fundamental requirements in pulse transmission.

The necessity for distinguishing between the presence or absence of a pulse in position 5 of $S_1$ and $S_2$ in Figure 6-10 is importantly dependent on a design that minimizes the effect of the presence of an unwanted signal in position 5 due to the pulse in position 4. This is accomplished by relating, in the design, the system transmission characteristic and pulse repetition rate so that the next pulse position (position 5) corresponds to the crossover of pulse number 4 at time $1/(2f_1)$ as illustrated in Figure 6-13.

This example illustrates the way in which the Fourier transform pair can be used. If an input signal which is a given function of time is assumed, the signal (as a function of time) at the output of a network can be found if the transmission characteristic of the network is known. The results may be expressed in very general functional terms in order to display the nature of a problem, or specific formulas may be used to obtain specific numerical results. In any particular case, finding the solution may be easy (as in Example 6-2) or may involve laborious or sophisticated mathematical manipulation of the specific functions involved in the problem. The basic idea remains the same.

Another class of transmission problems involves circuits having bandpass characteristics. Such problems are often difficult to solve directly but are amenable to solution by the methods of Fourier analysis using an equivalent low-pass circuit arrangement such as that in Example 6-2 [5, 6].

### REFERENCES

1. Scott, R. E. *Linear Circuits* (Reading, Mass.: Addison-Wesley Publishing Company, Inc., 1960).

2. *Reference Data for Radio Engineers* (Indianapolis: Howard W. Sams and Company, Inc., Sixth Edition, 1975), p. 44-12.

3. Franks, L. E. *Signal Theory* (Englewood Cliffs, N. J.: Prentice-Hall, Inc., 1969).

4. Bogert, B. P. *et al.* (Special issue on the fast Fourier transform) *IEEE Transactions on Audio and Electroacoustics*, Vol. AU-15 (June 1967).

5. Technical Staff of Bell Telephone Laboratories. *Transmission Systems for Communications*, Revised Third Edition (Winston-Salem, N. C.: Western Electric Company, Inc., 1964), Appendix A.

6. Sunde, E. D. "Theoretical Fundamentals of Pulse Transmission," I and II, *Bell System Tech. J.*, Vol. 33 (May 1954) pp. 721-788 and (July 1954) pp. 987-1010.

7. Watson, G. N. and E. T. Whittaker. *A Course of Modern Analysis* (Cambridge: University Press, 1940).

8. Wylie, C. R., Jr. *Advanced Engineering Mathematics* (New York: McGraw-Hill Book Company, Inc., 1951).

9. Van Valkenburg, M. E. *Network Analysis* (Englewood Cliffs, N. J.: Prentice-Hall, Inc., 1964).

10. Everitt, W. L. and G. E. Anner. *Communication Engineering* (New York: McGraw-Hill Book Company, Inc., 1956).

Chapter 7

# Negative Feedback Amplifiers

Detailed knowledge of feedback principles is needed only by those involved in the design and development of active transmission circuits. However, the high performance of modern transmission equipment is so dependent on the use of negative feedback that it appears desirable to provide some appreciation of why feedback is used, what it accomplishes, how it operates in electronic circuits, what some of the design limitations are, and what limitations exist in its application. With the design of feedback amplifiers used as the basis for discussion, feedback mechanisms and the interactions among them may be covered as background for an understanding of the interdependence of system and amplifier, or repeater, performance.

Negative feedback is commonly used in transmission systems for communications because it acts to suppress unwanted changes in amplifier gain and substantially reduces harmonic distortion and interchannel modulation noise. It also facilitates the design of amplifiers having much better broadband return loss characteristics than can be achieved without feedback.

## 7-1 THE PRINCIPLE OF NEGATIVE FEEDBACK

In its simplest form, a negative feedback amplifier can be regarded as a combination of an ordinary amplifier (the $\mu$ circuit) and a passive network (the $\beta$ circuit) ; by means of the latter, a portion of the output signal of the amplifier is combined out of phase with its input signal as illustrated in Figure 7-1. Ideally, this phase difference is 180 degrees and hence the term *negative feedback*.

The gain of a feedback amplifier may be written

$$\frac{e_2}{e_1} = \frac{\mu}{1 - \mu\beta} \quad . \tag{7-1}$$

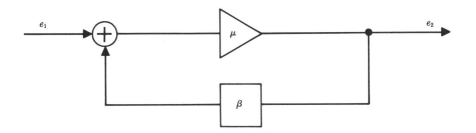

Figure 7-1. Feedback amplifier configuration.

Without feedback ($\beta = 0$), the gain would be simply $e_2/e_1 = \mu$. Thus, one effect of feedback is the reduction of gain by the term $1/(1 - \mu\beta)$.

In general, the $\mu$ gain is very much larger than unity. As a result, an approximation may be derived from Equation (7-1) as follows:

$$\frac{e_2}{e_1} = \frac{\mu}{1 - \mu\beta} = \frac{1}{(1/\mu) - \beta} \approx -\frac{1}{\beta}; \qquad (7\text{-}2)$$

that is, the gain of a feedback amplifier is approximately proportional to $\beta$-circuit loss and is independent of $\mu$-circuit gain.

These characteristics result in feedback amplifiers having attributes that far outweigh the disadvantage of reduced amplifier gain; consequently, in modern design, negative feedback is used in nearly all electronic amplifiers. It is especially valuable in amplifiers used in transmission systems where many amplifiers are connected in tandem. Here, without feedback the cumulative effect of small imperfections in individual amplifiers would be intolerable.

## 7-2 APPLICATIONS OF FEEDBACK

The design of transmission systems involves finding simultaneous solutions to problems of bandwidth, repeater spacing, and signal-to-noise performance. These in turn are related to channel capacity, transmission loss in the medium and the achievability of compensating gain, the cumulation of interferences such as thermal and intermodulation noise, and the provision of adequate signal load carrying capacity. The design of amplifiers to meet such requirements is made possible by feedback. It is incorporated in amplifiers of line repeaters used in analog and digital cable systems as well as in the amplifiers that are found in all types of terminal and station equipment.

One other important transmission system application is the use of feedback in dynamic backward-acting regulator and equalizer circuits. Such circuits utilize one or more single-frequency signals, called pilots, which are applied to a transmission system at the transmitting terminal at precise and carefully controlled frequencies and amplitudes. Immediately following a point of regulation, the pilot signal is picked off the line, rectified, and compared with a reference voltage. The error signal, i.e., the difference between the rectified pilot and the reference, is fed back to the input of a regulating amplifier through a network. The response of this network to the error signal changes the transmission gain in a direction and by an amount to correct the pilot amplitude at the output of the regulator. By the use of several pilots appropriately positioned in the signal spectrum, complex gain/frequency corrections are made across the entire signal band, resulting in dynamic equalization of the high-frequency line.

## 7-3  BENEFITS OF FEEDBACK

Once a system design is chosen, any departure from the ideal represents a penalty in performance. Departures in system gain result in increases in thermal noise if the gain is less than the design value or in intermodulation noise if the gain is greater than desired. Furthermore, in the latter situation the system may become overloaded. In addition to the performance penalties, such gain departures carry a cost penalty because they must be compensated by some form of equalization to correct the gain/frequency or delay/frequency characteristic, or both, to within tolerable limits over the transmission band.

Equation (7-2) shows that the gain of a feedback amplifier is nearly independent of the $\mu$ circuit. Thus, departures from the ideal gain/frequency characteristic (i.e., departures from design values) that are caused by changes in the $\mu$ circuit are effectively reduced by feedback. These changes may be caused by manufacturing, aging, and temperature-induced variations in $\mu$-circuit components, which include the active devices. Gain variations caused by power supply fluctuations are also reduced.

The nonlinear input/output characteristics of all active devices are another source of impairment in broadband electronic circuits. This type of impairment, often referred to as harmonic distortion or inter-modulation noise, is also reduced by the use of negative feedback. If no other benefits accrued from using feedback, this alone would justify application in analog cable transmission systems and in FM terminal equipment of microwave radio systems.

Additional feedback benefits accrue in the resolution of problems involving amplifier input and output impedances. Usually it is required that these impedances, or at least their absolute values, match the impedances of the circuits to which they connect. In nonfeedback amplifiers it is nearly always difficult to meet this requirement because the desired impedances are incompatible with the impedances of the devices used in the amplifiers. Circuit compromises often must be made to achieve an acceptable impedance match. In feedback amplifiers, however, the provision of feedback increases the flexibility of the design choices that can be made, and it is usually possible to achieve a better impedance match over a wide bandwidth by using a feedback amplifier than by using a nonfeedback amplifier.

### Example 7-1: Feedback Effects

This simple example illustrates how a $\mu$-gain change of about 0.8 dB may be suppressed by feedback to an amplifier gain change of approximately 0.1 dB.

Let the overall gain of an amplifier be 10 dB; that is,

$$20 \log \frac{e_2}{e_1} = 10 \, ; \quad \frac{e_2}{e_1} \approx 3.16.$$

From Equation (7-2),

$$\frac{e_2}{e_1} = \frac{\mu}{1 - \mu\beta} \approx 3.16.$$

Assume the $\mu$ gain (without feedback) is 20 log $\mu = 30$ dB; then

$\mu = 31.6$

and, by substitution,

$\beta = 0.284.$

Now, let the $\mu$ gain increase from 30 dB to 30.8 dB; that is, $\mu$ increases by about 10 percent from 31.6 to 34.8.

Then the overall amplifier gain is

$$\frac{e_2}{e_1} = \frac{34.8}{1 - 34.8\,(-0.284)} = 3.2$$

and

$$20 \log \frac{e_2}{e_1} = 20 \log 3.2 = 10.09 \text{ dB.}$$

Thus a 10 percent change in $\mu$-circuit gain is held to about a 1.3 percent change in overall gain (0.09 dB).

The fact that the amplifier gain increased as the $\mu$ gain increased is due to the phase relationships implied by the simple substitutions made. In complex feedback structures, the amplifier gain might increase or decrease over limited portions of the band and within a limited range of the $\mu$-gain change.

## 7-4  CIRCUIT CONFIGURATIONS

The principal circuit configurations useful in feedback circuits can be classified most easily in terms of the way in which the $\mu$ and $\beta$ circuits are connected to each other and to the external interconnections at amplifier input and output. The variety of connections that can be made cannot be clearly demonstrated by a simple drawing such as that of Figure 7-1. The actual situation is that shown broadly by Figure 7-2 in which the $\mu$, $\beta$, input, and output circuits are interconnected by means of six-terminal networks. The classification of feedback circuits then depends on the forms which these six-terminal networks assume.

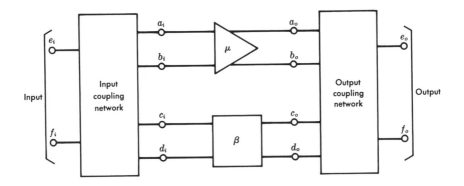

Figure 7-2. Feedback amplifier representation.

Illustrations of some of the more common feedback amplifier structures are given in Figures 7-3 through 7-6. Where appropriate, the network terminals are identified in accordance with the notation used in Figure 7-2. The $\mu$ circuits commonly have one, two, or three stages of gain; an unlimited number of network configurations may be found in the passive networks shown in the figures. To avoid complexity here, the internal network configurations are generally omitted in the figures.

### Series and Shunt Feedback

The configuration of Figure 7-3 is called series feedback because, as seen from the input and output terminals, the $\mu$ and $\beta$ circuits are in series. The $\beta$ circuit, shown here as a $\pi$ arrangement of three impedances ($A$, $B$, and $C$) may be much simpler or much more complex than that illustrated. The effective line terminals ($e_i$, $f_i$, $e_o$, and $f_o$) are shown at the high sides of the transformers since the transformer characteristics in this case may be added directly to those of the connecting circuits.

Figure 7-4 shows how feedback may be provided by means of shunt connections. The $\beta$ circuit, here represented as a T network of the three impedances, may again take on any of an unlimited

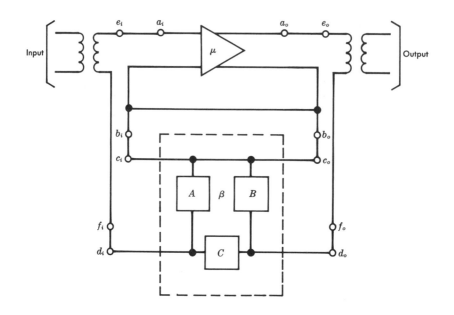

Figure 7-3. Series feedback amplifier.

number of configurations. Note that the connecting terminals (input and output), $\beta$ network, and $\mu$ network are all in parallel.

Series and shunt feedback designs are simple and they are convenient for many applications. The feedback phenomenon tends to change the effective input and output impedances of the amplifier to very high or very low values. As a result, it is possible to build out these impedances conveniently by the use of discrete components to achieve a good impedance match to the connecting network or transmission line. A disadvantage is that the line or connecting impedances form a part of the $\mu\beta$ loop. As a result, variations in the line impedance, sometimes large and impossible to control, affect the $\mu\beta$ characteristic; in some cases, the effect may be great enough to cause amplifier instability.

## Bridge-Type Feedback

These difficulties may be mitigated by using bridge-type feedback circuits. Many variations of these circuits exist, but the configuration

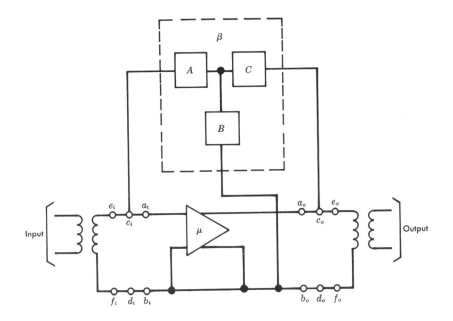

Figure 7-4. Shunt feedback amplifier.

that is most commonly used, especially for broadband repeaters in analog cable systems, is the high-side hybrid feedback arrangement illustrated in Figure 7-5. Several network branches must be added in this configuration to provide hybrid balance and input and output impedance control. These branches are designated $Z_n$ and $Z_1$ in Figure 7-5. The advantages of this circuit include the achievement of minimum noise and improved intermodulation performance while controlling both the input and output impedances.

Figures 7-3 through 7-5 show symmetrical arrangements at each end of the amplifier. This has been done only to simplify the illustrations. The number of configurations is increased greatly by combining different types of connections at input and output. Furthermore, circuit advantages can sometimes be realized by providing multiple loop configurations. An example of such a configuration is given in Figure 7-6. Here, a feedback amplifier with a series feedback network $Z_{\beta 1}$, similar to that of Figure 7-3, is shown with local shunt feedback $Z_{\beta 2}$ around the last stage of a three-stage configuration in the $\mu$ path. The impedances $Z_{i1}$ and $Z_{i2}$ are interstage networks in the $\mu$ path.

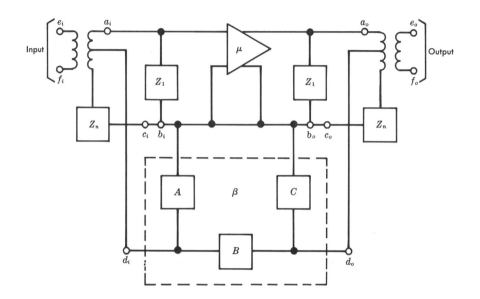

Figure 7-5. Amplifier with high-side hybrid feedback.

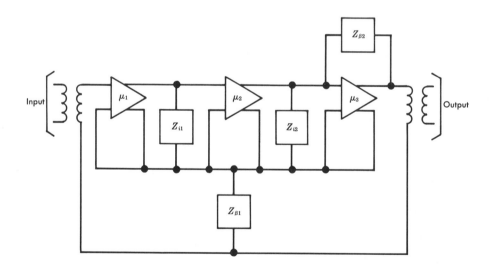

Figure 7-6. Three-stage series feedback amplifier with local shunt feedback on last stage.

## 7-5　DESIGN CONSIDERATIONS

It is not possible to review here the entire procedure followed in designing a feedback amplifier nor is it desirable to do so. However, some important relationships and design limitations are discussed in order to provide an improved understanding of how transmission systems operate and how system performance is related to the design of the individual amplifier.

### Gain and Feedback

The shape and magnitude of the gain/frequency characteristic are basic design considerations. The closeness of the gain/frequency characteristic to the desired characteristic may be determined by the degree of circuit complexity that can be tolerated; however, the better the match, the better will be the ultimate transmission characteristic of the system. Equipment size and power dissipation may also be important considerations in making this first set of compromises in amplifier design.

**Characteristic Shaping.** The characteristics of feedback amplifiers are all complex functions of frequency which are importantly related to the transmission characteristics of all of the networks making up the complete amplifier and its external terminations.

In many applications, it is desirable to design the amplifier to a flat gain, one that is equal over the entire transmitted band. In the case of line repeaters for analog cable systems, it is usually desirable to have the gain of the amplifier sections of the repeaters match the loss of the cable section over the band of interest. In either case, the desired flat or shaped gain/frequency characteristic is produced primarily by proper design of the $\beta$-circuit network since the gain is approximately equal to $-1/\beta$ as shown in Equation (7-2). Some gain shaping may also be provided in those networks that are outside the $\mu\beta$ loop, such as the coupling networks shown in many of the figures as simple transformers.

To achieve optimum signal-to-noise performance, it is also desirable in many cases to shape the feedback/frequency characteristic of an amplifier. For example, it is possible to increase low-frequency feedback at the expense of high-frequency feedback. This can be accomplished by careful designs of all networks in the $\mu\beta$ loop, using frequency-dependent reactive components, since the feedback is, by definition, proportional to $1/(1-\mu\beta)$.

**Gain and Phase Margins.** The selection of a circuit configuration and the amount of feedback to be provided depend on the magnitudes of the gain and bandwidth required and on the characteristics of available active devices. These considerations include the linearity of the device input/output characteristics, the noise figure of the input device, and the need for minimizing variations in circuit parameters due to device aging and ambient temperature changes.

As shown in Equation (7-2), the insertion gain of a feedback amplifier is

$$\frac{e_2}{e_1} = \frac{\mu}{1-\mu\beta} \approx -\frac{1}{\beta}.$$

The total gain around the feedback loop is defined as $\mu\beta$, where $\mu$ is the total gain provided by the active devices (and their related $\mu$-circuit networks) and $\beta$ is the loss of the network that connects the output back to the input. From these relationships, the loop gain in dB is

$$20 \log \mu\beta = 20 \log \mu + 20 \log \beta \approx 20 \log \mu - g_R \qquad (7\text{-}3)$$

where $g_R$ is the insertion gain of the complete closed-loop amplifier in dB. It is approximately equal to $-20 \log \beta$. Thus,

$$20 \log \mu \approx 20 \log \mu\beta + g_R \qquad \text{dB.} \qquad (7\text{-}4)$$

That is, the sum of the loop gain and insertion gain cannot exceed the total gain available in the $\mu$ circuit. It is therefore impossible to get loop gain in excess of the difference between the $\mu$ gain and the desired insertion gain. When the desired loop gain is greater, the design is said to be *gain limited*.

Most broadband amplifier designs, however, are not gain limited; the need for adequate stability margins is usually controlling. In the gain expression $\mu/(1-\mu\beta)$, the denominator may become zero, depending on phase relationships, when $\mu\beta = 1$. If $\mu\beta$ is equal to unity at any frequency, in-band or out-of-band, the amplifier may become unstable and break into spontaneous oscillation at that frequency if the phase of $\mu\beta$ is unfavorable. If it were possible to hold $|\mu\beta| \gg 1$

for all frequencies, this would not be a problem, but every active device has some frequency above which its gain decreases monotonically. The rate of decrease may be enhanced by circuit stray inductance or capacitance. Thus, there is always a frequency at which $|\mu\beta| = 1$.

Two criteria must be satisfied to guarantee a stable amplifier; the phase must be greater than 0 degrees where $|\mu\beta|$ passes through 0 dB, and $|\mu\beta|$ must represent several dB of loss where the phase passes through 0 degrees. These criteria are known as the *phase and gain margins* in an amplifier design. If an amplifier has such margins, it is said to meet the Nyquist stability criteria. Such margins are illustrated in Figure 7-7 where the characteristics are plotted on an arbitrary, normalized frequency scale. A phase margin of about 30 degrees and gain margin of about 10 dB, as illustrated, allow for variations in device characteristics which result from manufacturing processes, aging, and temperature effects.

The achievement of adequate phase and gain margins sets an upper limit on the achievable in-band feedback. When this limit is lower than that set solely by gain considerations, the design is said to be *stability limited.*

Ideally, maximum stability margins would result if the phase of the $\mu\beta$ characteristic could be held at 180 degrees. Then, the gain

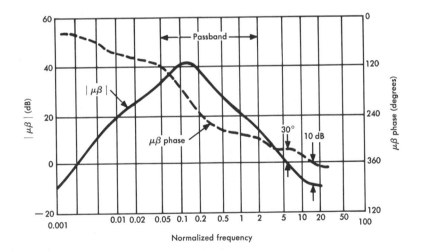

Figure 7-7. Typical feedback amplifier characteristics.

expression could be written as $1/(1+|\mu\beta|)$. Within the transmission band the phase is often controlled to approach this condition. However, out-of-band phase changes due to phase shifts inherently associated with any gain/frequency characteristic, such as the gain cutoff mentioned earlier. Furthermore, for very high frequencies the propagation time around the feedback loop contributes additional phase shift which can be minimized, but not eliminated, by careful design.

## Nonlinear Distortion and Overload

In addition to the related considerations of gain and achievable feedback, the related combination of overload, gain, and nonlinear distortion must be considered in feedback amplifier design. These can be studied by first examining the phenomenon of nonlinear distortion and its reduction by feedback and then relating these to the problems of gain and overload.

**Nonlinear Distortion.** The generation of intermodulation products caused by nonlinear input/output characteristics of transistors is a very complex phenomenon. The analysis here is oversimplified in order to illustrate how products are generated, how feedback tends to suppress them, and how gain and overload are affected.

The nonlinear input/output voltage relationships of an amplifier may be represented by the expression

$$e_o = a_0 e_i^0 + a_1 e_i^1 + a_2 e_i^2 + a_3 e_i^3 + \ldots \qquad , \qquad (7\text{-}5)$$

where $e_o$ and $e_i$ are the output and input signal voltages, and the $a$ coefficients provide magnitude values of various wanted and unwanted components in the output signal. If the input signal has many frequency components, Equation (7-5) may be used to study the intermodulation phenomenon by assuming

$$e_i = A \cos \alpha t + B \cos \beta t + C \cos \gamma t.$$

When this value of $e_i$ is substituted in Equation (7-5), the expression can be expanded by trigonometric identities. The output voltage then contains an infinite number of terms consisting of various combinations of input signal components; the magnitudes are represented by the coefficients $A$, $B$, and $C$ of the input signal and $a_0$, $a_1$, etc., of the input/output expression. Fortunately, in most applications, the magnitudes of terms in Equation (7-5) having exponents of the fourth power and higher are so small they usually may be ignored.

To demonstrate the nonlinear phenomenon and the effects of feedback, a few specific terms of the output voltage, extracted from expansion of Equation (7-5) after substituting the expression for $e_i$, may be examined. The terms of interest are

$$e_1 = a_1 A \cos \alpha t \qquad\qquad (7\text{-}6)$$

$$e_2 = a_2 AB \cos (\alpha+\beta)t \qquad\qquad (7\text{-}7)$$

$$e_3 = \frac{3}{2} a_3 ABC \cos (\alpha+\beta-\gamma)t. \qquad\qquad (7\text{-}8)$$

The first term, Equation (7-6), is a component of the output which corresponds exactly with the first term of the input signal ($A \cos \alpha t$) except for the coefficient $a_1$. This coefficient may be regarded as a measure of the gain of the amplifier, $g_R$. As shown in Equation (7-4), the value of $g_R$, and therefore the value of $a_1$, is a function of the feedback, $\mu\beta$.

Equation (7-7) represents an intermodulation distortion component derived from the second-order term of Equation (7-5). The coefficient of this term involves magnitudes $A$ and $B$ of the two intermodulating input signal components and the coefficient $a_2$ of Equation (7-5). The value of $a_2$ is a function of the feedback, $\mu\beta$; to a first approximation, the value of $a_2$ is reduced in direct proportion to the amount of feedback provided.

Equation (7-8) represents an intermodulation distortion component derived from the expansion of the third-order term of Equation (7-5). The coefficient involves the magnitudes $A$, $B$, and $C$ of the three intermodulating input signal components and the coefficient $a_3$ of Equation (7-5). The value of $a_3$ is also reduced by feedback but not by as simple a relationship as $a_1$ and $a_2$. Second-order modulation components, fed back to the input, mix with fundamental signal components to produce products that appear at the output as third-order products. The result is that the reduction of third-order intermodulation is not quite as effective as the reduction of second-order intermodulation.

There are many more terms in the three-frequency expansion of Equation (7-5) [2]. The distribution of the intermodulation products across the band, the frequency characteristics of the transmitted signal and the amplifier gain, the modulation coefficients, the feedback, and other phenomena make the calculation of intermodulation noise in system design a problem that is most tractable when solved by a digital computer. In contrast with computation, amplifier and system performance is more easily determined by measurements involving a noise loading technique.

The above discussion of nonlinear distortion is predominantly qualitative. For design purpose, the factors above are often manipulated in such a way as to define $20 \log M_2$ and $20 \log M_3$ as the ratios, expressed in decibels, of the second harmonic ($M_2$) or the third harmonic ($M_3$) to a 0-dBm fundamental at the output of a repeater. These modulation coefficients prove useful in analog cable system design. They are, of course, related to the $a$ coefficients of Equation (7-5).

**Overload.** The coefficients $20 \log M_2$ and $20 \log M_3$ are essentially constant over most of the signal amplitude range of interest (though they may be functions of frequency). However, as overload is approached, departures from constant values of $20 \log M_2$ and $20 \log M_3$ are observed as are departures from normally constant gain. These observations lead to a number of definitions of overload in a feedback amplifier. Typical characteristics are plotted in Figure 7-8 for departures of $20 \log M_3$ and gain from their nominal values as functions of the signal power at the output of a repeater. Three definitions of overload are discussed briefly below; two are related to departure of $20 \log M_3$ from a constant value, and one is related to the departure of gain from constant value.

*Definition 1:* By this definition, the overload point is that value of output signal power at which $20 \log M_3$, the third-order modulation coefficient, increases by 0.5 dB relative to its nominal constant value. This is identified as point $P_{R1}$ at 20 dBm in Figure 7-8. This definition, appropriate for use in systems limited by intermodulation, is conservative in the sense that only a slight performance impairment results from exceeding the limit by a small amount. A relatively small amount of overload margin would be allowed in a design based on this definition.

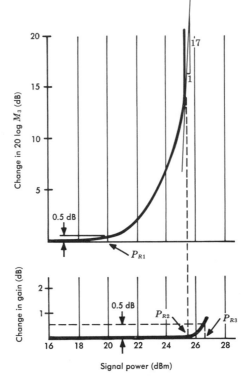

Figure 7-8. Overload point definitions as applied to a typical amplifier.

*Definition 2:* In this case the overload point is defined as that value of output signal power at which the third-harmonic power increases by 20 dB for a 1-dB increase in signal power; this corresponds to a 17-dB increase in 20 log $M_3$. Since under these conditions very serious transmission impairment may result, a more generous overload margin must be provided. This definition of overload is recommended by the CCITT.* Its use is justified by the statistics of system performance interactions in long analog cable systems and by the amplitude/frequency statistics of a broadband signal; together, these statistics are used to show a very low probability of overload. The overload point is illustrated by point $P_{R2}$ in Figure 7-8 at a signal power of about 25.5 dBm.

*Definition 3:* The overload phenomenon may be related to changes in amplifier gain, whereby the overload point is defined as the signal

---

* Comité Consultatif International Telegraphique et Telephonique, Recommendation G.222, II$^d$ Plenary Assembly, (New Delhi: December 8-16, 1960).

power at the output at which the amplifier gain departs from its nominal value by 0.5 dB, as illustrated by point $P_{R3}$ on the lower portion of Figure 7-8. For this illustration, the overload point is about 26.5 dBm. The use of this definition may be appropriate when intermodulation distortion is not a major consideration.

The range of values of defined amplifier overload points is fairly wide, for example, 6.5 dB in Figure 7-8. However, in the event that the overload point is exceeded under definition 2 or 3, performance degradation is so severe that wider system margins must be provided in most cases. Thus, the actual operating value of load might well be approximately the same no matter which definition is used.

## Noise and Terminations

It is desirable to introduce the subject of thermal noise generation in networks and systems here in order to relate the phenomenon to amplifier design and, thus,·to overall system performance [3].

It can be shown that the available noise power of a thermal noise source is directly proportional to the product of the bandwidth of the system or detector and the absolute temperature of the source. This relation can be expresed as

$$p_a = kTB \text{ watts,} \tag{7-9}$$

where $k$ is Boltzmann's constant ($1.3805 \times 10^{-23}$ joule per Kelvin), $T$ is the absolute temperature in Kelvins (290 K is taken as room temperature), and $B$ is the bandwidth in hertz. Available noise power may also be expressed as

$$P_a = -174 + 10 \log B \qquad \text{dBm.} \tag{7-10}$$

The noise figure for a two-port network is defined as follows: "The noise figure at a specified input frequency is the ratio of (1) the total noise power per unit bandwidth at a corresponding output frequency available at the output when the noise temperature of the input source is standard (290 K) to (2) that portion of this output power engendered at the input frequency by the input source" [4]. The noise figure, when applied according to this definition to a narrow band, $\Delta B$, is called a *spot noise figure*. The spot noise figure may vary as a function of frequency.

Alternately, the spot noise figure, $n_F$, can be expressed in terms of signal-to-noise ratios. Such an expression may be written

$$n_F = \frac{p_{si}/p_{ni}}{p_{so}/p_{no}} \qquad (7\text{-}11)$$

where $p$ represents power and the subscripts are $s$ for signal, $n$ for noise, $i$ for input, and $o$ for output. Here, the noise figure is defined as the ratio $(p_{si}/p_{ni})$ of the available signal-to-noise power ratio at the input of the two-port network to the available signal-to-noise power ratio $(p_{so}/p_{no})$ at the output of the two-port when the temperature of the noise source is standard $(T = 290$ K$)$.

The value of $p_{ni}$ can be determined, by substitution in Equation (7-9), as $p_{ni} = kT\Delta B$. The ratio $p_{si}/p_{so}$ is the gain, $g_a(f)$, of the network. The substitution of these values in Equation (7-11) yields

$$n_F = \frac{p_{no}}{g_a(f)\,kT\Delta B} \qquad . \qquad (7\text{-}12)$$

Examination of Equation (7-12) shows that the noise figure of an amplifier is importantly related to the thermal noise generated at the input (where the signal is at its lowest amplitude), to the gain of the amplifier, and to any sources of noise picked up within the amplifier that make the output noise greater than the input noise amplified by $g_a(f)$. These internal noise sources are to some extent subject to control by circuit design techniques. The dominant source, however, is usually at the amplifier input. Here, the noise source is outside the $\mu\beta$ loop and, as a result, the noise figure is not improved by feedback.

The selection of components and the design of the input circuits of amplifiers for minimum noise figure is important in transmission system design. The cumulation of noise in tandem-connected amplifiers is directly related to the nominal noise figure of each and to ten times the logarithm of the number of amplifiers in tandem. Thus, when the number of amplifiers has been set by repeater spacing, gain, and bandwidth considerations, the noise performance is controlled by the individual noise figures of the amplifiers.

As mentioned, the design of feedback amplifiers and their classification into a variety of types depend on the forms which the six-terminal coupling networks take and the manner in which $\beta$ and $\mu$ circuits and external circuit connections are made. At the input, the design must simultaneously (1) satisfy return loss requirements by providing a termination to properly match the amplifier input impedance to the line impedance, (2) minimize the noise figure of the first-stage device by suitably matching its input impedance to the driving point impedance, and (3) meet feedback and gain-shaping requirements. At the output, the design must again satisfy impedance matching and feedback requirements and, in addition, must minimize penalties in nonlinear and overload performance that might result from improper last-stage terminations. In general, these combinations of requirements can best be met by the use of hybrid feedback connections, described previously and illustrated in Figure 7-5.

## Summary

The design and application of negative feedback amplifiers in transmission systems has been described in terms of three sets of interrelated parameters: (1) gain and feedback, (2) nonlinear distortion and overload, and (3) noise and terminations. These relationships are neither unique nor independent of one another. All must be considered simultaneously in the design process.

Design criteria that are involved when a new design is to be undertaken include the bandwidth, the gain, the lowest amplitude the signal may be allowed to reach without picking up excessive noise, the highest permissible signal amplitude that will not exceed overload or intermodulation limits, gain and feedback shaping, device bias conditions, and many others. Only the most important have been touched on in this chapter.

### REFERENCES

1. Bode, H. W. *Network Analysis and Feedback Amplifier Design* (Princeton, N.J.: D. Van Nostrand and Company, Inc., 1945).

2. Technical Staff of Bell Telephone Laboratories. *Transmission Systems for Communications*, Fourth Edition (Winston-Salem, N.C.: Western Electric Company, Inc., 1970), p. 241.

3. Technical Staff of Bell Telephone Laboratories. *Transmission Systems for Communications*, Fourth Edition (Winston-Salem, N.C.: Western Electric Company, Inc., 1970), Chapters 8 and 16.

4. American Standards Association. *Definition of Electrical Terms*, ASA-C42.65, 1957.

5. Thomas, D. E. "High Frequency Transistor Amplifiers," *Bell System Tech. J.*, Vol. 38 (Nov. 1959).

6. Hakim, S. S. *Junction Transistor Circuit Analysis* (New York: John Wiley and Sons, Inc., 1962).

7. Blecher, F. H. "Design Principles for Single Loop Feedback Amplifiers," *Trans. IRE*, Vol. CT-4 (Sept. 1957).

Chapter 8

# Modulation

Communication signals must usually be transmitted via some medium separating the transmitter from the receiver. Since the information to be sent is rarely in the best form for direct transmission, efficiency of transmission requires that it be processed in some manner before being transmitted. *Modulation may be defined as that process whereby a signal is converted from its original form into one more suitable for transmission over the medium between the transmitter and receiver* [1]. The process may shift the signal frequencies to facilitate transmission or to change the bandwidth occupancy, or it may materially alter the form of the signal to optimize noise or distortion performance. At the receiver this process is reversed by methods called demodulation.

Satisfactory transmission and demodulation of modulated signals depend on the introduction by the medium of no more than a specified amount of distortion. The effects of distortion in the medium may be quite different for different modulation modes. If maximum distortion values are exceeded, signal impairments at the receiver are excessive. Distortions that must be considered are of many types. They include amplitude distortion, which results from the variation of transmission loss with frequency, and phase distortion (often expressed as delay distortion), which results from the departure from linear of the phase/frequency characteristic of the channel. Other forms of signal impairment which may result in imperfect signal demodulation include nonlinear channel input/output characteristics, frequency offset, amplitude and phase jump, echoes, and noise. These impairments are treated in later chapters. They are treated in this chapter only where they result directly from the modulation or demodulation process.

The modulation process can be represented mathematically by an equation which, in its most general form, can be used to express any of several forms of modulation. The several forms include amplitude modulation, angle modulation (frequency or phase), and pulse modulation. While other expressions are more representative of the various forms of pulse modulation, there is one form of the equation that lends itself particularly to studies of amplitude and angle modulation,

$$M(t) = a(t) \cos[\omega_c t + \phi(t)].  \qquad (8\text{-}1)$$

Here $a(t)$ represents the amplitude of the sinusoidal carrier, and $\cos[\omega_c t + \phi(t)]$ is the carrier and its instantaneous phase angle. An amplitude-modulated system is one in which $\phi(t)$ is a constant, and $a(t)$ is functionally related to the modulating signal. An angle-modulated system results when $a(t)$ is held constant and $\phi(t)$ is made to bear a functional relationship to the modulating signal. It is appropriate to discuss each of these two types separately and in some detail.

All three general types of modulation (amplitude, angle, and pulse) are used extensively in Bell System equipment. For example, L-type mutiplex equipment and N-type carrier systems employ several forms of amplitude modulation, most microwave radio systems employ angle modulation for the high-frequency signal transmitted between transmitting and receiving antennas, and T-type carrier systems employ pulse modulation and time division multiplex techniques to form the high-frequency line signal.

## 8-1 PROPERTIES OF AMPLITUDE-MODULATED (AM) SIGNALS

Equation (8-1) can be modified to represent amplitude modulation by making $\phi(t)$ a constant. For convenience, let $\phi(t) = 0$ to obtain

$$M(t) = a(t) \cos \omega_c t,  \qquad (8\text{-}2)$$

where the carrier is at the frequency $f_c = \omega_c/2\pi$ and where $a(t)$ is the modulation signal which is a function of time. Since the modulated wave, $M(t)$, is the product of $a(t)$ and a carrier wave, the process is often called product modulation.

A general expression for $a(t)$ may be written as

$$a(t) = [a_0 + mv(t)].  \qquad (8\text{-}3)$$

If Equation (8-3) is normalized by letting the dc component, $a_0$, equal 1, the coefficient $m$ is defined as the modulation index and is equal to unity for 100 percent modulation.

Now, let $v(t)$ represent a signal containing two components at different frequencies, $f_m$ and $f_n$, having amplitudes of $a_m$ and $a_n$, respectively. Then,

$$mv(t) = m(a_m \cos \omega_m t + a_n \cos \omega_n t)$$

and, by substitution in Equation (8-3),

$$a(t) = a_0 + m(a_m \cos \omega_m t + a_n \cos \omega_n t). \tag{8-4}$$

By substitution in Equation (8-2) and by trigonometric expansion, the modulated signal becomes

$$M(t) = [a_0 + m(a_m \cos \omega_m t + a_n \cos \omega_n t)] \cos \omega_c t$$

$$= a_0 \cos \omega_c t$$

$$+ \frac{m}{2} [a_m \cos(\omega_c - \omega_m)t + a_n \cos(\omega_c - \omega_n)t]$$

$$+ \frac{m}{2} [a_m \cos(\omega_c + \omega_m)t + a_n \cos(\omega_c + \omega_n)t]. \tag{8-5}$$

If in Equation (8-5) the coefficients $a_0$ and $a_n$ are zero and in addition $a_m$ and $m$ both equal unity, the resulting modulated wave expressed by Equation (8-5) reduces to

$$M(t) = \frac{1}{2} \cos(\omega_c - \omega_m)t + \frac{1}{2} \cos(\omega_c + \omega_m)t. \tag{8-6}$$

Equation (8-6) contains no component at the original carrier frequency, $f_c$, but only a side frequency on either side of the carrier and spaced $f_m$ hertz from the carrier frequency as shown in Figure 8-1. The terms in Equation (8-6) containing $(\omega_c - \omega_m)$ and $(\omega_c + \omega_m)$ are known as the lower and upper sidebands (LSB and USB), respectively. The resultant wave of Equation (8-6) represents a form of modulation known as double-sideband suppressed-carrier (DSBSC). This form is characterized by a zero-amplitude dc component in the

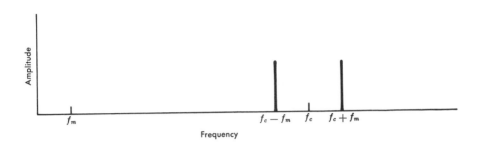

Figure 8-1. Product modulator — single-frequency modulating signal.

modulating signal and, as a result, a modulated signal having no component at the carrier frequency.

Consider next the resultant form of Equation (8-5) if $a_0 = 0$ and $m$, $a_n$, and $a_m$ are all unity. Then the modulated wave is

$$M(t) = \frac{1}{2} \left[ \cos(\omega_c - \omega_m) t + \cos(\omega_c - \omega_n) t \right]$$

$$+ \frac{1}{2} \left[ \cos(\omega_c + \omega_m) t + \cos(\omega_c + \omega_n) t \right]. \qquad (8\text{-}7)$$

The result is as if the two modulating frequency components at $f_m$ and $f_n$ were modulated independently and then added linearly. Thus, superposition holds, the product modulation process is quasi-linear, and it may be inferred that product modulation translates the baseband signal in frequency and reflects it symmetrically about the carrier frequency without distortion.* The result is illustrated in Figure 8-2(a), which shows the two-frequency case, and in Figure 8-2(b), which shows the more general case of a modulating wave having a spectrum from $f_a$ to $f_b$ where $f_b < f_c/2$. Note that if $f_b > f_c/2$, the baseband and lower sideband signals overlap. Ambiguity or distortion, which can occur in the recovered signal, may be avoided in design by choosing frequencies to make $f_b < f_c/2$.

---

*Note that while the mathematical analysis for product modulation is linear, the physical realization of the process often involves the use of nonlinear devices. The mode of operation in these cases still results in a quasi-linear process output.

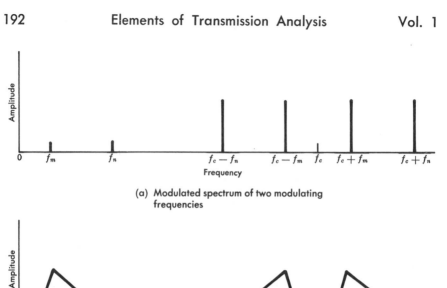

(a) Modulated spectrum of two modulating
    frequencies

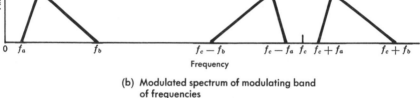

(b) Modulated spectrum of modulating band
    of frequencies

Figure 8-2. Product modulator frequency spectrum — complex modulating signal.

If $a(t)$ is given a strong dc component, i.e., $a_0 \neq 0$, the function $a(t)$ may be restricted to values of one sign only (for example, positive values only). Then, a carrier component in the output wave would result as shown by the first term in Equation (8-5). The resultant wave is known as a double sideband with transmitted carrier signal (DSBTC).

If either sideband in the DSBSC spectrum, Figure 8-2(b), is rejected by a filter or other means, the result is a single-sideband (SSB) wave. Basically, single-sideband modulation is simply frequency translation, with or without the inversion obtainable by selecting the lower rather than the upper sideband. Sideband suppression by filtering is most common. When this is done, the carrier component is usually effectively suppressed with the unwanted sideband.

Up to this point, three types of amplitude-modulated signals have been mentioned: double sideband with transmitted carrier (DSBTC), double-sideband suppressed-carrier (DSBSC), and single sideband (SSB). Subsequently, the properties of these three signals are further examined, and finally a fourth type, known as vestigial sideband (VSB), is considered.

## Double Sideband with Transmitted Carrier

Double-sidedband modulation with transmitted carrier provides a basis for discussing various forms of amplitude modulation. Consider a baseband signal (e.g., a complex wave with a continuous but band-limited frequency spectrum) with a time function represented by $v(t)$ and, for simplicity, a maximum amplitude of unity. The modulating function, $a(t)$, can be forced positive at all times by letting $a_0 \geqq 1$ in Equation (8-3). This ensures that there are no phase reversals in the carrier component.

For a single-frequency modulating wave, $a_n$ equals zero in Equation (8-5) and, letting $a_0 = 1$ and $a_m = 1$, the modulated wave is

$$M(t) = \cos \omega_c t + \frac{m}{2} \cos (\omega_c - \omega_m) t + \frac{m}{2} \cos (\omega_c + \omega_m) t. \qquad (8-8)$$

In many instances the use of exponential notation for periodic functions has advantages over the trigonometric notation which has been used thus far in this chapter. A particularly useful application is in the phasor representation of modulated waves as an aid in understanding the various modulation processes. A sinusoidal carrier, $\cos \omega_c t$, can be written

$$\text{Re}\left[ e^{j\omega_c t} \right] \quad ,$$

where Re represents the real part of the complex quantity and

$$e^{j\omega_c t} = \cos \omega_c t + j \sin \omega_c t.$$

The exponential $e^{j\omega_c t}$ is a counterclockwise rotating phasor of unit length in the complex plane, and its real part is its projection on the real axis. This phasor is shown for three values of time in Figure 8-3.

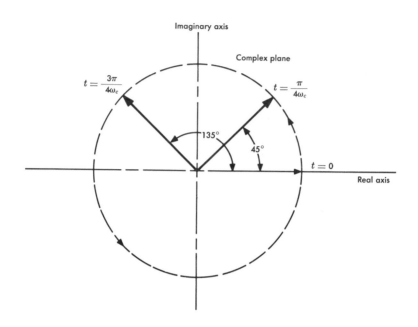

Figure 8-3. Phasor diagram of $e^{j\omega_c t}$.

Now consider the amplitude-modulated wave of Equation (8-8).
This can be written in exponential notation as

$$M(t) = \text{Re}\left[ e^{j\omega_c t} + \frac{m}{2} e^{j(\omega_c - \omega_m)t} + \frac{m}{2} e^{j(\omega_c + \omega_m)t} \right]$$

$$= \text{Re}\left[ e^{j\omega_c t}\left( 1 + \frac{m}{2} e^{j\omega_m t} + \frac{m}{2} e^{-j\omega_m t} \right) \right].$$

In this form the carrier phasor is multiplied by the sum of a
stationary vector and two rotating vectors of equal size which rotate
in opposite directions. As may be seen in Figure 8-4, the sum of
these three vectors is always real and, consequently, acts only to
modify the length of the real part of the rotating carrier phasor.
This produces amplitude modulation as expected.

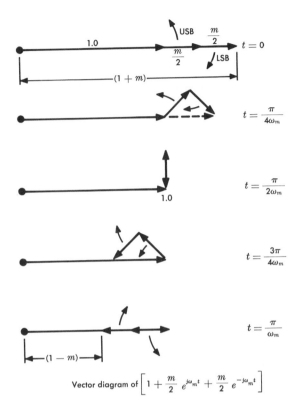

Vector diagram of $\left[1 + \dfrac{m}{2}\,e^{j\omega_m t} + \dfrac{m}{2}\,e^{-j\omega_m t}\right]$

Figure 8-4. Amplitude modulation — index of modulation $= m$.

At this point the average power in the carrier and in the sideband frequencies should be considered. For a unit amplitude carrier and a circuit impedance such that average carrier power is 1 watt, the power in each side frequency is $m^2/4$ watts; thus, the total sideband power is $m^2/2$ watts. Thus, for 100 percent modulation, only one-third of the total power is in the information-bearing sidebands. The sidebands get an even smaller share of the total power when the modulating function is a speech signal which has a higher peak-to-rms ratio than a sinusoid has. The sideband power must be reduced to a few percent of the total power to prevent occasional peaks from over-modulating the carrier.

While the DSBTC signal is sensitive to certain types of transmission phase distortion, it is not impaired by a transmission phase characteristic that is linear with the frequency. The basic requirement for no impairment is that the transmission characteristic have odd symmetry of phase about the carrier frequency.

An interesting degradation occurs under certain extreme transmission phase conditions. Suppose that the lower sideband frequency vector in Figure 8-4 is shifted clockwise by $\theta$ degrees, and the upper sideband frequency is shifted clockwise by $180 - \theta$ degrees. The resulting signal, Figure 8-5, consists of a carrier phasor with the sideband frequency vectors adding at right angles. The resultant vector represents a phase-modulated wave whose amplitude modulation has been largely cancelled, or washed out. A low-index DSBTC signal so distorted is indistinguishable from a low-index phase-modulated signal.

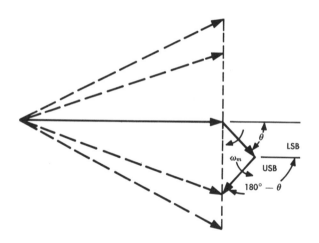

Figure 8-5. Result of certain extreme phase distortion of DSBTC signal to produce phase modulation.

The condition of a lower sideband vector shifted by $\theta$ degrees and the upper sideband vector shifted by $(180-\theta)$ degrees, of course, represents a worst case. Any change in phase relationship between the two sideband vectors and the carrier, other than in odd symmetry, causes partial washout and some phase modulation. Among other things, the index of modulation is, in effect, reduced.

## Double Sideband Suppressed Carrier

The DSBSC signal requires the same transmission bandwidth as DSBTC, but the power efficiency is improved by the suppression of the carrier. This requires reintroduction of a carrier at the receiving terminal, which must be done with extreme phase accuracy to avoid the type of washout distortion just discussed. Examination of Figure 8-4 shows that a $\theta$-degree phase error of the inserted carrier results in the effective amplitude modulation being reduced by the factor cos $\theta$. In the extreme, this effect can be seen by shifting only the stationary unit phasor (the carrier) of Figure 8-4 by 90 degrees to obtain the washout result of Figure 8-5. If the phase error $\theta$ is $\Delta\omega_e t$ radians and the baseband signal is a single-frequency sinusoid, the demodulated signal consists of two sinusoids separated by twice the error frequency of the inserted carrier, $\Delta f_e$ hertz.

The difficulty of accurately reinserting the carrier is the greatest disadvantage of DSBSC and is probably the reason this form has not seen more use. However, the transmitted sidebands contain the information required to establish the exact frequency and, except for a 180-degree ambiguity, the phase of the required demodulating carrier. This is so by virtue of symmetry about the carrier frequency, even with a random modulating wave. One means of establishing the carrier at $f_c$ is to square the DSBSC wave, filter the component present at frequency $2f_c$, and electrically divide the frequency in half [2]. It should be noted that a carrier thus derived disappears in the absence of modulation.

## Single Sideband

The single-sideband signal is not subject to the demodulation washout effect discussed in connection with the DSB signals. In fact, the local carrier at the receiving terminal is sometimes allowed to have a slight frequency error. This produces a frequency shift in each demodulated baseband component. If the error is kept within 1 or 2 Hz, the system is adequate for high quality telephone circuits. However, the single-sideband method of transmission with a fixed or rotating phase error in demodulation does not preserve the baseband waveform at all. This may be seen in Figure 8-6 by considering the phasor representing the upper sideband signal as arising from a single baseband frequency component at $f_m$. The dashed line represents the reference carrier phasor about which the sideband rotates with a relative angular velocity, $\omega_m$.

If a strong carrier of reference phase is added to the received sideband (as could be done in the receiving terminal just ahead of an envelope detector), the envelope of the resultant wave is sinusoidal and peaks when the sideband phasor aligns itself with the carrier. An envelope detector would produce, in the proper phase, a sinusoidal wave of frequency $f_m$.

Figure 8-6.   Upper sideband and reference carrier phasors for SSB signal.

If the phase of the added carrier is advanced 90 degrees, the peaks in the demodulated wave occur 90 degrees later; as a result, the baseband signal is retarded by 90 degrees Although this does not distort the waveform of the single-frequency wave considered, each frequency component in a complex baseband wave would be retarded 90 degrees causing gross waveform distortion as illustrated in Figure 8-7 where the baseband fundamental and the third harmonic are both shifted 90 degrees. Although an envelope detector is assumed here, similar results would follow from analyzing product detection of the SSB signal if the demodulating carrier were shifted relative to the required value, i.e., relative to the real or virtual carrier of the transmitted signal.

Single-sideband signals inherently contain quadrature components, a source of distortion that can cause serious impairment where faithful recovery of the (time-domain) baseband waveform is necessary for satisfactory transmission quality. An SSB signal can be represented as two DSB signal pairs superimposed as in Figure 8-8. One DSB pair has its resultant in phase with the carrier; the other has its resultant at right angles, or in quadrature. The inherent quadrature components and their related desired components are sometimes further shifted by a form of channel distortion called intercept distortion. Whether the distortion is inherent (quadrature distortion) or added (intercept distortion) its reduction or elimination from the demodulated signal is dependent on the signal format and on the design of the demodulator. The desired condition can be approached by adding a strong, or exalted, local carrier to the signal

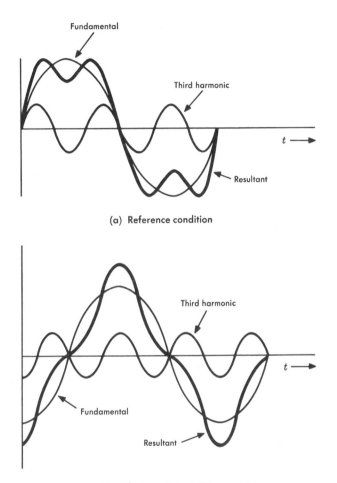

(a) Reference condition

(b) 90° phase shift of all frequencies

Figure 8-7. Waveform distortion due to 90° reference carrier phase error causing 90° lag of all frequencies.

and then using an envelope detector. This approach, illustrated in Figure 8-9, shows that the angle $\theta$ (a measure of unwanted phase modulation) is reduced with exalted carrier as in Figure 8-9(b) relative to its value in Figure 8-9(a). However, the index of modulation is seen to be also reduced. When it is possible to establish the

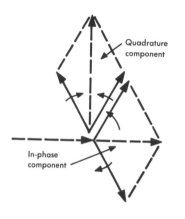

Figure 8-8. Analysis of SSB signal into in-phase and quadrature components.

correct phase of the transmitted or virtual (suppressed) carrier, a more effective way to eliminate quadrature distortion is to use product detection.

Since voice transmission is very tolerant of quadrature distortion, the design of early carrier systems allowed reintroduction of the carrier with a frequency error. The resulting severe quadrature distortion renders these systems unsuitable for transmission of accurate baseband waveforms and makes these systems theoretically unfit for data pulse transmission. Also, many data signals contain very low-frequency or even dc components. An SSB system will not transmit these components since practical filters cannot be built to suppress all of the unwanted sideband without cutting into the carrier frequency and the equivalent low frequencies of the wanted sideband.

A common technique used in carrying data traffic on SSB channels is to modulate a subcarrier in the data terminal, using angle modula-

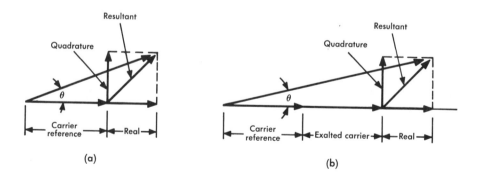

(a)             (b)

Figure 8-9. Quadrature distortion and reduction of phase modulation by exalted carrier.

tion or types of amplitude modulation which permit transmission of dc components. This also solves the quadrature distortion problem, since the subcarrier is transmitted and used in the ultimate demodulation in the receiving data terminal. Since the data subcarrier and the data sidebands travel the same path, the former provides the proper reference information for demodulating the latter, even in the presence of frequency shift. Of course, the baseband channel must be adequately equalized for delay and attenuation.

Single sideband is the modulation technique usually used for the frequency division multiplexing of multiple message channels prior to transmission over broadband facilities. Actually, SSB techniques are often used for interim frequency translations in the multiplex terminal for purposes of convenient filtering [3]. The bandwidth of the signal, measured in octaves, may be increased or decreased by such translations.

## Vestigial Sideband

Vestigial-sideband (VSB) modulation is a modification of DSB in which part of the frequency spectrum is suppressed. It can be produced by passing a DSB wave through a filter to remove part of one sideband as shown in Figure 8-10. The demodulation of such a wave results in addition of the lower and upper sideband components to form the baseband signal. To preserve the baseband frequency spectrum, it is necessary for the filter cutoff characteristic to be made symmetrical about the carrier frequency. This results in the spectrum of the sideband vestige effectively complementing the attenuated portion of the desired sideband. For the same reason and to avoid quadrature distortion, the phase must exhibit odd symmetry about the carrier frequency. As long as the cutoff is symmetrical about the carrier, it can be gradual (approaching DSB conditions) or sharp (approaching SSB conditions) or anywhere between these extremes.

The desired transmission characteristics may be shared among the transmitting and receiving terminals and the transmission medium. The apportioning of the characteristic is determined by economics and signal-to-noise considerations.

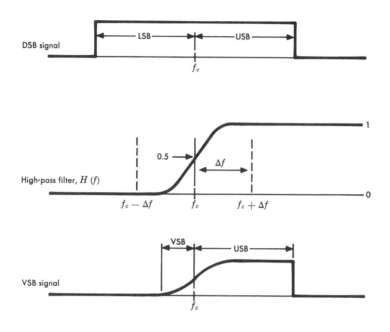

Figure 8-10. Generation of VSB wave. For no distortion,
$1 - H(f_c + \Delta f) = H(f_c - \Delta f)$.

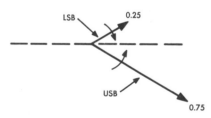

Figure 8-11. VSB phasors for intermediate modulating frequency.

The VSB signal is similar to DSBTC for low baseband frequencies and to SSB for high baseband frequencies. In the cutoff region, the behavior is as shown in Figure 8-11. The upper and lower sideband component vectors add to unity when they peak along the reference carrier line and, if properly demodulated, they produce the same baseband signal as an SSB signal of unit amplitude.

Transmission by VSB conserves bandwidth almost as efficiently as SSB, while retaining the excellent low-frequency baseband characteristics of DSB. Although the ideal SSB signal should allow the sideband spectrum to extend all the way to the carrier frequency, practical limitations on filters and phase distortion make it impractical. Thus, VSB has become standard for television and similar signals where good phase characteristics and transmission of low-frequency components are important but the bandwidth required for DSB transmission is unavailable or uneconomical. It requires somewhat more bandwidth than SSB and has the additional disadvantage that the transmitted carrier, only partially suppressed, may add significantly to signal loading.

## 8-2  PROPERTIES OF ANGLE-MODULATED SIGNALS

Equation (8-1), with $a(t)$ held constant, may be rewritten

$$M(t) = A_c \cos [\omega_c t + \phi(t)] \qquad (8-9)$$

where $\phi(t)$ is the angle modulation in radians. If angle modulation is used to transmit information, it is necessary that $\phi(t)$ be a prescribed function of the modulating signal. For example, if $v(t)$ is the modulating signal, the angle modulation $\phi(t)$ can be expressed as some function of $v(t)$.

Many varieties of angle modulation are possible depending on the selection of the functional relationship between the angle and the modulating wave. Two of these are important enough to have the individual names of phase modulation (PM) and frequency modulation (FM).

### Phase Modulation and Frequency Modulation

The difference between phase and frequency modulation can be understood by first defining four terms with reference to Equation (8-9):

$$\text{Instantaneous phase} = \omega_c t + \phi(t) \qquad \text{rad,} \qquad (8\text{-}10)$$

$$\text{Instantaneous phase deviation} = \phi(t) \qquad \text{rad,} \qquad (8\text{-}11)$$

$$\text{Instantaneous frequency*} = \frac{d}{dt} [\omega_c t + \phi(t)]$$

$$= \omega_c + \phi'(t) \qquad \text{rad/sec,} \qquad (8\text{-}12)$$

$$\text{Instantaneous frequency deviation} = \phi'(t) \qquad \text{rad/sec.} \qquad (8\text{-}13)$$

*The instantaneous frequency of an angle-modulated carrier is defined as the first time derivative of the instantaneous phase.

*Phase modulation* can then be defined as angle modulation in which the instantaneous phase deviation, $\phi(t)$, is proportional to the modulating signal voltage, $v(t)$. Similarly, *frequency modulation* is angle modulation in which the instantaneous frequency deviation, $\phi'(t)$, is proportional to the modulating signal voltage, $v(t)$. Mathematically, these statements become, for phase modulation,

$$\phi(t) = kv(t) \qquad \text{rad} \tag{8-14}$$

and, for frequency modulation,

$$\phi'(t) = k_1 v(t) \qquad \text{rad/sec} \tag{8-15}$$

from which

$$\phi(t) = k_1 \int v(t)\,dt \qquad \text{rad} \tag{8-16}$$

where $k$ and $k_1$ are constants.

These results are summarized in Figure 8-12. This figure also illustrates phase-modulated and frequency-modulated waves which occur when the modulating wave is a single sinusoid.

| TYPE OF MODULATION | MODULATING SIGNAL | ANGLE-MODULATED CARRIER |
|---|---|---|
| (a) Phase | $v(t)$ | $M(t) = A_c \cos\left[\omega_c t + kv(t)\right]$ |
| (b) Frequency | $v(t)$ | $M(t) = A_c \cos\left[\omega_c t + k_1 \int v(t)\,dt\right]$ |
| (c) Phase | $A_m \cos \omega_m t$ | $M(t) = A_c \cos\left(\omega_c t + kA_m \cos \omega_m t\right)$ |
| (d) Frequency | $-A_m \sin \omega_m t$ | $M(t) = A_c \cos\left(\omega_c t + \dfrac{k_1 A_m}{\omega_m} \cos \omega_m t\right)$ |
| (e) Frequency | $A_m \cos \omega_m t$ | $M(t) = A_c \cos\left(\omega_c t + \dfrac{k_1 A_m}{\omega_m} \sin \omega_m t\right)$ |

Figure 8-12. Equations for phase- and frequency-modulated carriers.

Figure 8-13 illustrates amplitude, phase, and frequency modulation of a carrier by a single sinusoid. The similarity of waveforms of the PM and FM waves shows that for angle-modulated waves it is necessary to know the modulation function; that is, the waveform alone cannot be used to distinguish between PM and FM. Similarly, it is not apparent from Equation (8-9) whether an FM or a PM wave is represented. It could be either. A knowledge of the modulation function, however, permits correct identification. If $\phi(t) = kv(t)$, it is phase modulation, and if $\phi'(t) = k_1 v(t)$, it is frequency modulation.

Comparison of (c) and (d) in Figure 8-12 shows that the expression for a carrier which is phase or frequency modulated by a sinusoidal-type signal can be written in the general form of

$$M(t) = A_c \cos (\omega_c t + X \cos \omega_m t) \tag{8-17}$$

where

$$X = kA_m \qquad \text{rad for PM} \tag{8-18}$$

and

$$X = \frac{k_1 A_m}{\omega_m} \qquad \text{rad for FM} \tag{8-19}$$

Here $X$ is the peak phase deviation in radians and is called the index of modulation. For PM the index of modulation is a constant, independent of the frequency of the modulating wave; for FM it is inversely proportional to the frequency of the modulating wave. Note that in the FM case the modulation index can also be expressed as the peak frequency deviation, $k_1 A_m$, divided by the modulating signal frequency, $\omega_m$. The terms *high index* and *low index* of modulation are often used. It is difficult to define a sharp division; however, in general, *low index* is used when the peak phase deviation is less than 1 radian. It is shown later that the frequency spectrum of the modulated wave is dependent on the index of modulation.

•When the modulation function consists of a single sinusoid, it is evident from Equation (8-17) that the phase angle of the carrier varies from its unmodulated value in a simple sinusoidal fashion,

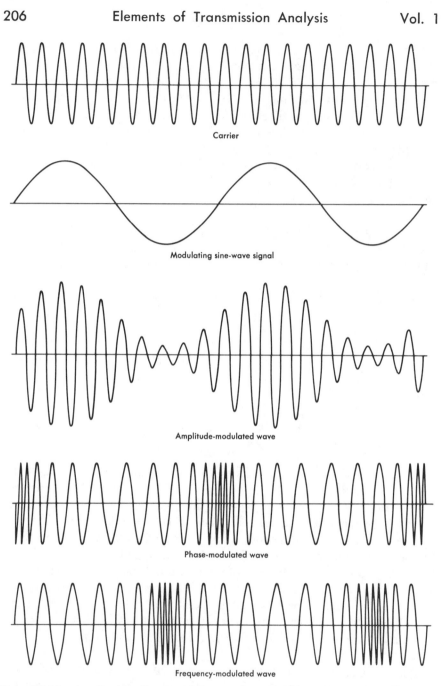

Figure 8-13. Amplitude, phase, and frequency modulation of a sine-wave carrier by a sine-wave signal.

with the peak phase deviation being equal to $X$. The phase deviation can also be expressed in terms of the mean square phase deviation, $D_\phi$, which for this case is $X^2/2$. Similarly, the frequency deviation of a sinusoidally modulated carrier can be expressed either in terms of the peak frequency deviation, $k_1 A_m$ rad/sec $= k_1 A_m/2\pi$ Hz, or the mean square frequency deviation, $D_f$, which is $k_1^2 A_m^2/8\pi^2$ Hz$^2$.

Where a large number of speech signals comprise the complex modulating function, the modulated signal closely approximates a random signal having a Gaussian spectral density function. Hence, from the statistics of the modulated signal, it is possible to define the value of instantaneous voltage that would be exceeded only a specified percentage of the time. Since instantaneous frequency deviation is proportional to instantaneous voltage, it follows that this voltage defines the value of instantaneous frequency deviation that is exceeded only the specified percentage of the time. It is customary to define the peak frequency deviation produced by the complex message load as the deviation exceeded 0.001 percent of the time. The peak deviation determines the required bandwidth.

### Phasor Representation

A wave angle-modulated by sinusoids can be represented by phasors as was done for the AM waves. Generally, the angle-modulated case is more complex as can be seen by expanding Equation (8-17) into a Bessel series of sinusoids. In the special case of very low index ($X$ less than 1/2 radian), all terms after the first can be ignored, and the phasor diagram is very similar to that for an AM wave except for the phase relationship of the sidebands relative to the carrier. In the PM case, the sidebands are phased to change the angle, rather than the amplitude, of the carrier as illustrated in Figure 8-14. A close examination of the phasor diagrams shows that one sideband of the PM wave is 180 degrees out of phase with the corresponding sideband in the AM wave. This can be seen by comparing Figure 8-4 at $t = 0$, for example, with Figure 8-14 at $t = \pi/2\omega_m$. In fact, it was pointed out in the AM discussion that if the inserted carrier of a DSBSC signal has a phase error of 90 degrees, severe washout occurs and the previously amplitude-modulated wave has very little amplitude modulation but considerable phase (or angle) modulation. The approximate phasor diagram for a low-index angle-modulated system

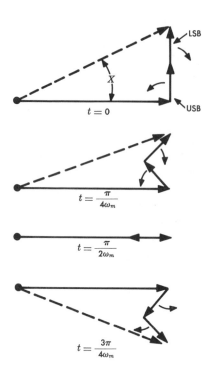

Figure 8-14. Phase modulation — low index.

modulated by a single-frequency sinusoid at $f_m$ is shown in Figure 8-14 for several values of time. The resultant vector has an amplitude close to unity at all times and an index, or maximum phase deviation, of $X$ radians. A true angle-modulated wave would include higher order terms and would have no amplitude variation. If $X$ is small enough, these terms are often ignored.

Several interesting conclusions may be observed by comparing the low-index angle-modulated wave with the AM signal shown in Figure 8-4. Both types of modulation are similar in the sense that they both contain the carrier and the same first-order sideband frequency components. In fact, for the low-index case, the amplitudes of the first-order sidebands are approximately the same when the indices are equal $(X = m)$. The important difference is the phase of the sideband components. It may be expected, therefore, that in the transmission of an FM or PM wave the phase characteristic of the transmission path is extremely important, and certain phase irregularities could easily convert phase-modulation components into amplitude-modulation components.

## Average Power of an Angle-Modulated Wave

The average power of an FM or PM wave is independent of the modulating signal and is equal to the average power of the carrier when the modulation is zero. Hence, the modulation process takes power from the carrier and distributes it among the many sidebands but does not alter the average power present. This may be demonstrated by assuming a voltage of the form of Equation (8-9), squaring, and dividing by a resistance, $R$, to obtain the instantaneous power,

$$P(t) = \frac{M^2(t)}{R}$$

$$= \frac{A_c^2}{R} \cos^2 [\omega_c t + \phi(t)]$$

$$= \frac{A_c^2}{R} \left\{ \frac{1}{2} + \frac{1}{2} \cos [2\omega_c t + 2\phi(t)] \right\}. \qquad (8\text{-}20)$$

The second term can be assumed to consist of a large number of sinusoidal sideband components about a carrier frequency of $2f_c$ Hz; therefore, the average value of the second term of Equation (8-20) is zero. Thus, the average power is given by the zero frequency term

$$P_{avg} = \frac{A_c^2}{2R}. \qquad (8\text{-}21)$$

This, of course, is the same as the average power in the absence of modulation.

## Bandwidth Required for Angle-Modulated Waves

For the low-index case, where the peak phase deviation is less than 1 radian, most of the signal information of an angle-modulated wave is carried by the first-order sidebands. It follows that the bandwidth required is at least twice the frequency of the highest frequency component of interest in the modulating signal. This would permit the transmission of the entire first-order sideband.

For the high-index signal a different method called the quasi-stationary approach must be used [4]. In this approach, the assumption is made that the modulating waveform is changing very slowly so that static response can be used. For example, assume that a 1-volt baseband signal causes a 1-MHz frequency deviation of the carrier. This corresponds to $k_1 = 2\pi \times 10^6$ radians per volt-sec. Then, if the modulating signal has a 1-volt peak, the peak frequency deviation is 1 MHz. Thus, it is obvious that *if the rate of change of frequency is very small* the bandwidth is determined by the peak-to-peak frequency deviation. It was mathematically proven by J. R. Carson in 1922 that frequency modulation could not be accommodated in a narrower band than amplitude modulation, but might actually require a wider band [5]. The quasi-stationary approach for large index indicates that the minimum bandwidth required is equal to the peak-to-peak (or twice the peak) frequency deviation.

Thus, for low-index systems ($X < 1$) the minimum bandwidth is given by $2f_T$, where $f_T$ is the highest frequency in the modulating signal. For high-index systems ($X > 10$), the minimum bandwidth is given by $2\Delta F$, where $\Delta F$ is the peak frequency deviation. It would be desirable to have an estimate of the bandwidth for all angle-modulated systems regardless of index. A general rule (first stated by J. R. Carson in an unpublished memorandum dated August 28, 1939) is that the minimum bandwidth required for the transmission of an angle-modulated signal is equal to two times the sum of the peak frequency deviation and the highest modulating frequency to be transmitted. Thus,

$$B_W = 2(f_T + \Delta F) \qquad \text{Hz.} \qquad (8\text{-}22)$$

This rule (called Carson's rule) gives results which agree quite well with the bandwidths actually used in the Bell System. It should be realized, however, that this is only an approximate rule and that the actual bandwidth required is to some extent a function of the waveform of the modulating signal and the quality of transmission desired.

## 8-3 PROPERTIES OF PULSE MODULATION

In pulse-modulation systems the unmodulated carrier is usually a series of regularly recurrent pulses. Modulation results from varying some parameter of the transmitted pulses, such as the amplitude, duration, or timing. If the baseband signal is a continuous waveform, it is broken up by the discrete nature of the pulses. In considering the feasibility of pulse modulation, it must be recognized that the continuous transmission of information describing the modulating function is unnecessary, provided the modulating function is bandlimited and the pulses occur often enough. The necessary conditions are expressed by the sampling principle, as subsequently discussed.

It is usually convenient to specify the signalling speed or pulse rate in *bauds*. A baud is defined as the unit of modulation rate corresponding to a rate of one unit interval per second; i.e., baud $= 1/T$ where $T$ is the minimum signalling interval in seconds. When the duration of signalling elements in a pulse stream is constant, the baud rate is equal to the number of signalling elements or symbols per second. Thus, the baud denotes pulses per second in a manner analogous to hertz denoting cycles per second. Note that all possible pulses are counted whether or not a pulse is sent, since no pulse is usually also a valid symbol. Since there is no restriction on the allowed amplitudes of the pulses, a baud can contain any arbitrary information rate in bits per second. Unfortunately, *bits per second* is often used incorrectly to specify a digital transmission rate in bauds. For binary symbols of equal time duration, the information rate in bits per second is equal to the signalling speed in bauds if there is no redundancy. In general, the relation between information rate and signalling rate depends upon the coding scheme employed.

### Sampling

In any physically realizable transmission system, the message or modulating function is limited to a finite frequency band. Such a bandlimited function is continuous with time and limited in its possible range of excursions in a small time interval. Thus, it is only necessary to specify the amplitude of the function at discrete time intervals in order to specify it exactly. The basic principle discussed here is called the sampling theorem, which in a restricted form states [6]:

> If a message that is a magnitude-time function is sampled instantaneously at regular intervals and at a rate at least twice the highest significant message frequency, then the samples contain all of the information of the original message.

The application of the sampling theorem reduces the problem of transmitting a continuously varying message to one of transmitting information representing a discrete number of amplitude samples per given time interval. For example, a message bandlimited to $f_T$ hertz is completely specified by the amplitudes at any set of points in time spaced $T$ seconds apart, where $T = 1/2f_T$ [7]. Hence, to transmit a bandlimited message, it is only necessary to transmit $2f_T$ independent values per second. The time interval, $T$, is often referred to as the Nyquist interval.

The process of sampling can be thought of as the product modulation of a message function and a set of impulses, as shown in Figure 8-15. The message function of time, $v(t)$, is multiplied by a train of impulses, $c(t)$, to produce a series of amplitude-modulated pulses, $s(t)$. If the spectrum (i.e., the Fourier transform) of $v(t)$ is given by $F(f)$ as shown in Figure 8-15, the spectrum of the sampled wave, $s(t)$, is then shown by $S(f)$ in the figure. The output spectrum, $S(f)$, is periodic on the frequency scale with period $f_s$, the sampling frequency. It is important to note that a pair of sidebands has been produced around $f_s$, $2f_s$, and so on through each harmonic of the sampling frequency. This figure also shows the need for $f_s > 2f_T$, so that the sidebands do not overlap. Note also that all sidebands around all harmonics of the sampling frequency have the same amplitude. This is a result of the fact that the frequency spectrum of an impulse is flat with frequency. In a practical case, of course, finite width pulses would have to be used for the sampling function, and the spectrum of the sampled signal would fall off with frequency as the spectrum of the sampling function does.

The amplitude-modulated pulse signal that results from sampling the input message may be transmitted to the receiver in any form that is convenient or desirable from a transmission standpoint. At the receiver the incoming signal, which may no longer resemble the impulse train, must be operated on to re-create the original pulse amplitude-modulated sample values in their original time sequence at a rate of $2f_T$ samples per second. To reconstruct the message, it is necessary to generate from each sample a proportional impulse and to pass this regularly spaced series of impulses through an ideal low-pass filter having a cutoff frequency $f_T$. Examination of the spectrum of $S(f)$ in Figure 8-15 makes the feasibility of this obvious. Except for an overall time delay and possibly a constant of proportionality, the output of this filter would then be identical to the original message. Ideally, then, it is possible to transmit information exactly, given the instantaneous amplitude of the message at intervals spaced not further than $1/2f_T$ seconds apart.

## Pulse Amplitude Modulation

In pulse amplitude modulation (PAM), the amplitude of a pulse carrier is varied in accordance with the value of the modulating wave as shown in Figure 8-16(c). It is convenient to look upon

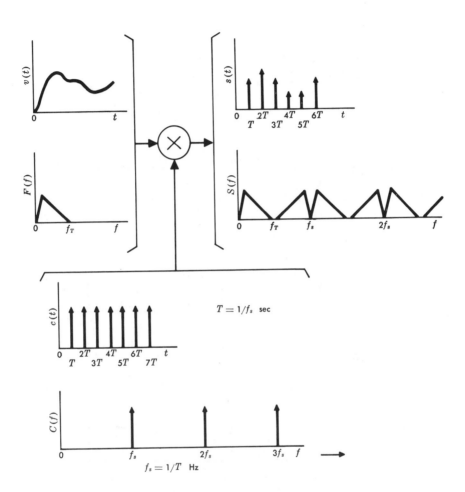

Figure 8-15. Sampling with an impulse modulator.

PAM as modulation in which the value of each instantaneous sample of the modulating wave is caused to modulate the amplitude of a pulse. Signal processing in time division multiplex terminals often begins with PAM, although further processing usually takes place before the signal is launched onto a transmission system.

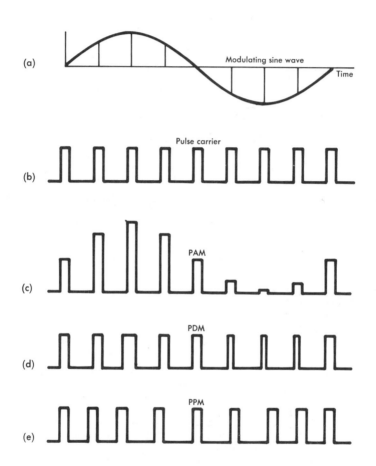

Figure 8-16. Examples of pulse-modulation systems.

## Pulse Duration Modulation

Pulse duration modulation (PDM), sometimes referred to as pulse length modulation or pulse width modulation, is a particular form of pulse time modulation. It is modulation of a pulse carrier in which the value of each instantaneous sample of a continuously varying modulating wave is caused to produce a pulse of proportional duration, as shown in Figure 8-16(d). The modulating wave may vary the time of occurrence of the leading edge, the trailing edge, or both edges of the pulse. In any case, the message to be transmitted is composed of sample values at discrete times, and each value must be uniquely defined by the duration of a modulated pulse.

In PDM, long pulses expend considerable power during the pulse while bearing no additional information. If this unused power is subtracted from PDM so that only transitions are preserved, another type of pulse modulation, called pulse position modulation, results. The power saved represents the fundamental advantage of pulse position modulation over PDM.

## Pulse Position Modulation

A particular form of pulse time modulation, in which the value of each instantaneous sample of a modulating wave varies the position of a pulse relative to its unmodulated time of occurrence, is pulse position modulation (PPM). This is illustrated in Figure 8-16(e). The variation in relative position may be related to the modulating wave in any predetermined unique manner. Practical applications of PPM systems have been on a modest scale, even though their instrumentation can be extremely simple.

If either PDM or PPM is used to time division multiplex several channels, the maximum modulating signal must not cause a pulse to enter adjacent allotted time intervals. In telephone systems with high peak-to-rms ratios, this requirement leads to a very wasteful use of time space. In fact, almost all of the time available for modulation is wasted because many of the busy channels may be expected to be inactive and most of the rest will be carrying small signal power. Consequently, although PPM is more efficient than PDM, both fall short of the theoretical ideal when used for multiplexing ordinary telephone channels.

## Pulse Code Modulation

A favored form of pulse modulation is that known as pulse code modulation (PCM). This mode of signal processing may take any of several forms; each requires the successive steps of sampling, quantizing, and coding. If the input signal is analog in nature, the sampling is usually a sampling of the signal amplitude; if the signal is digital, the sampling process may take the form of time sampling to determine the time of occurrence of transitions from one signal state to another. The process of sampling, common to pulse modulation generally, was described previously. Following is a discussion of quantizing and several forms of coding.

**Quantizing.** Instead of attempting the impossible task of transmitting the exact amplitude of a sampled signal, suppose only certain discrete amplitudes of sample size are allowed. Then, when the message is sampled in a PAM system, the discrete amplitude nearest the true amplitude is sent. When received and amplified, this signal sample has an amplitude slightly different from any of the specified discrete steps because of the disturbances encountered in transmission. But if the noise and distortion are not too great, it is possible to tell accurately which discrete amplitude of the signal was transmitted. Then the signal can be reformed, or a new signal created which has the amplitude originally sent.

Representing the message by a discrete and therefore limited number of signal amplitudes is called *quantizing*. It inherently introduces an initial error in the amplitude of the samples, giving rise to quantization noise. But once the message information is in a quantized state, it can be relayed for any distance without further loss in quality, provided only that the added noise in the signal received at each repeater is not too great to prevent correct recognition of the particular amplitude each given signal is intended to represent. If the received signal lies between $a$ and $b$ and is closer to $b$, it is surmised that $b$ was sent. If the noise is small enough, there are no errors. Note, therefore, that in quantized signal transmission the maximum noise is determined by the number of bits in the code; while in analog signal transmission, it is controlled by the repeater spacing, the characteristics of the medium, and the amplitude of the transmitted signal.

**Coding.** A quantized sample can be sent as a single pulse having certain possible discrete amplitudes or certain discrete positions with respect to a reference position. If, however, many discrete sample amplitudes are required (100 for example), it is difficult to design circuits that can distinguish between amplitudes. It is much less difficult to design a circuit that can determine whether or not a pulse is present. If several pulses are used as a code group to describe the amplitude of a single sample, each pulse can be present (1) or absent (0). For instance, if three pulse positions are used, then a code can be devised to represent the eight different amplitudes shown in Figure 8-17. These codes are, in fact, just the numbers (amplitudes) at the left written in binary notation. In general, a code group of $n$ on-off pulses can be used to represent $2^n$ amplitudes. For example, 7 binary pulses yield 128 sample levels.

| AMPLITUDE REPRESENTED | CODE |
|:---:|:---:|
| 0 | 000 |
| 1 | 001 |
| 2 | 010 |
| 3 | 011 |
| 4 | 100 |
| 5 | 101 |
| 6 | 110 |
| 7 | 111 |

Figure 8-17. Binary code representation of sample amplitudes.

It is possible, of course, to code the amplitude in terms of a number of pulses which have discrete amplitudes of 0, 1, and 2 (ternary, or base 3) or 0, 1, 2, and 3 (quaternary, or base 4), etc., instead of the pulses with amplitudes 0 and 1 (binary, or base 2). If ten levels are allowed for each pulse, then each pulse in a code group is simply a digit or an ordinary decimal number expressing the amplitude of the sample. If $n$ is the number of pulses and $b$ is the base, the number of quantizing levels the code can express is $b^n$. To decode this code group, it is necessary to generate a pulse which is the linear sum of all pulses in the group, each pulse of which is multiplied by its place value $(1, b, b^2, b^3 \ldots)$ in the code.

**Differential Pulse Code Modulation.** This form of pulse modulation has two major potential advantages that can sometimes be used advantageously in particular design situations. First, it can sometimes result in a lower digital rate than straight PCM coding and yet give equivalent transmission performance. Second, the sampling, quantizing, and coding of a signal can be accomplished without the use of large amounts of common equipment. Thus, in situations where large numbers of signals need not be processed simultaneously, it may be more economical than conventional PCM.

Many forms of differential PCM exist [8]. One, known as delta modulation, samples the analog signal at a high rate and codes the samples in terms of the *change* of signal amplitude from sample to sample. The digital rate must be higher than the sampling rate given previously (sampling at a rate at least twice the highest message frequency) because of distortion that might be introduced when the rate of change of signal amplitude is high. The combined sampling and coding process, however, may still result in a lower net digital rate.

### REFERENCES

1. Panter, P. F. *Modulation, Noise, and Spectral Analysis* (New York: McGraw-Hill Book Company, Inc., 1965), p. 1.

2. Rieke, J. W. and R. S. Graham. "The L3 Coaxial System — Television Terminals," *Bell System Tech. J.*, Vol. 32 (July 1953), pp. 915-942.

3. Blecher, F. H. and F. J. Hallenbeck. "The Transistorized A5 Channel Bank for Broadband Systems," *Bell System Tech. J.*, Vol. 41 (Jan. 1952), pp. 321-359.

4. Rowe, H. E. *Signals and Noise in Communication Systems* (Princeton: D. Van Nostrand Company, 1965), pp. 103 and 119-124.

5. Carson, J. R. "Notes on the Theory of Modulation," *Proc. IRE* (Feb. 1922).

6. Black, H. S. *Modulation Theory* (Princeton: D. Van Nostrand Company, 1953), p. 37.

7. Oliver, B. M., J. R. Pierce, and C. E. Shannon. "The Philosophy of PCM," *Proc. IRE*, Vol. 36 (Nov. 1948), pp. 1324-1331.

8. Technical Staff of Bell Telephone Laboratories. *Transmission Systems for Communications*, Fourth Edition (Winston-Salem, N. C.: Western Electric Company, Inc., 1970), pp. 592-597 and Chapter 5.

Chapter 9

# Probability and Statistics

The parameters in most engineering problems are not unique or deterministic; that is, they can assume a range of values. If extreme or worst case values of the important parameters are used, solutions to such problems are seldom economical, frequently inaccurate, and sometimes not even realizable. Probabilistic solutions must be sought; that is, the nature of the distribution of parameters must be studied and understood, appropriate values must be found to represent the parameters in question, and answers must be found that adequately represent the range of values that the solutions can take on as a result of the range of values of the important parameters. The tools for finding economic solutions to such problems are provided by the related subjects, *probability* and *statistics*.

While the use of extreme values of parameters often leads to impractical solutions to problems, the use of other parameter values (nominal, mean, or average) may also lead to equally impractical solutions. It is important to consider overall distributions of values; in some cases only the average is important, but in other cases extreme values (the tails of the distributions) may have to be taken into account. Sometimes, the extreme cases are solved by legislating against them. For example, telephone loops may be laid out by assuming the use of a single gauge of wire in the cables used in the loop plant. If this were done, the losses of loops longer than some specific value would exceed the loss that can give satisfactory service. Loops having such excess loss are avoided by applying loop design rules that require the addition of gain devices, the use of loading coils, or the use of heavier gauge wire when the loop length exceeds the limit. Losses, however, are still functions of all the parameters mentioned (wire gauge, distance, loading, and gain). If the rules are written so that no possible connection could have excessive loss, the solution would be uneconomical; if too many connections have excess loss, grade

219

of service suffers. Thus, the problem is to find an economical compromise which can provide an overall satisfactory grade of service.

Since the Bell System is so large and complex, it is impractical to measure the values of all similar parameters (noise and loss on all trunks, for example) in order to determine the performance of any part of the plant or to describe the characteristics of any part of the plant. Instead, the plant is described on the basis of statistical parameters using only a few key numbers, such as one to represent some central or average value and one to represent the dispersion or spread of the data. Estimates of such numbers can be determined by measuring only a properly chosen sample of the total universe of values.

Probability theory provides a mathematical basis for the evaluation and manipulation of statistical data. The theory treats events that may occur singly or in combination as a result of interacting phenomena which themselves may be occurring sequentially or simultaneously.

Following a classical process of deductive reasoning, the theory of probability [1] evolved from a number of postulates which were based on experimental observations. The postulates were tested and, where necessary, modified to fit observed data. Finally, clearly defined axioms evolved, and the entire theory was built upon these axioms. Probability theory provides the means for expressing or describing a set of observations more efficiently than by enumerating all numbers in the set. The unknowns are expressed as functions of a random variable; these functions, which describe the domain and range of the unknown, are derived by a mapping process. This process, together with some of the terminology and symbology that are unique to probability theory, must be described.

The *mean* (or *expected value*), the *standard deviation,* and the *variance* are the principal parameters used in expressions for discrete and continuous functions of a random variable. Methods of summing random variables are available and a number of different types of distributions may be used to represent communications phenomena of various characteristics. Each is represented by a different distribution function. Where functional relationships are not known, statisical analyses are often used.

## 9-1   ELEMENTS OF PROBABILITY THEORY

Probability theory is applied to the study of and relationships among *sets* of observations or data. The largest set, consisting of all the observations or all the data, is known as the *universe*, the *domain,* or the *sample space. Subsets,* which are made up of certain interrelated elements defined according to some specified criteria, are all contained in the sample space. The interrelations among subsets, often referred to simply as sets, are conveniently displayed for study in a sketch called a *Venn diagram* in which the sample space is displayed as a square. Subsets are depicted as geometrical figures within the square; within each figure are located all the elements of that subset.

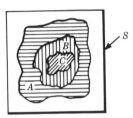

Figure 9-1 is an example of a Venn diagram illustrating the relationships among sets *A, B,* and *C* and the sample space, *S,* of which they are parts. Examination of Figure 9-1 shows that *C* is a subset

Figure 9-1. Venn diagram of three subsets.

of *B, B* is a subset of *A,* and *A* is a subset of *S.* It follows that *C* is a subset of *A* and that *B* and *C* are subsets of *S.* The above statements regarding subsets may be written as follows:

$$C \subset B, B \subset A, A \subset S, C \subset A, B \subset S, C \subset S,$$

where the symbol $\subset$ is used to indicate that every element of the subset shown at the closed end of the symbol is also an element of the larger set shown at the open end of the symbol. Thus, $C \subset B$ (*C* is contained in *B*) may also be written $B \supset C$ (*B* contains *C*).

### Axioms

A number of axioms form the basis of probability theory. These are

(1) *The probability of an event, A, is the ratio of the outcomes favorable to A to the total number of outcomes, n,* where it is assumed that all *n* outcomes are equally likely. Here the

total number of events represents the sample space, and the event $A$ represents the subset of the sample space which satisfies some specific criterion. The axiom may be expressed $P(A) = n_A/n$.

(2) *Probability is a positive real number between 0 and 1 inclusive;* i.e.,

$$0 \leq P \leq 1.$$

In the physical world, negative probability has no meaning and nothing can occur more than 100 percent of the time.

(3) *The probability of an impossible event is zero.* Note that the rule does not imply the converse; i.e., a probability of zero does not mean that an event is impossible. (The impossible event is sometimes called the *empty set,* or *null set,* one that contains no elements.)

(4) *The probability of a certain event is unity.* By certain event is meant one that is certain to occur at every trial. It is the set represented by the sample space. Again, the converse is not necessarily true; i.e., a probability of unity does not necessarily mean that the event is certain.

(5) *The probability that at least one of two events occurs is the sum of the individual probabilities of each event minus the probability of their simultaneous occurrence.*

(6) *The probability of the simultaneous occurrence of two events is the product of the probability of one event and the conditional probability of the second event given the first.*

## Set Operations

Many relationships among the sets (or subsets) of a sample space may be established for the purpose of performing mathematical operations. Such set operations include those of union, intersection, and complement, each of which requires the introduction of additional commonly used symbology.

(1) The *union* of two sets, written $A \cup B$, is defined as the set whose elements are all the elements either in $A$ or in $B$ or in both. The union of $A$ and $B$ is illustrated in the Venn diagram of Figure 9-2(a).

 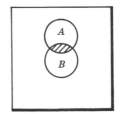

(a) Union $A \cup B$             (b) Intersection $A \cap B$

Figure 9-2. Union and intersection of sets.

(2) The *intersection* of two sets, written $A \cap B$, is defined as the set whose elements are common to set $A$ and set $B$, as illustrated in Figure 9-2(b).

(3) The *complement* of set $A$ is the set consisting of all the elements of the sample space that are not in $A$. The complement is identified by the use of the prime symbol. It is illustrated in Figure 9-3 as $A'$.

Consider now a hypothetical experiment where the totality of results makes up a sample space, $S$, and involves two events, $A$ and $B$. In developing the probabilities associated with these two events, it is convenient to use the above symbology to indicate various compound events, i.e., those involving union or intersection. The total number of possible outcomes (elements of the sample space) is taken as $n$. Any of the $n$ outcomes is assumed to be equally probable. Compound

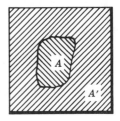

Figure 9-3. Complement sets.

| EVENT | NO. OF OUTCOMES | PROBABILITY |
|-------|-----------------|-------------|
| $A \cap B'$ | $n_1$ | $n_1/n$ |
| $A' \cap B$ | $n_2$ | $n_2/n$ |
| $A \cap B$ | $n_3$ | $n_3/n$ |
| $A' \cap B'$ | $n_4$ | $n_4/n$ |

Figure 9-4. Compound events.

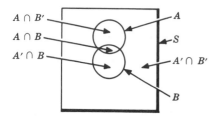

Figure 9-5. Venn diagram of compound events.

events involving $A$ and $B$ can be summarized as in Figure 9-4 and as illustrated by the Venn diagram of Figure 9-5. Each area in Figure 9-5 illustrates the events shown in the first column of Figure 9-4, one of which occurred after each performance of the experiment.

A number of probability relations can be defined and related, by observation, to Figures 9-4 and 9-5. The probability of $A$, without regard for the occurrence of another event, is

$$P(A) = P(A \cap B') + P(A \cap B) = (n_1 + n_3)/n. \qquad (9-1)$$

Similarly, the probability of event $B$ is

$$P(B) = P(B \cap A') + P(B \cap A) = (n_2 + n_3)/n. \qquad (9-2)$$

The probability of either $A$ or $B$ or both is

$$P(A \cup B) = P(A) + P(B) - P(A \cap B) = (n_1 + n_2 + n_3)/n. \quad (9-3)$$

The probability of the event which is the intersection of $A$ and $B$ may be written

$$P(A \cap B) = P(A) \, P(B \mid A) \qquad (9-4)$$

or

$$P(A \cap B) = P(B) \, P(A \mid B). \qquad (9-5)$$

The expressions $P(B \mid A)$ and $P(A \mid B)$ are known as conditional probabilities. These may be read "the conditional probability of $B$, given $A$" and "the conditional probability of $A$, given $B$," respectively. These conditional probabilities may then be determined as

$$P(B \mid A) = \frac{P(A \cap B)}{P(A)} = n_3/(n_1+n_3) \qquad (9\text{-}6)$$

and

$$P(A \mid B) = \frac{P(A \cap B)}{P(B)} = n_3/(n_2+n_3). \qquad (9\text{-}7)$$

Note also that the probability of $A$ and $B$ occurring simultaneously may then be determined

$$P(A \cap B) = n_3/n. \qquad (9\text{-}8)$$

Some additional definitions and conclusions may now be presented. If events $A$ and $B$ cannot occur simultaneously, they are mutually exclusive, or *disjoint;* the probability of their simultaneous occurrence is the probability of the empty set, $\phi$; that is,

$$P(A \cap B)_{\text{disjoint}} = P(\phi) = 0.$$

If this conclusion is combined with Axiom 5, it may be stated that if two events are mutually exclusive, the probability of at least one of them is the sum of their individual probabilities; that is,

$$P(A \cup B)_{\text{disjoint}} = P(A) + P(B).$$

If the occurrence of an event in no way depends on the occurrence of a second event, the two are *independent.* Mathematically, $A$ and $B$ are independent if

$$P(A \mid B) = P(A)$$

or if

$$P(B \mid A) = P(B).$$

Then from Equation (9-4) or (9-5), it is seen that $P(A \cap B) = P(A) \ P(B)$. Note that this does not mean that $P(A \cap B) = 0$. The fact that two events are independent means that there is no functional relationship between their probabilities of occurrence. The expression $P(A \cap B) = 0$ says that $A$ can never occur when $B$ does, a functional relationship of mutual exclusion.

If events $A$ and $B$ can occur simultaneously, then a certain fraction of events $B$ have event $A$ associated with them. If this fraction is the same as the fraction of all possible events that have event $A$ associated with them, then the events $A$ and $B$ are independent. Symbolically, independence implies that

$$P(A \mid B) = \frac{P(A \cap B)}{P(B)} = \frac{P(A) \ P(B)}{P(B)} = P(A). \qquad (9\text{-}9)$$

This can be demonstrated by combining the definition of independence with Axiom 6. If $A$ and $B$ are statistically independent, the probability of their simultaneous occurrence is the product of their individual probabilities, that is,

$$P(A \cap B)_{\text{independent}} = P(A) \ P(B). \qquad (9\text{-}10)$$

Note that Equation (9-10) is symmetric in $A$ and $B$. This implies that if $A$ is independent of $B$, then $B$ is independent of $A$. This need be true only in the statistical sense. It is important to recognize the difference between statistical dependence and causal dependence. From the causal viewpoint, subscriber complaints are dependent on noisy trunks, but noisy trunks are not dependent on subscriber complaints. Statistical analysis would merely show a dependence or correlation between the two without any indication as to which is the cause and which is the effect.

Much statistical work is simplified if it can be assumed that events are either mutually exclusive or independent. Where events are mutually exclusive, the probability of at least one of the events is the simple sum of the probabilities of the mutually exclusive events. The probability of the simultaneous occurrence of independent events may be found as the product of the probabilities of the independent events.

*Example 9-1:*

This example concerns a group of 1000 trunks between two cities. All these trunks are measured for loss and noise. It is found that 925 trunks meet the noise objective, 875 trunks meet the loss objective, and 850 trunks meet both objectives. If a connection is established between the two cities and if there is an equal probability that any trunk may be used, what relationships can be evaluated from the foregoing set operations in regard to calls between the two cities?

Various events and their probabilities may now be tabulated, as in Figure 9-4; symbolic and numerical values are both given in the table below. A Venn diagram of the relationships among the subsets of trunks is given in Figure 9-6.

| EVENT | NOTE | NO. OF OCCURRENCES | PROBABILITY OR RELATIVE FREQUENCY |
|---|---|---|---|
| $S$ | 1 | $n = 1000$ | $n/n = 1000/1000 = 1$ |
| $A$ | 2 | $n_1 = 925$ | $n_1/n = 925/1000 = 0.925$ |
| $B$ | 2 | $n_2 = 875$ | $n_2/n = 875/1000 = 0.875$ |
| $A'$ | 3 | $n_3 = 75$ | $n_3/n = 75/1000 = 0.075$ |
| $B'$ | 3 | $n_4 = 125$ | $n_4/n = 125/1000 = 0.125$ |
| $A \cap B$ | 4 | $n_5 = 850$ | $n_5/n = 850/1000 = 0.85$ |
| $A \cap B'$ | 4 | $n_6 = 75$ | $n_6/n = 75/1000 = 0.075$ |
| $A' \cap B$ | 4 | $n_7 = 25$ | $n_7/n = 25/1000 = 0.025$ |
| $A' \cap B'$ | 4 | $n_8 = 50$ | $n_8/n = 50/1000 = 0.05$ |
| $A \mid B$ | 5 | $n_9 = 850$ | $n_5/(n_7+n_5) = 850/875 = 0.971$ |
| $B \mid A$ | 6 | $n_{10} = 850$ | $n_5/(n_6+n_5) = 850/925 = 0.919$ |
| $A \cup B$ | 7 | $n_5+n_6+n_7 = 950$ | $(n_5+n_6+n_7)/n = 950/1000 = 0.95$ |

*Notes:*

1. $S$ is the sample space, 1000 trunks.

2. $A$ and $B$ are two subsets, the trunks which meet the noise objective and the loss objective, respectively.

3. $A'$ and $B'$ are the complements of $A$ and $B$.

4. These are the four mutually exclusive events which make up the sample space, $S$.

5. $A \mid B$, the event $A$ given $B$, consists of those trunks meeting the noise objective among the trunks which meet the loss objective.

6. $B \mid A$, the event $B$ given $A$, contains the trunks meeting the loss objective among those that meet the noise objective.

7. $A \cup B$ represents all the trunks that meet the noise objective, the loss objective, or both.

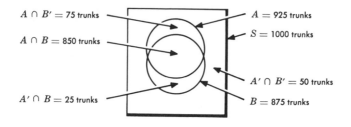

Figure 9-6. Venn diagram for a set of trunks.

## 9-2 DISCRETE AND CONTINUOUS FUNCTIONS

An important objective in working with statistics and with probability theory is a more efficient way of describing a set of observations than by enumerating all of the numbers in the set. A common problem is that of characterizing a set of measurements which are supposed to be similar or identical but which are not. The random variable is a function which may be discrete, as the trunks in Example 9-1, where the trunks either met objectives or they did not. The random variable may also be continuous. In Example 9-1, the data may have related to actual measurements of loss and noise, and the random variable might have represented the distribution of these measurements, i.e., the number of trunks showing noise or loss values in some recognizable measurement system such as dB of loss or dBrnc0 of noise.

### Mapping

Consider a sample space made up of elements designated as $\rho_i$. By a process called mapping, the elements of the space (or domain) can be expressed in terms of a random variable, $X$, which is plotted along an axis. The mapping process is illustrated in Figure 9-7 where $X(\rho_i) = x_i$. By virtue of the rule of correspondence, each element, $\rho_i$, maps into one and only one value, $x_i$, although it is possible for more than one $\rho_i$ to map into the same $x_i$. While every element $\rho_i$ must map into some value, $x_i$, it is not necessary that every $x_i$ be an image of an element, $\rho_i$.

Theoretically, the variable $X(\rho_i)$ may take any value from $-\infty$ to $+\infty$ as indicated in Figure 9-7. It is generally true, however, that the mapping process establishes a restricted range of $x$ between minimum and maximum values. This is also illustrated in Figure 9-7.

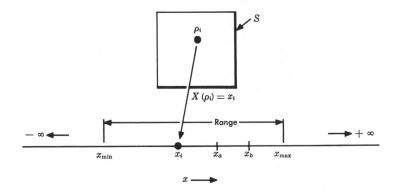

Figure 9-7. Mapping.

If the elements of a sample space exhibit characteristics that involve two parameters, the mapping becomes a two-dimensional process as illustrated by Figure 9-8 where the trunks of Example 9-1 are mapped onto the $x$-$x$ and $y$-$y$ axes. The various events then map into areas in the $x$-$y$ plane of Figure 9-8.

As mentioned previously, the random variable, $X$ or $Y$, may be continuous or discrete. In either case, the treatment and manipulation of data depend on the ability to express these variables by suitable functional relationships, such as the cumulative distribution function or the probability density function.

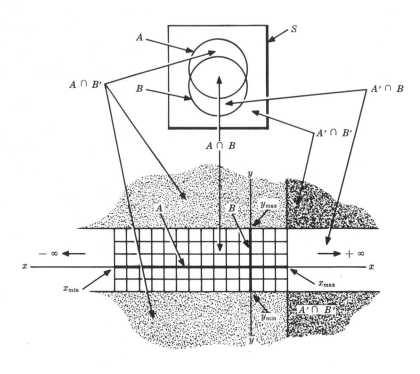

Figure 9-8. Mapping in two dimensions.

## Cumulative Distribution Function

A real random variable (r.v.) is a real function whose domain is the sample space, $S$, and whose range is the real line (the $x$ axis). The r.v. also satisfies the conditions (1) that the set $\{\rho_i : X(\rho_i) \leq x_a\}$* is an event for any real number, $x_i$, and (2) that the probability $P\{\rho_i : X(\rho_i) = \pm\infty\}$ is equal to zero. The function describing the probability distribution of the random variable is called the cumulative distribution function (c.d.f.) and may be written [2]

$$F_X(x_i) = P\{\rho_i : X(\rho_i) \leq x_i\}. \tag{9-11}$$

*In expressions such as this, the braces define the *set* and the colon is read *such that*.

This equation states that the c.d.f. is a function equal to the probability that the variable, $X$ (representing the elements, $\rho_i$, of the sample space), is equal to or less than the value $x_i$. For present purposes, this equation must meet the following conditions:

(1) It is a real function of a real number.

(2) It is right-continuous; that is, the value of the function $F_X(x)$ at any point is *equal to or less than* the given value [$X(\rho_i) \leqq x_i$ in Equation (9-11)].

(3) It is single-valued, monotonic, nondecreasing.

(4) $\lim\limits_{x \to -\infty} F_X(x) = 0$;        $\lim\limits_{x \to \infty} F_X(x) = 1$.

The random variable, $X$, may be continuous or discrete or mixed. When $X$ is continuous, the c.d.f. is continuous. When $X$ is discrete, the c.d.f. is not continuous and, when plotted, appears as a set of steps. These relationships show that $X(\rho_i)$ may take on values from $-\infty$ to $+\infty$; the c.d.f. correspondingly takes on values from 0 to 1 for $X(\rho_i) = -\infty$ to $X(\rho_i) = +\infty$.

If an estimation of a continuous c.d.f. is plotted as in Figure 9-9, the curve looks like an uneven staircase having flat treads and discontinuities in place of vertical risers. As the number of observations increases and the granularity of readings becomes finer, the treads and risers become smaller. A continuous c.d.f. is a smooth curve as illustrated in Figure 9-10.

It should be noted that the plot of a discrete c.d.f. would also look like Figure 9-9.

## Probability Density Function

The derivative of the c.d.f. is defined as the probability density function (p.d.f.). It may be written

$$f_X(x) = dF_X(x)/dx. \tag{9-12}$$

The function is illustrated in Figure 9-11.

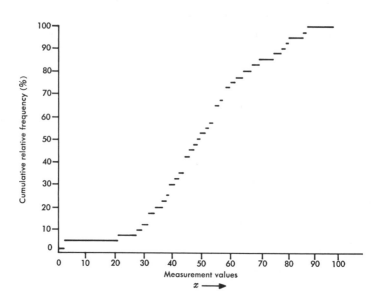

Figure 9-9.  Approximation to a continuous c.d.f.

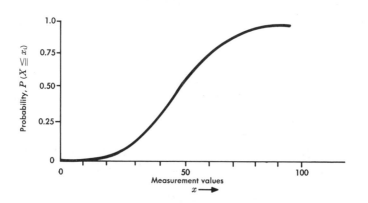

Figure 9-10.  A continuous c.d.f.

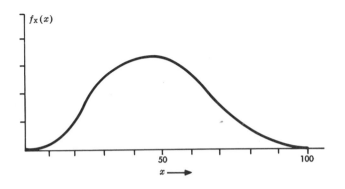

Figure 9-11. Probability density function.

## 9-3 THE PRINCIPAL PARAMETERS

*Expected value, variance,* and *standard deviation* are terms that define the important and useful characteristics of random variables. The definitions of other parameters, sometimes useful in statistical studies but seldom used in probability theory, are given later. When available data permit the use of approximations or estimates of the random variable, estimates of expected value, variance, and standard deviation may also be used for statistical analysis.

### Expected Value

The expected, or mean, value of a random variable, $X$, may be estimated from repeated trials of an appropriate experiment by

$$\overline{X} \approx \frac{\sum_{i=1}^{n} x_i}{n} \tag{9-13}$$

where $\sum_{i=1}^{n} x_i$ is the sum of the values $x_i$, and $n$ is the total number of values of $x$. Here, $x_i$ is the numerical value that the random variable, $X$, takes for the $i^{\text{th}}$ trial.

If the random variable is discrete, the expected value may be found by

$$E[X] = \overline{X} = \sum_{x_i=1}^{n} x_i \, P\{X = x_i\}, \qquad (9\text{-}14)$$

where $x_i$ represents the discrete values assumed by the variable, $X$. If the random variable is continuous, the expected value is found by

$$E[X] = \overline{X} = \int_{-\infty}^{+\infty} x \, f_X\,(x) \; dx. \qquad (9\text{-}15)$$

The term *expectation* has been extended to include the expectation of any function of $X$, provided $X$ has a probability function. The expectation of $g(X)$ is

$$E[g(X)] = \int_{-\infty}^{+\infty} g(x) \, f_X(x) \; dx,$$

where $f_X(x)$ is the probability density function. Of particular interest is $g(X) = X^2$, the mean squared value, or

$$E[X^2] = \int_{-\infty}^{+\infty} x^2 \, f_X(x) \; dx = \overline{X^2}. \qquad (9\text{-}16)$$

## Variance

The expected value of a random variable gives no information regarding the variation or range of values that may be assumed by a random variable. The most useful measure of this parameter is the *variance*, defined as the expectation of the square of the deviations of observations from their mean, or expected, value. The expression for the variance, which may be derived from the function of the random variable, is

$$\sigma_X^2 \approx \int_{-\infty}^{+\infty} (x - \overline{X})^2 \, f_X(x) \; dx.$$

By substituting Equations (9-15) and (9-16) and noting

that $\int\limits_{-\infty}^{+\infty} f_X(x)\,dx = 1$ (since the probability of the entire sample

must be unity), the above expression may be written

$$\sigma_X^2 = \int\limits_{-\infty}^{+\infty} x^2\,f_X(x)\,dx - 2\overline{X}\int\limits_{-\infty}^{+\infty} x\,f_X(x)\,dx + \overline{X}^2 \int\limits_{-\infty}^{+\infty} f_X(x)\,dx$$

$$= \overline{X^2} - 2\overline{X}^2 + \overline{X}^2$$

$$= \overline{X^2} - \overline{X}^2. \tag{9-17}$$

The variance may be estimated, from repeated trials of an experiment,
by

$$\sigma_X^2 \approx \frac{\left(\sum\limits_{i=1}^{n} x_i^2\right) - \left(\sum\limits_{i=1}^{n} x_i\right)^2}{n}. \tag{9-18}$$

Since expectation is a sum or integral, it obeys the same laws
as sums or integrals. The expectation of a constant is that constant.
The expectation of a constant times a random variable is the constant
times the expectation of the random variable. The expectation of a
sum is the sum of the expectations. The mean or any other statistical
average is a constant and not a random variable.

## Standard Deviation

The square root of the variance is often a convenient parameter
to use as a measure of variation or dispersion. It is called the
*standard deviation.* For the approximation given in Equation (9-18),
it is

$$\sigma_X \approx \sqrt{\frac{\left(\sum\limits_{i=1}^{n} x_i^2\right) - \left(\sum\limits_{i=1}^{n} x_i\right)^2}{n}}. \tag{9-19}$$

The exact expression for the standard deviation is found from
Equation (9-17),

$$\sigma_X = \sqrt{\overline{X^2} - \overline{X}^2}. \tag{9-20}$$

*Example 9-2:*

The approximation to the continuous cumulative distribution
function of Figure 9-9 is a plot of the available data concerning

the sample space. From the data determine the expected value, the variance, and the standard deviation. The first three columns in the accompanying table represent the data from which the figure was constructed; the last two are computed values which are summed.

The multiplier, $n$, in the last two columns reflects the fact that all $x_i$ points are not different; $n$ is the number of readings of each value (column 2).

| VALUE (ABSCISSA) | DATA | | $nx_i$ | $nx_i^2$ |
| --- | --- | --- | --- | --- |
| | $n$ | CUM. $n$ | | |
| 3 | 1 | 1 | 3 | 9 |
| 21 | 4 | 5 | 84 | 1764 |
| 27 | 3 | 8 | 81 | 2187 |
| 29 | 2 | 10 | 58 | 1682 |
| 31 | 2 | 12 | 62 | 1922 |
| 33 | 6 | 18 | 198 | 6534 |
| 36 | 2 | 20 | 72 | 2592 |
| 37 | 2 | 22 | 74 | 2738 |
| 39 | 2 | 24 | 78 | 3042 |
| 42 | 6 | 30 | 252 | 10,584 |
| 43 | 2 | 32 | 86 | 3698 |
| 45 | 3 | 35 | 135 | 6075 |
| 46 | 7 | 42 | 322 | 14,812 |
| 47 | 3 | 45 | 141 | 6627 |
| 49 | 3 | 48 | 147 | 7203 |
| 50 | 2 | 50 | 100 | 5000 |
| 52 | 2 | 52 | 104 | 5408 |
| 53 | 3 | 55 | 159 | 8427 |
| 54 | 2 | 57 | 108 | 5832 |
| 56 | 8 | 65 | 448 | 25,088 |
| 58 | 3 | 68 | 174 | 10,092 |
| 60 | 4 | 72 | 240 | 14,400 |
| 61 | 4 | 76 | 244 | 14,884 |
| 64 | 2 | 78 | 128 | 8192 |
| 67 | 2 | 80 | 134 | 8978 |
| 69 | 3 | 83 | 207 | 14,283 |
| 74 | 3 | 86 | 222 | 16,428 |
| 77 | 2 | 88 | 154 | 11,858 |
| 79 | 2 | 90 | 158 | 12,482 |
| 80 | 2 | 92 | 160 | 12,800 |
| 87 | 3 | 95 | 261 | 22,707 |
| 88 | 2 | 97 | 176 | 15,488 |
| 98 | 3 | 100 | 294 | 28,812 |
| | | | 5264 | 312,628 |

From the above table, the expected value may be computed by Equation (9-13) as

$$\overline{X} \approx \frac{\sum\limits_{i=1}^{n} x_i}{n} \approx \frac{5264}{100} \approx 52.6.$$

The variance may be computed from Equation (9-18) as

$$\sigma_X^2 \approx \frac{\left(\sum\limits_{i=1}^{n} x_i^2\right) - \left(\sum\limits_{i=1}^{n} x_i\right)^2}{n}$$

$$\approx \frac{312{,}628}{100} - \left(\frac{5264}{100}\right)^2 \approx 355.$$

The standard deviation, from Equation (9-19), is

$$\sigma_X \approx \sqrt{355} \approx 18.8.$$

## 9-4  SUMS OF RANDOM VARIABLES

In statistical analysis and in applications of probability theory, it is possible to make use of certain relationships between several sample spaces or between a sample space and subsets of that sample space. One example of many such useful relationships is the summing of random variables.

If two independent random variables are known, a new random variable may be derived by adding together repetitively one member from each of the two original random variables. The mean value of the random variable is the sum of the mean values of the original two; that is,

$$(\overline{X+Y}) = \overline{X} + \overline{Y}. \tag{9-21}$$

The variance of the derived random variable is the sum of the original variances. This may be written

$$\sigma_{(X+Y)}^2 = = \sigma_X^2 + \sigma_Y^2. \tag{9-22}$$

These relationships for the random variable derived from the sum of the two independent random variables are valid provided the values of all means and variances are finite. It is also assumed in the derivation of Equations (9-21) and (9-22) that, in addition to the first two random variables being independent, there is equal probability of one member of one random variable combining with any member of the other. The equations may be extended to apply to any number of variables, provided the universes are all independent.

If the two random variables are subtracted, the means subtract but the variances add.

*Example 9-3:*

In this example, it is assumed that telephone connections may be established between switching machines in two cities, A and C, by way of a switching machine in city B. The trunks between A and B have a mean loss of 2.7 dB and a standard deviation of 0.7 dB. The trunks between B and C have a mean loss of 1.6 dB and a standard deviation of 0.3 dB. When connections are established from A to C, there is in each link (AB and BC) equal likelihood of connection via any trunk in the group. Determine the mean loss of connections from A to C and the standard deviation of the distribution of loss between A and C.

The standard deviation of the distribution of overall losses may be found from Equation (9-22). It is

$$\sigma_{AC} = \sqrt{\sigma_{AB}^2 + \sigma_{BC}^2}$$

$$= \sqrt{0.7^2 + 0.3^2} = 0.76 \text{ dB.}$$

The mean value of the derived random variable (the mean loss from A to C) is found from Equation (9-21) to be

$$\overline{X}_{AC} = \overline{X}_{AB} + \overline{X}_{BC}$$

$$= 2.7 + 1.6 = 4.3 \text{ dB.}$$

*Example 9-4:*

Assume the distribution of A to C trunk losses determined in Example 9-3, i.e., $\overline{X}_{AC} = 4.3$ dB and $\sigma_{AC} = 0.76$ dB. Assume further that the distribution of talker volumes at $A$ is given by $\overline{X}_{\text{vol } A} = -15$ vu, and $\sigma_{\text{vol } A} = 2$ vu. The mean of the distribution of volumes at $C$ may be determined by

$$\overline{X}_{\text{vol } C} = \overline{X}_{\text{vol } A} - \overline{X}_{AC}$$

$$= -15 - 4.3 = -19.3 \text{ vu.}$$

The standard deviation of volumes at C is given by

$$\sigma_{\text{vol } C} = \sqrt{\sigma^2_{\text{vol } A} + \sigma^2_{AC}} = \sqrt{2^2 + 0.76^2}$$

$$= 2.14 \text{ vu.}$$

This type of computation, involving the difference between mean values, is applicable to the determination of grade of service.

## 9-5 DISTRIBUTION FUNCTIONS

A number of different distribution functions of random variables are used to represent various phenomena in the field of telecommunications. Each may be expressed mathematically and graphically to illustrate its applicability and general characteristics.

### Gaussian or Normal Distribution

A random variable is said to be normally distributed if its density function is a Gaussian curve, i.e., if the function can be written in the form

$$f_{\mathbf{X}}(x) = A e^{-\alpha x^2}, \quad \alpha > 0.$$

The density functions of many random variables are found to take this form and may be expressed by

$$f_X(x) = \frac{1}{\sigma_X \sqrt{2\pi}} \, e^{-(x-X)^2 2\sigma_X^2} \, , \quad -\infty < x < +\infty \quad (9\text{-}23)$$

where $e$ is the base of natural logarithms. If it is assumed that $X = 0$ and $\sigma_X = 1$, Equation (9-23) represents the unit (standard form) normal density function. It may be written

$$f_X(x) = \frac{1}{\sqrt{2\pi}} \, e^{-x^2/2}. \quad (9\text{-}24)$$

The corresponding unit normal cumulative distribution function is

$$F_X(x) = \frac{1}{\sqrt{2\pi}} \int_{-\infty}^{x} e^{-u^2/2} \, du \quad (9\text{-}25)$$

where $u$ is the variable dummy of integration. To illustrate these functions, Equations (9-24) and (9-25) are plotted as Figures 9-12 and 9-13.

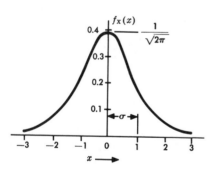

Figure 9-12. Unit normal density function.

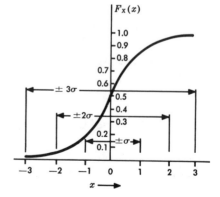

Figure 9-13. Unit normal cumulative distribution function.

The density function of the normal distribution is written in rather simple form as shown in Equation (9-23). The cumulative distribution function, which is its integral, cannot be written in closed form. Its values have been computed by numerical techniques with considerable precision. Values are given in Figure 9-14 for the unit normal cumulative distribution function, Equation (9-25). Per-

| $x$ | $F(x)$ | $x$ | $F(x)$ | $x$ | $F(x)$ |
|---|---|---|---|---|---|
| −4.0 | 0.00003 | −0.9 | 0.1841 | 1.1 | 0.8643 |
| −3.301 | .0005 | −0.842 | .2000 | 1.2 | .8849 |
| −3.090 | .0010 | −0.8 | .2119 | 1.282 | .9000 |
| −3.0 | .0013 | −0.7 | .2420 | 1.3 | .9032 |
| −2.9 | .0019 | −0.674 | .2500 | 1.4 | .9192 |
| −2.881 | .0020 | −0.6 | .2741 | 1.5 | .9332 |
| −2.8 | .0026 | −0.524 | .3000 | 1.6 | .9452 |
| −2.749 | .0030 | −0.5 | .3085 | 1.645 | .9500 |
| −2.7 | .0035 | −0.4 | .3446 | 1.7 | .9554 |
| −2.652 | .0040 | −0.385 | .3500 | 1.8 | .9641 |
| −2.6 | .0047 | −0.3 | .3821 | 1.9 | .9713 |
| −2.576 | .0050 | −0.253 | .4000 | 1.960 | .9750 |
| −2.5 | .0062 | −0.2 | .4207 | 2.0 | .9772 |
| −2.4 | .0082 | −0.126 | .4500 | 2.1 | .9821 |
| −2.326 | .0100 | −0.1 | .4602 | 2.2 | .9861 |
| −2.3 | .0107 | 0 | .5000 | 2.3 | .9893 |
| −2.2 | .0139 | 0.1 | .5398 | 2.326 | .9900 |
| −2.1 | .0179 | 0.126 | .5500 | 2.4 | .9918 |
| −2.0 | .0228 | 0.2 | .5793 | 2.5 | .9938 |
| −1.960 | .0250 | 0.253 | .6000 | 2.576 | .9950 |
| −1.9 | .0287 | 0.3 | .6179 | 2.6 | .9953 |
| −1.8 | .0359 | 0.385 | .6500 | 2.652 | .9960 |
| −1.7 | .0446 | 0.4 | .6554 | 2.7 | .9965 |
| −1.645 | .0500 | 0.5 | .6915 | 2.749 | .9970 |
| −1.6 | .0548 | 0.524 | .7000 | 2.8 | .9974 |
| −1.5 | .0668 | 0.6 | .7257 | 2.881 | .9980 |
| −1.4 | .0808 | 0.674 | .7500 | 2.9 | .9981 |
| −1.3 | .0968 | 0.7 | .7580 | 3.0 | .9987 |
| −1.282 | .1000 | 0.8 | .7881 | 3.090 | .9990 |
| −1.2 | .1151 | 0.842 | .8000 | 3.301 | .9995 |
| −1.1 | .1357 | 0.9 | .8159 | 4.0 | .99997 |
| −1.036 | .1500 | 1.0 | .8413 | | |
| −1.0 | .1587 | 1.036 | .8500 | | |

Figure 9-14. Normal probability distribution function values.

centages of the normal distribution that lie within and outside certain symmetric limits of the normal density function are illustrated in Figure 9-15.

It is often useful to plot the cumulative distribution function from collected data. For the normal distribution this gives an S-shaped

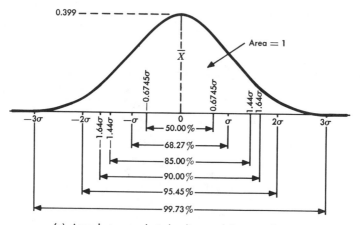

(a) Areas between selected ordinates of the normal curve

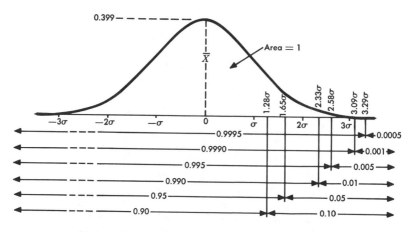

(b) Areas beyond selected ordinates of the normal curve

Figure 9-15. Areas between and beyond selected ordinates of the normal curve.

curve, called an ogive, such as that illustrated in Figure 9-13. By suitable distortion of the cumulative probability scale, the ogive can be made to appear as a straight line. Commercially available graph paper, called arithmetic probability paper, having just such a distorted scale has been designed for use with normal distributions. When a set of observations has been plotted on such paper, it is a simple matter to estimate the mean by reading the 50 percent point, and the standard deviation by reading the values at the 16 percent and 84 percent points, which are separated by approximately $2\sigma$.

It can be shown that (1) with certain constraints, if $n$ samples are drawn from a sample space, the mean values of the samples constitute a random variable whose density, $f_{\bar{x}}(x)$, is concentrated near its mean and (2) as $n$ increases, $f_{\bar{x}}(x)$ tends to a normal density curve regardless of the shape of the densities of the samples of $n$. The constraints are that $n$ must be large (usually greater than 10) and that the standard deviation of the random variable must be finite. This is the *central limit theorem*.

## Poisson Distribution

The Poisson distribution is a discrete probability distribution function which takes the form

$$F_X(x) = \frac{\lambda^x e^{-\lambda}}{x!}, \quad x = 0, 1, 2 \dots , \qquad (9\text{-}26)$$
$$\lambda > 0.$$

The corresponding probability density function is a sequence of impulses expressed

$$f_X(x) = e^{-\lambda} \sum_{x=0}^{\infty} \frac{\lambda^x}{x!}. \qquad (9\text{-}27)$$

In these equations, $\lambda$ is a constant. The derivation of the Poisson distribution is based on the assumptions that the number of observations, $n$, is large (usually greater than 50), that the probability of success, $p$, is small (less than $0.075n$), and that the product of the two, $np$, is a constant. Among the properties of the Poisson distribu-

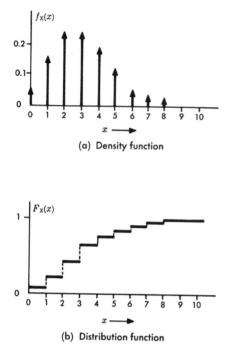

(a) Density function

(b) Distribution function

Figure 9-16. Poisson distribution.

tion are the facts that $\lambda$ is equal to the mean value and that the variance, $\sigma^2$, is also equal to $\lambda$. The Poisson distribution is illustrated in Figure 9-16 for $\lambda = 3$.

This distribution is useful in studying the control of defects in a manufacturing process, the occurrence of accidents or rare disease, and the congestion of traffic, including telephone traffic. It has also been used to represent the statistics of discontinuities in a transmission medium due to certain manufacturing processes and to damage caused by rocks falling upon the cable during installation.

## Binomial Distribution

A combination of $n$ different objects taken $x$ at a time is called a *selection* of $x$ out of $n$ with no attention given to the order of arrangement. The number of combinations of such a selection is denoted by $\begin{pmatrix} n \\ x \end{pmatrix}$. It is defined as

$$\begin{pmatrix} n \\ x \end{pmatrix} = \frac{n!}{x!\,(n-x)\,!}$$

If $p$ is the probability of sucess in any single trial and $q = 1-p$ is the probability of failure, then the probability of success for $x$ times out of $n$ trials is given by

$$P(x) = \begin{pmatrix} n \\ x \end{pmatrix} p^x\, q^{n-x} = \frac{n!}{x!\,(n-x)\,!}\, p^x\, q^{n-x}. \tag{9-28}$$

This is known as the binomial distribution. Its density function may be written

$$f_X(x) = \sum_{x=0}^{n} \binom{n}{x} p^x q^{n-x}. \quad (9\text{-}29)$$

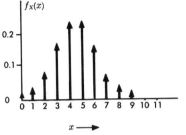

This function is a sequence of impulses as illustrated by Figure 9-17, where $n = 9$ and $p = q = 1/2$.

The mean value of the binomial distribution is equal to $np$ and the variance is $\sigma^2 = npq$.

Figure 9-17. Density function for binomial distribution.

### Example 9-5:

Consider the 1000 trunks of Example 9-1. Recall that 850 of these trunks meet both noise and loss objectives. Thus, 150 trunks fail to meet the loss or the noise objective or both. In five consecutive connections using these trunks, where equal probability of using any trunk is assumed, (1) what is the probability that all five connections will be satisfactory with respect to both noise and loss and (2) what is the probability that two out of five calls will be unsatisfactory?

(1) $p = \dfrac{850}{1000} = 0.85$

$q = 1 - p = 0.15$

$n = 5$ trials

$x = 5$ successful trials.

Using Equation (9-28),

$$P(x) = \frac{5!}{5!(5-5)!} (0.85^5)(0.15^{5-5}).$$

Since it can be shown that $0! = 1$ and $x^0 = 1$, $P(x) = 0.85^5 = 0.44$.

(2) $p = 0.15$

$q = 0.85$

$n = 5$

$x = 2.$

Again using Equation (9-28),

$$P(x) = \frac{5!}{2!(5-2)!} \, (0.15^2) \, (0.85^{5-2})$$

$$= 0.14.$$

## Binomial-Poisson-Normal Relationships

The three distributions described so far are related to one another. If $np$ and $nq$ are both greater than 5, the binomial distribution can be closely approximated by a normal distribution with standardized variable,

$$z = \frac{x - np}{\sqrt{npq}}.$$

The unit normal density function, Equation (9-24), may then be written

$$f_Z(z) = \frac{1}{\sqrt{2\pi}} \, e^{-z^2/2}.$$

If, in the binomial distribution, $n$ is large and the probability, $p$, of an event is close to zero ($q = 1-p$ is nearly 1), the event is called a *rare event*. In practice, an event can be considered rare if $n \geq 50$ and if $np < 5$. In such a case, the binomial distribution is very closely approximated by the Poisson distribution with $\lambda = np$. For this case the Poisson density function, Equation (9-27), may be written

$$f_X(x) = e^{-np} \sum_{x=0}^{\infty} \frac{(np)^x}{x!}$$

## Log-Normal Distribution

Here the random variable is normally distributed when expressed in logarithmic units, for example, decibels. Commercially available graph paper is designed so that a log-normal distribution plots as a straight line.

The log-normal distribution is often encountered in transmission work. In some cases, where the phenomena to be analyzed are multiplicative, the treatment of log-normal distributions is straight-forward because in logarithmic form the phenomena are additive and so may be treated as any other random variable in which additive combinations are under consideration. An example is the evaluation of the overall gain or loss of a circuit containing many tandem-connected components, each of which may be represented by a random variable whose distribution is log-normal. A transmission system having a number of transmission line sections and a number of amplifiers in tandem can be so analyzed.

In some cases, the phenomena are individually log-normal but are combined in such a way that the antilogarithms must be considered as the random variables. An example is found in the analysis of signal voltages of combinations of talker signals in multichannel telephone transmission systems. Here, the individual talker distributions are log-normal. The distributions, however, combine by voltage (not log-voltage) to produce a total signal which must be characterized with sufficient accuracy to evaluate the probability of system overload. The analysis, which must be made by graphical or mathematical approximations, has been applied to load-rating theory for transmission systems [3].

## Uniform Distribution

This distribution, sometimes called a rectangular distribution from the shape of the density function, is represented by the density function

$$f_X(x) = \frac{1}{x_b - x_a}, \quad x_a \leqq x \leqq x_b \tag{9-30}$$

$$= 0, \text{ elsewhere.}$$

This function and the corresponding distribution function are shown in Figure 9-18.

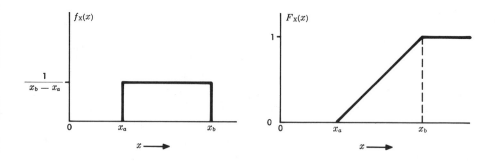

Figure 9-18. Density and cumulative distribution functions of a uniform, or rectangular, distribution.

*Example 9-6:*

Given a manufacturing process for an amplifier having 6-dB gain with acceptance limits of ± 0.25 dB and given that the random variable (the gain) is uniformly distributed between the two limits, what is the probability that the gain, $G$, is between 5.9 and 6.1 dB?

From Equation (9-12), it can be shown [1] that

$$F_X(x_2) - F_X(x_1) = \int_{x_1}^{x_2} f_X(x)\,dx$$

and that

$$P\{5.9 \leqq G \leqq 6.1\} = \int_{5.9}^{6.1} f_G(x)\,dx.$$

From Equation (9-30),

$$f_G(x) = \frac{1}{x_b - x_a} = \frac{1}{0.5}.$$

Then

$$P\{5.9 \leqq G \leqq 6.1\} = \frac{1}{0.5} \int_{5.9}^{6.1} dx$$

$$= \frac{0.2}{0.5} = 0.4.$$

Thus, about 40 percent of all amplifiers of this type will have gain values between 5.9 and 6.1 dB.

### Rayleigh Distribution

The density function for the Rayleigh distribution may be written

$$f_X(x) = \frac{x}{\sigma^2} e^{-x^2/2\sigma^2} , x \geqq 0 \qquad (9\text{-}31)$$

$$= 0, \qquad\qquad x < 0 \quad .$$

This density function, illustrated by Figure 9-19, is often used to approximate microwave fading phenomena.

## 9-6   STATISTICS

The subject of statistics covers the treatment and analysis of data and the relationships between the data and samples taken from the data. Statistics also includes methods of evaluating the confidence in the accuracy of the relationships inferred from data samples.

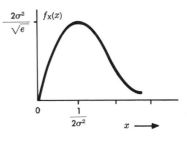

### Central Values and Dispersions

For many purposes, statements of the central value, $\overline{X}$, and the dispersion, $\sigma$, provide an adequate summary of a set of observations.

Figure 9-19. Rayleigh density function.

Estimates of central values and dispersions, based on experimental outcomes, are frequently used instead of functional relationships which are often not known.

There are a number of ways of expressing both the central value and the dispersion of a random variable. Since the mean and variance are most easily treated, it is frequently convenient to transform other measures of central values or dispersions to the mean and variance. A specific reason for this convenience is that the relationships of the mean and variance of a subset of samples to the mean and variance of the sample are simple and essentially independent of the nature of the density and distribution functions representing the sample space.

**Central Values.** A central value may be regarded as an average, where the word *average* is used in its broadest sense. Following is a list of expressions for the central value of a set of observations that might be used in various circumstances:

(1) The *median* is a central value of the random variable defined such that, in a set of observations, half the observations have values greater than the median and half less than the median. If a number of discrete observations are arranged in order of magnitude, the median is the middle one if there is an odd number of observations. If there is an even number of observations, the median is the arithmetic average of the two middle observations.

(2) The *midrange* is one-half the sum of the largest and smallest of the observations.

(3) The *mode* is the most common value of the variable. It is an estimate of the value of $x$ at the maximum of the density function, Equation (9-12). If the density function has two or more maxima, the distribution is described as bimodal or multimodal.

(4) The *geometric mean* is the $n$th root of the magnitude of the product of all $n$ observations.

(5) The *root mean square* (rms) is the square root of the arithmetic mean of the squares of the observations.

(6) The *arithmetic mean* is the measure having the greatest utility in probability theory. It is sometimes simply called the mean

value or the average, where here *average* has a narrower connotation than used earlier. Arithmetic mean is an estimate of mathematical expectation. As shown previously in Equation (9-13), the estimate of the mean may be written

$$\overline{X} \approx \frac{\sum\limits_{i=1}^{n} x_i}{n}$$

where $\sum\limits_{i=1}^{n} x_i$ is the sum of the values of the observations, $x_i$, and $n$ is the total number of observations.

**Dispersions.** A complete description of the dispersion might consist of a tabulation of all deviations. The deviation of any observation, in turn, is the magnitude of the difference between that observation and some stated central value of the observations. Some central value of deviations may be defined as a measure of dispersion. Several commonly used expressions and definitions for dispersions are given in the following:

(1) The *range* is simply the difference between the smallest and largest observations in the sample.

(2) The *mean deviation* is the arithmetic mean of absolute deviations about the mean central value. It is seldom used.

(3) The *standard deviation* is the measure of dispersion which is used as the basis for most of the mathematical treatment of dispersion values in probability theory and in statistical analysis. It is sometimes called the rms deviation. It is given the Greek letter $\sigma$ as its symbol. The square of the standard deviation is the variance.

## Histogram

Sometimes it is desirable to display graphically the number of observations that fall in certain small ranges or intervals. The entire range of observations is divided into cells, and the number of observations falling in each cell is listed. The upper bound of each cell is included in the cell. A graphical representation, shown in Figure 9-20, may be prepared by showing values of $x$ as the abscissa and constructing at each cell a rectangle having an area proportional to the number of observations in the cell. The resulting diagram is called a histogram. If the ordinate is expressed in terms of the fraction or percentage of total observations, the histogram is also called a relative frequency diagram. This is illustrated by the right-hand ordinate scale in Figure 9-20. The illustration is a plot of the data of Example 9-2 and of the c.d.f. illustrated by Figure 9-9.

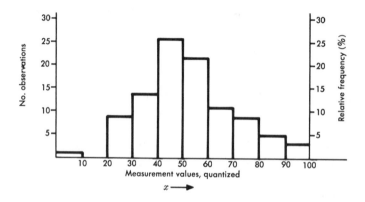

Figure 9-20. Histogram.

## Sampling

Sampling involves the measurement of some elements of a universe or sample space. By proper choice of sampling procedure, parameters describing the samples can be used to establish relationships between the parameters of the samples and the parameters of the universe from which they were drawn. Thus, sampling is useful in the estimation of the parameters of the sample space.

Sampling theory is also useful in the determination of the significance of differences between two samples. Tests of significance and decision theory depend on sampling theory [4].

To assure the validity of the results of a sampling procedure, samples must be chosen so that they are representative of the universe. One such process is called random sampling. This may be accomplished physically, for example, by drawing the samples from a bowl in which the universe is represented by properly identified slips of paper or other representative elements. The bowl is agitated before each drawing of a sample to guarantee random selection. Tables of random numbers are also available and can sometimes be used to advantage.

If a sample is drawn from the universe, recorded, and then placed back into the universe before the next is drawn, the process is called sampling with replacement. If it is not returned, the process is called sampling without replacement. Both processes are used, the choice depending on circumstances. The process of determining the method of sampling is involved in the design of the experiment.

As previously discussed, the sum of random variables may be expressed in terms of probability theory as the addition of two or more functions of the random variable. Summing may also be a statistical process. A sample may be drawn from one sample space and another sample from the same sample space or from a second one. The two values are added and recorded, and the samples are returned and mixed. The process is repeated many times, and from the recorded data the mean and variance may be computed for the summed data by Equations (9-21) and (9-22). Such a process is often necessary when the parameters of the two original sample spaces are not known.

If one value is drawn from each of several similar universes or if several values are drawn from a single universe, independence among the members of the sample is maintained by sampling with replacement. The relationships among the samples can then be used to estimate the parameters of the original universe(s). Such a sample is called a random sample. The size of the sample is designated by the number of members or values, $n$.

The mean of the sample is the sum of the sample values divided by $n$. Thus, the mean of the sample means is equal to the mean of the universe. The variance of the sample means is equal to the variance of the universe divided by the sample size.

**Estimation.** Estimation involves the drawing of inferences about a universe or sample space from measurements of a sample drawn from the sample space. Estimation is sometimes broken down into point estimation and interval estimation. In point estimation a particular parameter, such as the mean of the unknown universe, is sought. In interval estimation two values are sought between which some fraction (such as 99 percent) of the unknown universe is believed to lie. In some respects, interval estimation is simply two problems in point estimation.

In choosing the methods of estimation, the sample and the set of observations based on it may contain extraneous material which does not belong but which may have a serious effect on the estimate. Estimation may have as one of its objectives the identification and elimination of such invalid data.

In experimentation and production, measurements are made to be used as samples from the potential universe of measurements that might be made. The universe as a whole is nonexistent; its parameters are not and cannot be known, and there is no way to determine what they really are. However, it is expected that, if many measurements are made and the results expressed statistically, the computed values will very nearly represent "true values" for the universe. Since the universe is in fact nonexistent and since the "true values" cannot really be defined, there is no way to define a best estimation. It is necessary to rely on statistical analyses to determine that results are consistent and unbiased.

This discussion of estimation has been presented to give realization that, while there are similarities between the methods of estimation and prediction, there are significant differences, too. Space does not permit a more thorough discussion of estimation but one more point must be stressed, that of confidence limits.

**Confidence Limits.** In addition to making estimates of sample space parameters, it is often desirable to express a measure of the limits of confidence in the values. If there is a sample of $n$ observations having a mean value of $\bar{x}_n$ taken from a sample space having a standard deviation of $\sigma$, the mean of the sample space may be said to have a value $\bar{x}$ lying between limits of $a\sigma/\sqrt{n}$ and $-a\sigma/\sqrt{n}$ about $\bar{x}$. This interval is called a confidence interval and its two end values,

$\bar{x} \pm a\sigma/\sqrt{n}$, are the confidence limits. If by using this estimation procedure to set the interval it is expected that the right answer is obtained 99.7 percent of the time (the area under the normal density curve between $\pm 3\sigma$ points, i.e., $a = 3$), the limits are called 99.7 percent confidence limits. Methods are available for determining limits for various levels of confidence for different types of distributions [4].

### REFERENCES

1. Papoulis, A. *Probability, Random Variables, and Stochastic Processes* (New York: McGraw-Hill Book Company, Inc., 1965).

2. Thomas, J. B. *An Introduction to Statistical Communication Theory* (New York: John Wiley and Sons, Inc., 1969).

3. Holbrook, B. D. and J. T. Dixon. "Load Rating Theory for Multi-Channel Amplifiers," *Bell System Tech. J.*, Vol. 18 (Oct. 1939), pp. 624-644.

4. Spiegel, M. R. "Theory and Problems of Statistics," *Schaum's Outline Series* (New York: McGraw-Hill Book Company, Inc., 1961).

5. Fry, T. C. *Probability and Its Engineering Uses* (Princeton, N. J.: D. Van Nostrand Company, Inc., 1928).

6. Feller, W. *An Introduction to Probability Theory and Its Applications*, Vol. I (New York: John Wiley and Sons, Inc., 1957).

7. Wortham, A. W. and T. E. Smith. *Practical Statistics in Experimental Design* (Columbus, Ohio: Charles E. Merrill Books, Inc., 1959).

Chapter 10

# Information Theory

The significance and impact of information theory on the conception, design, and understanding of communication systems have been very large in the years since the publication in 1948 of Shannon's first paper on the subject, later published in book form [1]. While the subject has its roots and genesis in abstract mathematical thinking, its importance is so great that it cannot be bypassed or overlooked here on the excuse that its thorough understanding requires a full knowledge of underlying mathematical principles which are beyond the scope of this book and assumed level of academic background of its readers.

The transmission and storage of information—by human speech, letters, newspapers, machine data, television, and countless other means—are among the most commonplace and most important aspects of modern life. The processes have at least three major facets: syntactic, semantic, and pragmatic.

The syntactic aspects of information involve the number of possible symbols, words, or other elements of information, together with the constraints imposed by the rules of the language or coding system being used. Syntactics also involves the study of the information-carrying capabilities of communications channels and the design of coding systems for efficient information transmission with high reliability.

In communications engineering, the technical problems of the syntactic aspects of information are of primary concern. While this may appear to restrict the engineering role to one that is relatively superficial, it must be recognized that the semantic and pragmatic aspects of information transmission may be seriously degraded if excessive syntactic errors are introduced. Therefore, the importance of these other aspects of information transmission must be appreciated while the technical problems of transmitting and storing information are being solved.

The semantic aspects of information often involve the ultimate recipient of the information. The understanding of a message depends on whether the person receiving it has the deciphering key or understands the language. The problems of semantics generally have little to do with the properties of the communication channel per se.

The pragmatic aspects of information involve the value or utility of information. This is even more a function of the ultimate recipient than semantics. The pragmatic content of information depends strongly on time. For example, in a production management system, the information on production, sales, inventories, distribution, etc., is made available at regular intervals. If the information is late, its value may be significantly decreased; indeed it may be worthless to the recipient.

Ultimately, the value of any information system is dependent on all three aspects of storage and transmission of information. The user's willingness to pay for a system is a function of its practical utility, and a more complex and expensive system can be justified only by the increased utility of faster response times or greater accuracy.

The purposes here are (1) to present a brief historical sketch of the mathematical background to Shannon's work, (2) to provide some appreciation for the subject in terms of what is meant by *information* and its important relationships to probability theory, (3) to present enough mathematical background to illustrate the importance and power of information theory, and (4) to present the fundamental theorems of information theory and discuss their relationship to transmission system design and operating problems.

## 10-1   THE HISTORICAL BASIS OF INFORMATION THEORY

The basis upon which most modern communication theory is built has an extensive, implicit background in the work of Fourier. Early in the nineteenth century he demonstrated the great utility of sinusoidal oscillations as building blocks for representing complex phenomena and, by his studies on heat flow, he revealed the nature of factors governing response time in physical systems. This led to the modern description of communications systems in terms of available bandwidth, which, in turn, is related to the impulse response of the system when signals more complicated than sine waves are impressed. The signal itself is regarded as having a spectrum defining the relative importance of different frequencies in its composition and a bandwidth determined by the frequency range. If the bandwidth of the system is less than that of the signal, imperfect transmission occurs.

During the late 1920s, Nyquist and Hartley made significant contributions [2, 3, 4]. Hartley's work quantified the relationship between signalling speed, channel bandwidth, and channel time of availability. Nyquist's analyses led to conclusions that are now well-known throughout the communications industry as Nyquist's criteria for pulse transmission [3]. They apply to the suppression of intersymbol interference in a bandlimited medium. The criteria may be stated as follows:

(1) Theoretically error-free transmission of information may be achieved if the signalling rate of the transmitted signal is properly related to the impulse response of the channel as discussed in Chapter 6. These conditions are met if the time of occurrence of any pulse corresponds to the zero amplitude crossings of pulses received during any other time interval. When the proper conditions are met, the maximum signalling rate is $2f_1$ bauds, where $f_1$ is the cutoff frequency of the channel expressed in terms of a low-pass filter characteristic.

(2) Equally valid error-free transmission of information may be accomplished if the channel characteristic produces zero amplitude crossings of the received pulses at intervals corresponding to those halfway between adjacent signal impulses. In this case, the receiving circuits are adjusted to detect transitions in the signal at intervals corresponding to those times halfway between adjacent signal impulses. (At the cost of doubling the band, criteria 1 and 2 can be met simultaneously by providing a certain channel characteristic, namely the so-called raised cosine characteristic. This channel characteristic is often used because it provides margin for departures from ideal in filter design, in the timing circuits needed to perform the detection function, and in protection against external sources of interference.)

(3) The third criterion for error-free transmission is that the area under a received signal pulse should be proportional to the corresponding impressed signal pulse value. The response to each impulse, therefore, has zero area for every signalling interval except its own.

Nyquist and Hartley were concerned with maximum efficiency (highest speed) of transmission of telegraph signals in a bandlimited system; the rate of transmission must take into account performance limitations due to intersymbol interference and interrelated channel and signal characteristics. Insofar as external sources of interference were concerned, they assumed an ideal, noise-free transmission medium. As a result of his work, Nyquist's name has found a place in the technical vocabulary in such terms as *Nyquist bandwidth, Nyquist rate,* and *Nyquist interval.*

Applying their research efforts to a generalized channel and to considerations of performance in the presence of noise and interference from external sources, Wiener and Shannon made significant contributions to communications theory during the 1940s and 1950s [1, 5, 6]. Shannon, particularly, is credited with initiating the science of information theory.

## 10-2  THE UNIT OF INFORMATION

To be useful, information must be expressed in some form of symbology that is known and understandable to both the originator and the recipient of a message. The symbology may be spoken or written English, French, or German; it may be the dots and dashes of Morse code; it may be the varying waveforms of a television video signal; it may be the 0s and 1s of a binary code, etc. Although *information* is popularly associated with the idea of knowledge, in information theory it is associated with the uncertainty in the content of a message and the resolution of that uncertainty upon receipt of the message.

If a message source forms messages as a set of distinct entities, such as Morse code symbols, the source is called a *discrete* source. If the messages form a set whose members can differ minutely, such as the acoustic waves at a telephone set or the light variation picked up by a television camera, the message source is said to be a *continuous* source. In either case, it is possible to express the information in terms of equivalent discrete symbols. If the message produces a continuous signal, the translation from a continuous to a discrete format is accomplished by the use of the sampling theorem and a process of quantization (see Chapter 8).

The simplest discrete format is binary; that is, the information is expressed in symbols that can attain one of two equally likely values. The unit used to express the binary format is the *bit* (*binary digit*). In binary terms, the number of information symbols generated may be expressed as

$$m = \log_2 n \text{ bits,} \tag{10-1}$$

where $m$, the amount of information, is a function of the logarithm of the number of outcomes, $n$, that may be attained by the message source. In Equation (10-1), the logarithm is taken to the base 2; this has been found to be the most convenient in solving theoretical communication problems because most practical system applications are binary.* Therefore, the unit most generally used in information theory is the bit.

Although *bit* was derived from *binary digit*, the two are really different and care should be exercised in their use. The bit is a measure of information, while a binary digit is a symbol used to convey that information. To illustrate the difference, consider a channel capable of transmitting 2400 arbitrarily chosen off-or-on pulses per second. Such a channel has an information capacity of 2400 bits per second but, if the channel is used to transmit a completely repetitive series of off-on pulses at 2400 per second, the actual rate of information transmission is 0 bits per second despite the fact that the channel is then transmitting 2400 binary digits per second. To say the channel is transmitting 2400 bits per second under these conditions is to misuse the word *bit*.

## 10-3  ENTROPY

It should be recognized now that the information contained in a message is a matter of probability. The message is a set of symbols taken from a larger set. If there is no uncertainty about what the

---

*Information can, of course, be measured in logarithmic units other than those to the base 2. If base 10 is used, the information is measured in decimal digits, or hartleys; if the base $e$ is used, information is in natural units, or nats.

message is (what set of symbols is expected by the recipient), the message contains no information. If there is uncertainty and by successful receipt and decoding of the message the uncertainty is resolved, an amount of information has been transmitted equal to that defined by Equation (10-1). Something more is needed, however, some measure of the uncertainty in a message before it is decoded. This measure is called *entropy*.

Since a coded message is chosen from among a set of code symbols, there are more choices and, therefore, more uncertainty in long messages than in short messages. For example, there are just two possible messages consisting of one binary digit (*0* or *1*), four messages consisting of two binary digits (*00, 01, 10,* or *11*), 16 consisting of four binary digits, and so on. The entropy in the message increases for each of these cases as the number of choices increases. If the freedom of choice and the uncertainty decrease, the entropy decreases.

If a message source is not synchronous, that is, not producing information symbols at a constant rate, it is said to have an entropy, $H$, of so many bits per symbol (letter, word, or message). However, if the source does produce symbols at a constant rate, its entropy, $H'$, is expressed as so many bits per second.

The simple expression in Equation (10-1) may be expanded as an expression for entropy by including a factor for the expectation of each possible outcome:

$$H = - (p_1 \log_2 p_1 + p_2 \log_2 p_2 + \ldots + p_n \log_2 p_n)$$

$$= - \sum_{i=1}^{n} p_i \log_2 p_i \text{ bits per symbol.} \qquad (10\text{-}2)$$

In this equation, there are $n$ independent symbols, or outcomes, whose probabilities of occurrence are $p_1, p_2 \ldots p_n$.

Equation (10-2) may be used to illustrate the effect on the entropy of a source when probability $p_1$ is changed. For the simple case in which there are just two choices ($X$ with probability $p_1$ and $Y$ with probability $p_2 = 1 - p_1$), the value of $H$ is plotted in Figure 10-1. Examination of the figure makes it clear that for this case the entropy,

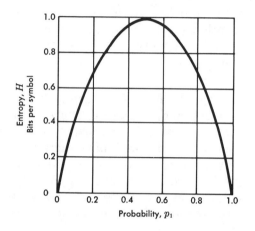

Figure 10-1. Entropy of a simple source.

$H$, is a maximum of one bit per symbol when $p_1 = p_2 = 1/2$ and that the entropy is zero when $p_1 = 0$ or 1 (no uncertainty in the message).

It may be shown that this situation is typical even when the number of choices is large. The entropy is a maximum when the probabilities of the various choices are about equal and is a small value when one of the choices has a probability near unity.

*Example 10-1:*

Given an honest coin. If it is tossed once, there are two equally probable outcomes, namely, head or tail. The entropy is computed by Equation (10-2) as

$$H = -[0.5(-1.0) + 0.5(-1.0)]$$

$$= 1.0 \text{ bit.}$$

If the coin has two heads, the probability of a head is unity. Then

$$H = -(1.0 \log_2 1.0)$$

$$= 0 \text{ bit.}$$

Thus, the tossing of a two-headed coin gives no information. The outcome has no uncertainty; it is always heads.

*Example 10-2:*

Given an honest die. One roll of such a die can result in any one of six equally probable outcomes. Thus, using Equation (10-2) again, it is seen that the entropy of the source (the die) is

$$H = -\log_2 \ (1/6)$$

$$= 2.58 \text{ bits.}$$

Now, assume the die is loaded; in this case the outcomes are not equally probable. Assume the following probabilities for the various possible outcomes:

| DIE FACE | PROBABILITY |
|----------|-------------|
| 1 | 0.4 |
| 2 | 0.2 |
| 3 | 0.1 |
| 4 | 0.1 |
| 5 | 0.1 |
| 6 | 0.1 |

$$H = - \ (0.4 \log_2 0.4 + 0.2 \log_2 0.2 + 0.4 \log_2 0.1)$$

$$= 2.32 \text{ bits.}$$

An example of a more complex relationship between probability of occurrence and entropy is illustrated by an evaluation of the information content of the written English language. As a first approximation, the occurrence of the 27 symbols representing the 26 letters of the alphabet and a space may be assumed to be equally probable. Such an approximation sets the upper bound at

$$H = \log_2 27 = 4.75 \text{ bits per symbol.}$$

This approximation, however, is inaccurate since the probabilities of occurrence are quite different for different letters. For example, in typical English text the letter $E$ occurs with a probability of about 0.13, while $Z$ occurs with a probability of only about 0.0008. By considering such probabilities and other refinements, the information content of English text is estimated to be about one bit per symbol.

## 10-4 THE COMMUNICATION SYSTEM

The term *information* has been defined in terms of logarithmic units (bits), and the measure of information has been defined as entropy in bits. The concept of information transfer from a source to a destination has been described as a probabilistic phenomenon involving the probabilities of message generation by the source and the resolution of the uncertainties at the destination. These concepts may now be more specifically related to the communication system.

The general communication system is commonly represented in one simplified form by a sketch such as the block diagram of Figure 10-2. The system is made up of a source of information and a destination for the information. Between the source and destination are a transmitter, a channel to carry the information, and a receiver. The function of the transmitter is to process the message from the source into a form suitable for transmission over the channel, a process frequently referred to as *channel coding*. The receiver reverses this process to restore the signal to its original form so that the message can be delivered to the destination in suitable form. This process is called *decoding*. The channel in such a system interconnects the physically separated transmitter and receiver. It is often assumed to be ideal, introducing no noise or distortion. This is, in fact, never achievable; every channel introduces some noise and distortion. It should also be recognized that there are noise sources that enter the system at places other than the channel; however, it is convenient to assume that all the perturbations on ideal signal transmission are

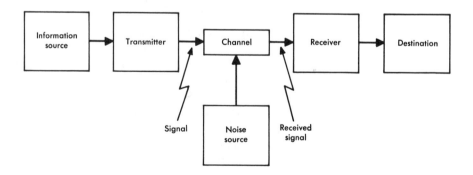

Figure 10-2. Block schematic of a general communication system.

introduced into the channel by a source external to the channel, as shown in Figure 10-2.

## Coding

The coding (and decoding) of messages in the transmitter (and receiver) of the communication system of Figure 10-2 may take any of a large number of forms and may be done for a number of different reasons. At the source the signal is often encoded for the purpose of increasing the entropy of the source. Morse's dot-dash coding of alphabetical symbols is an example; he assigned short dot-dash symbols to those letters most frequently found in English text and longer symbols to those less frequently encountered. Encoding is found in the transmitter where the purpose may be to increase the efficiency of transmission over the channel, i.e., to increase the rate of transmission of information. Encoding may also provide error detection, error correction, or both. A thorough review of coding principles and techniques is beyond the scope of this chapter; however, practical applications of coding techniques appear in Chapter 8 (Modulation) and in several chapters of Volume 2 in which terminals for specific systems are described.

## Noise

Noise contains components whose characteristics generally can be defined in probability terms and thus theoretically in terms of information. The presence of noise in a communication system adds bits of information and increases the uncertainty of the received signal; therefore, one might erroneously conclude that noise is beneficial. However, since the added noise perturbs the original set of choices, it introduces an undesirable uncertainty.

Figure 10-3 illustrates a simple case of how noise may introduce errors in transmission. The transmitted message, shown in Figure 10-3(a) as a series of 0s and 1s, is transformed into a series of plus (for 1) and minus (for 0) voltages in Figure 10-3(b). This signal is perturbed by an interfering noise depicted in Figure 10-3(c). The signal and noise voltages add as in Figure 10-3(d), and the receiver translates the received composite signal into the received message of Figure 10-3(e) which contains three errors.

The noise introduced in a communication system, such as that illustrated in Figure 10-2, may consist of any of a large number of kinds of interference introduced in the transmission path. Its

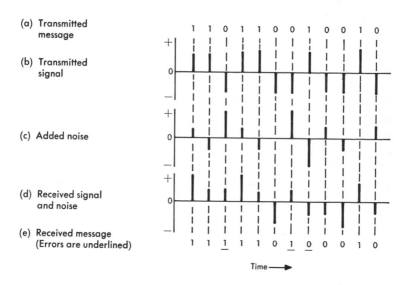

Figure 10-3. Effect of noise on transmission.

characteristic may be well-defined by some probabilistic expression; Gaussian or Poisson distributions are examples. Intersymbol interference, resulting from gain or envelope delay distortion (or both), may be regarded as noise. Signal-dependent distortion of the signal itself, due to nonlinear devices, may also be considered as noise. Other possible types of interference (noise) include frequency offset or frequency shift, sudden amplitude or phase hits due to external influences, echoes (perhaps due to impedance mismatches), or crosstalk due to some other signal being superimposed on the wanted signal by way of an unwanted path. Any or all of these interferences can produce transmission errors.

## 10-5  THE FUNDAMENTAL THEOREMS

The value and broad scope of information theory are expressed succinctly by Pierce: "To me the indubitably valuable content of information theory seems clear and simple. It embraces the ideas of the information rate or entropy of an ergodic message source, the information capacity of noiseless and noisy channels, and the efficient encoding of messages produced by the source, so as to approach errorless transmission at a rate approaching the channel capacity. The world of which information theory gives us an understanding of clear and present value is that of electrical communication systems and, especially, that of intelligently designing such systems" [7].

Shannon presents a large number of theorems, all of which pertain to the comments quoted above. There are three, however, that are basic and of sufficient importance to discuss here. These are (1) the fundamental theorem for the noiseless channel, (2) the fundamental theorem for the discrete channel with noise, and (3) the theorem for the channel capacity with an average power limitation.

### The Noiseless Channel

Let a source have an entropy of $H$ bits per symbol and a channel have a capacity of $C$ bits per second. Then it is possible to encode the output, $m'$, of the source in such a way as to transmit at the average rate of $C/H - \epsilon$ symbols per second over the channel, where $\epsilon$ is arbitrarily small. This may be written

$$m' = \frac{C}{H} - \epsilon \text{ symbols per second.} \qquad (10\text{-}3)$$

It is not possible to transmit at an average rate greater than $C/H$.

It is necessary to distinguish carefully between the $m$ of Equation (10-1) expressed in bits and the $m'$ of Equation (10-3) expressed in symbols per second. The quantity $m$ as used in Equation (10-1) is a measure of information produced by a source (in the simple binary case, the number of $1$s and $0$s). Then,

$$H' = m'H \text{ bits per second,} \qquad (10\text{-}4)$$

where $m'$ is the average number of symbols produced per second, $H$ is the entropy produced by the source in bits per symbol, and $H'$ is the entropy in bits per second.

For cases of interest, $H' \leqq C$ and the number of symbols per second is

$$m' = \frac{H'}{H} \leqq \frac{C}{H} \text{ symbols per second.} \qquad (10\text{-}5)$$

Equations (10-3) and (10-5) are equivalent since $\epsilon$ in Equation (10-3) is the amount by which $C/H$ exceeds $m'$ in Equation (10-5). Thus, a source may produce symbols at a rate of $m'$ symbols per second. The entropy may be such that the information produced is only $H$ bits per symbol or $H'$ bits per second. The theorem shows that as long as $H' \leqq C$, the source may be coded so that a rate of $C/H - \epsilon$ symbols per second may be transmitted over the channel of capacity $C$. A rate greater than $C/H$ cannot be achieved by any coding without error in transmission.

## The Discrete Channel with Noise

Consider a discrete channel with a capacity, $C$, and a discrete source with an entropy, $H'$. If $H' \leq C$, there exists a coding system such that the output of the source can be transmitted over the channel with an arbitrarily small frequency of errors (or an arbitrarily small equivocation). If $H' > C$, it is possible to encode the source so that the equivocation is less than $H'-C+\epsilon$ where $\epsilon$ is arbitrarily small. There is no method of encoding which gives an equivocation less than $H'-C$.

It seems strange to find a theorem relating to a "discrete channel with noise" that has no explicit mention of noise in its statement. However, this situation arises from the manner in which Shannon leads up to the theorem. Shannon defines the capacity, $C$, of the discrete channel with noise as

$$C = \text{Max} \ [H'(x) - H'_y(x)] \quad \text{bits per second.} \qquad (10\text{-}6)$$

In this equation, $C$ is given as the maximum value of source $x$ with entropy $H'(x)$ minus the conditional entropy $H'_y(x)$. The latter, in turn, is defined as the equivocation, which measures the average ambiguity of the received signal due to the presence of noise in the system. Thus, noise is included by implication.

## Channel Capacity with an Average Power Limitation

The capacity of a channel of band $W$ perturbed by white noise* of power $P_{noise}$ when the average transmitter power is limited to $P_{max}$ is given by

$$C = W \log_2 \frac{P_{max} + P_{noise}}{P_{noise}} \quad \text{bits per second.} \qquad (10\text{-}7)$$

---

*White noise has a flat or constant power spectral density.

Shannon explains, "This means that by sufficiently involved coding systems we can transmit binary digits at the rate $W \log_2 \dfrac{P_{max} + P_{noise}}{P_{noise}}$ bits per second, with arbitrarily small frequency of errors. It is not possible to transmit at a higher rate by any encoding system without definite positive frequency of errors" [1]. In Equation (10-7), $W$ is the bandwidth in hertz; $P_{max}$ and $P_{noise}$ are signal and noise powers that may be expressed in any consistent set of units (as a ratio, the units cancel out in the equation). As a final restriction, Shannon points out that to approximate this limiting rate of transmission the transmitted signals must approximate white noise in statistical properties. Coding, used to improve the transmission rate, is accomplished only at the expense of introducing delay and complexity. To achieve or approach the limiting rate may introduce sufficient delay in practice as to make the process impractical.

Other theorems of Shannon give the rate of information transmission for other sets of conditions. For example, the condition of peak power rather than average power limitation is covered. For noise other than white noise the transmission rate cannot be stated explicitly but can be bounded. The bounds are usually near enough to being equal that most practical problems can be solved satisfactorily.

## 10-6 CHANNEL SYMMETRY

It may be shown that the maximum rate of transmission of information (the capacity) can be determined for a symmetical channel by straightforward means but that the computation for an unsymmetical channel becomes complicated [7]. A symmetrical channel is one in which the probability, $p$, of a *0* from the source being received as a *0* is equal to the probability that a *1* from the source is received as a *1*. Thus, the probability that a transmitted *0* would be received as a *1* and the the probability that a *1* would be received as a *0* are both equal to $(1-p)$. Most practical problems involve symmetrical channels.

*Example 10-3:*

Given the symmetrical channel of Figure 10-4(a) having a transmitter, $x$, a receiver, $y$, and additive noise; given channel performance such that $p = 0.9$ [as shown in Figures 10-4(b) and 10-4(c)]; and given the statistics of the transmitter such that the probability of a 1 is 0.6 and of a 0 is 0.4. What is the entropy of the signal received at $y$?

From Equation (10-2), the entropy of the transmitter is

$$H(x) = -(0.6 \log_2 0.6 + 0.4 \log_2 0.4) = 0.97 \text{ bit per symbol.}$$

The rate of transmission, $R$, may be shown by an expression similar to Equation (10-6),

$$R = H'(x) - H'_y(x) \text{ bits per second.}$$

The equivocation, $H'_y(x)$, may be found by

$$H'_y(x) = -p \log_2 p - (1-p) \log_2 (1-p)$$

$$= -0.9 \log_2 0.9 - 0.1 \log_2 0.1 = 0.469 \text{ bit per second.}$$

Thus,

$$R = 0.97 - 0.469 = 0.501 \text{ bit per second.}$$

This is the entropy of the signal at $y$ for the situation in Figure 10-4(b).

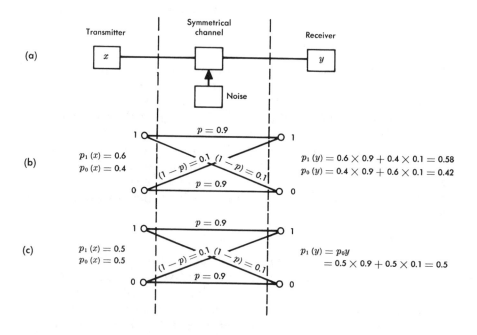

Figure 10-4. Transmission over a discrete symmetrical channel with noise.

Next, assume that the transmitter may be encoded differently so that $p_1(x) = p_0(x) = 0.5$ as illustrated in Figure 10-4(c). What is now the entropy of the signal received at $y$?

For this condition,

$$H'(x) = -0.5 \log_2 0.5 - 0.5 \log_2 0.5 = 1.0 \text{ bit per second.}$$

The channel is the same as in Figure 10-4(b). Therefore, the new rate is

$$R = 1 - 0.469 = 0.531 \text{ bit per second.}$$

Thus, the entropy of the signal received at $y$ has been increased by increasing the entropy of the transmitter.

For the channel assumed, one having $p = 0.9$, this can be shown to be the maximum rate and thus the channel capacity, $C$, of Equation (10-6).

It may be shown that if the symmetrical channel has performance such that $p = 0.99$, the maximum rate improves to 0.92 bit per second when the source entropy is unity.

REFERENCES

1. Shannon, C. E. and W. Weaver. *The Mathematical Theory of Communication* (Urbana, Ill.: The University of Illinois Press, 1963).

2. Nyquist, H. "Certain Factors Affecting Telegraph Speed," *Bell System Tech. J.*, Vol. 3 (April 1924).

3. Nyquist, H. "Certain Topics in Telegraph Transmission Theory," *A.I.E.E. Transactions*, Vol. 47 (April 1928).

4. Hartley, R. V. L. "Transmission of Information," *Bell System Tech. J.*, Vol. 7 (July 1928).

5. Wiener, N. *Cybernetics* (Cambridge, Mass.: Technology Press, 1948).

6. Wiener, N. *Extrapolation, Interpolation, and Smoothing of Stationary Time Series* (New York: John Wiley and Sons, Inc., 1949).

7. Pierce, J. R. *Symbols, Signals and Noise: The Nature and Process of Communication* (New York: Harper and Brothers, 1961).

8. Hyvärinen, L. P. *Information Theory for Systems Engineers* (New York: Springer-Verlag, 1968).

9. Raisbeck, G. *Information Theory—An Introduction for Scientists and Engineers* (Cambridge, Mass.: M.I.T. Press, 1963).

10. Bennett, W. R. and J. R. Davey. *Data Transmission* (New York: McGraw-Hill Book Company, Inc., 1965).

Chapter 11

# Engineering Economy

Solutions to engineering problems are usually considered complete only after economic analyses have been made of several alternative solutions and the results compared. This is true in the field of transmission as it is in any other field. Sometimes, a choice must be made on the basis of incomplete information and it becomes necessary to exercise engineering judgement in respect to the impact of intangible aspects of the problem. Bell System objectives are to provide *economically* the best possible service. To meet these objectives, engineering economy studies must be made to demonstrate the value of new systems, new services, and specific proposals for network expansion; service and performance improvements must also be evaluated.

Financial accounting and engineering are fields that appear to be quite remote from each other; however, there are numerous points of contact in the paths followed by the two professions. Both are concerned with the use of capital and expense funds. The major difference is that in financial accounting these funds are dealt with in retrospect by examining the results of expenditures while in engineering one of several alternate future courses of action must be selected to use available funds most effectively.

There are two broad categories of expenditure which consume most of the funds available to the Bell System. One is the cost of operating the business and maintaining the plant in service. These expenses are charged in the period in which they are accrued; they are planned, budgeted, controlled, and paid out of current revenues. The second is the capital required to construct the new plant needed to satisfy growing service demands. Capital expenditures are paid out of funds accumulated as retained earnings from current revenues, the sale of stocks and bonds, depreciation, deferred taxes, and investment tax credits. Funds are planned, budgeted, controlled, and spent in accordance with procedures generally categorized as the *construction program*.

Planning and implementing the construction program involve many factors that affect the choice of a course of action. Relative service and performance capability, operating conditions, maintenance complexities, revenues, and costs must all be considered. Costs are given considerable weight because they provide a tangible and quantitative measure of relative worth in terms that most people understand.

Many types of cost studies are made to determine the effects of some action on pricing policy, financial position, or accounting results. Such studies must often be made within the constraints of the Uniform System of Accounts prescribed by the Federal Communications Commission. Many are made after a course of action has been determined. Engineering economy studies are intended to show which of several plans is economically most attractive in fulfilling service requirements. Therefore, they are important aids in making decisions which cumulatively result in the formulation of the construction program. In the field of transmission engineering, as in many other areas, there are often several possible courses of action which may be feasible. Therefore, familiarity with the principles of engineering economy studies is necessary in fulfilling transmission engineering functions.

An engineering economy study may be made (1) to determine which of several plans or methods of doing a job will be the most economical over a given time interval; (2) to prepare cost estimates for studies of new and existing service offerings or special service arrangements; (3) to establish priorities for discretionary plant investment opportunities; and (4) to establish revenue and capital requirements over long periods of time as major projects are programmed and initiated [1]. Objectives such as these are often satisfied by studies in which engineering data are used as input information. The provision of the necessary data for such studies requires an understanding of basic engineering economy principles and often contributes valuable perspective on the total engineering problem.

## 11-1 TIME VALUE OF MONEY

Engineering economy studies deal wtih money to be spent or received in various amounts and at different times. The objective of such studies is to evaluate the money involved in the plans under consideration; therefore, it is essential to understand the basic rules that govern the comparison of money spent or received at different times. An understanding of the time value of money is implicit in the basic rules of economy studies and in the application of sound principles to the conduct of such studies. Simply, it must be understood that a dollar today is not equal to a dollar a year from now or a year ago. However, there are means of expressing such dollars in equivalent terms, i.e., means of expressing the *time value of money.*

### The Earning Power of Money

There are costs involved in the use of money that are derived from the potential earning power that money has as a commodity. These costs must be measured in terms of this earning power which is a continuous function of time that increases with the period of use.

The term *interest* is often used to represent the earning power of money. However, the term is most applicable to designate the return on borrowed money, i.e., debt. In engineering economy studies of the type to be considered here, it is common practice to use a composite of debt interest and equity return. The composite *return*, equivalent to the cost of all capital, is determined according to the percentage of each type in the capital structure.

The effect of return on the time value of money may be evaluated for a particular time relative to another (usually taken as a reference, $T = 0$) by the expression

$$\frac{D_{T=0}}{D_{T=1}} = \frac{1}{1+i} \tag{11-1}$$

where $D$ is the value of money at times denoted by the subscripts and $i$ is the rate of return for the period. Thus, if one dollar is needed one year from now $(T = 1)$ and if the rate of return, $i$, is 0.10, only 91 cents are required now $(T = 0)$.

The rate of return may be compounded at intervals of typically (though not necessarily) one year. Compounding involves the computation of the value of money on the basis of the return on the original amount plus the accrued return during the compounding interval. Thus, to determine the value of money where the return has been compounded over $n$ intervals

$$\frac{D_{T=0}}{D_{T=n}} = \frac{1}{(1+i)^n}$$

or

$$D_{T=n} = D_{T=0} (1+i)^n \quad . \tag{11-2}$$

The interpretation of Equations (11-1) and (11-2) are that if an amount $D$ is to be made available at a future time, $T = n$, a smaller amount of money may be made available now, $T = 0$, when it is invested with a rate of return, $i$.

## Equivalent Time-Value Expressions

In the conduct of engineering economy studies, several time-value equivalencies are used; the choice depends on the nature of the study. Sometimes it is desirable to express all costs in terms of equivalent amounts at a time arbitrarily chosen and defined as "the present." In other studies, it is desirable to express costs in terms of some selected time in the past or the future. In some cases, it is desirable to express the costs as an annuity, an amount of money that must be provided, in equal amounts and at equal intervals, to be equivalent to a single lump-sum amount at a specified time.

The expressions used are all equivalent to one another and may be converted from one form to another, as desired. The following are such commonly-used expressions.

Future worth of a present amount, a form similar to Equation (11-2),

$$F/P = (1+i)^n \quad . \tag{11-3}$$

Present worth of a future amount,

$$P/F = \frac{1}{(1+i)^n} \quad . \tag{11-4}$$

Future worth of an annuity,

$$F/A = \frac{(1+i)^n - 1}{i} \quad . \tag{11-5}$$

Annuity for a future amount,

$$A/F = \frac{i}{(1+i)^n - 1} \ . \tag{11-6}$$

Present worth of an annuity,

$$P/A = \frac{(1+i)^n - 1}{i(1+i)^n} \ . \tag{11-7}$$

Annuity from a present amount,

$$A/P = \frac{i(1+i)^n}{(1+i)^n - 1} \ . \tag{11-8}$$

The symbols used in Equations (11-3) through (11-8) are $i$ for return rate, $n$ for the number of time periods (compounding intervals), $F$ for the future worth of an amount, $P$ for the present worth, and $A$ for an annuity. Future worth and present worth are methods of expressing money values as a single amount at some particular time.

Application of Equations (11-3) through (11-8) is greatly facilitated by the use of tabulations found in many standard texts [1]. In addition, such equations may now easily be solved by the use of modern engineering calculators. An illustration of how these equations may be applied is given in Figure 11-1. Here, the value of $1,000 in year 8, point A, may be traced through various processes to the future and past and finally back to the same value, $1,000 in year 8.

## 11-2  ECONOMY STUDY PARAMETERS

The plant is a dynamic conglomeration of telecommunications gear which grows each year in size and complexity. Additions to the plant must be chosen with care to insure that continually increasing service requirements are met, that they are met economically, and that performance objectives are satisfied. The processes of anticipating needs, recommending new plant, and implementing a construction program are, in most telephone companies, engineering functions that begin long before the year of implementation.

To aid in the implementation of construction programs, engineering economy studies are made to determine which course of action is most attractive economically. Although initial and recurrent costs are the most important factors affecting engineering economy studies,

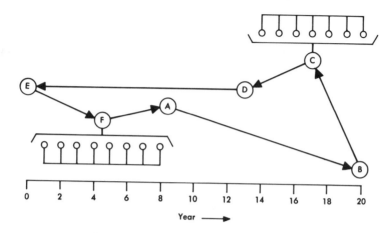

| CONVERSION | DESCRIPTION | FACTOR | EQUIVALENCE | EQUATION |
|---|---|---|---|---|
| A → B | Single amount in year 8 to single amount in year 20 | $F/P$ ($i = 0.1$, $n = 12$) | B = 3.1384A = \$3138.40 | 11-3 |
| B → C | Single amount in year 20 to equal amount in years 14 to 20 | $A/F$ ($i = 0.1$, $n = 7$) | C = 0.1054B = \$330.79/yr | 11-6 |
| C → D | Equal amounts in years 14 to 20 to single amount in year 13 | $P/A$ ($i = 0.1$, $n = 7$) | D = 4.8684C = \$1610.42 | 11-7 |
| D → E | Single amount in year 13 to single amount in year 0 | $P/F$ ($i = 0.1$, $n = 13$) | E = 0.2897D = \$466.54 | 11-4 |
| E → F | Single amount in year 0 to equal amounts in years 1 to 8 | $A/P$ ($i = 0.1$, $n = 8$) | F = 0.1874E = \$87.43/yr. | 11-8 |
| F → A | Equal amounts in years 1 to 8 to single amount in year 8 | $F/A$ ($i = 0.1$, $n = 8$) | A = 11.4359F ≈ \$1000 | 11-5 |

Figure 11-1. Time-value equivalence.

there are other vital parameters, such as life of plant, service requirements, inflation, the debt-to-equity ratio of the company, composite
cost of capital, and tax laws which must be taken into consideration.

An engineering economy study must take into account the interval
over which the problem is to be studied and its relation to the life of
the plant involved. If these time intervals are not the same, adjust-

ments must be made in the study program. The choice of interval is a matter of judgment; it may be relatively short, such as two or three years, or it may continue, at least in theory, indefinitely into the future. Since most telephone plant has long life, the study period is often taken as 10 to 20 years.

Long-range planning is undertaken in order to provide guidance for gross changes in plant makeup. Studies of traffic and private line growth patterns, shifts in population densities, emerging new services, and expected new system designs must be made and continually refined. The basis for long-term planning is a simplified 30- to 40-year customer services forecast with incremental five-year study periods. Adequate time must be allowed for the complex processes of evaluation, compromise, and final decision. Long-range studies must include some evaluation of the effects of the planning results in the period immediately following the selected study period. Planning decisions finally come into focus in the form of current planning processes about three years before implementation is scheduled. Construction programs are reviewed and adjusted during these three years and at least three times during the final year.

Provision must be made in the planning process for flexibility and changes. For example, traffic and private line forecasts might show a need five years in the future for a substantial number of new trunks between two cities 50 miles apart. Since cable pairs between the two cities are limited in number, a T1 Carrier System or systems, a new radio system, and a new multipair cable installation must all be considered. The results of long-range studies might show the new multipair cable to be most economical. Two years later, during the preparation of the first construction program for the proposed project, it might become evident that these two cities are along a major route that has developed a need for a large number of circuits and for which a new microwave radio system must be installed. The circuits needed for the original project may be provided much more economically by the new radio system. Such adjustments and changes are frequently made to achieve a more economical program.

Other unforeseen events may cause construction program changes. As plans are reviewed, changes in the economic climate may require upward or downward adjustments in the budget, which must be reflected in corresponding changes in the construction program. For example, emergency needs may have developed from massive plant damage caused by a hurricane or the unanticipated rapid growth of a new industrial park may impose a sudden and unexpected demand for new facilities. Such events are reviewed frequently by committees which are empowered to approve certain changes in the construction program in order to meet the unexpected need.

The economic analysis of engineering problems primarily involves consideration of a number of aspects of costs. Included are a variety of capital costs that are incurred because investors provide funds needed to acquire plant and operations costs that are incurred by the existence of the plant. Such things as the changing technology and inflation must also be considered for their effects on costs.

## Capital Costs

The costs associated with the acquisition of property are called capital costs; accounting procedures are used to monitor, control, and recover such costs. The property is an asset assigned for accounting purposes to a specific *plant account*. Capital costs related to engineering economy studies include the concept of the composite cost of money, the first-cost investment, and the sources of capital used to recover the cost of the investment. The process of capitalization of money invested in plant involves recurring annual costs having three elements: return, capital repayment, and income taxes.

**The Composite Cost of Money.** Money needed to pay the initial costs of plant investments is obtained from a number of sources each of which involves cost factors that must be evaluated for engineering economy studies. The two basic sources of money are called *debt capital* and *equity capital*. The ratio of the debt capital (obtained by borrowing) to the total capital (debt plus equity) carried on the books of a company is called the *debt ratio*. A discussion of a desirable or optimum debt ratio for a regulated industry, a subject of continuing scrutiny and study on the part of industry management and regulatory agencies, is beyond the scope of this text. However, an understanding of the relation of the debt ratio to the composite cost of money is necessary in making engineering economy analyses.

The composite cost of money, or return rate, may be expressed by an equation that relates the composite cost of money, $i$, to the debt ratio, $r$, the interest paid on the debt, $i_d$, and the return on equity (stock dividends and retained earnings), $i_e$. These are expressed

$$i = ri_d + (1 - r)i_e \quad .$$

$$(11\text{-}9)$$

Thus, if the debt ratio is 45 percent, the composite cost of money is

$$i = 0.45i_d + 0.55i_e \quad .$$

It must be recognized that the debt ratio, debt interest, and equity return may all change with time. However, such variations are usually not accounted for in engineering economy studies because they are not predictable and they generally tend to affect alternative study plans proportionately. Long-term forecasts of debt and equity costs must sometimes be changed to reflect changes in the corporate debt ratio.

**First Costs.** The amount of money required to build a new plant is called the *first cost*. The first cost of a project is the invested capital upon which the rate of return is initially calculated. (Later, the return rate is based on unrecovered investment.) Included are the costs of materials, transportation, labor and incidentals related to installation, supervision, tools, engineering, and a number of other miscellaneous items. These costs are accumulated during the construction interval and do not recur during the life of that plant item; however, they must be recovered during the life of the plant if the company operation is to be based on sound economic principles.

**Capital Recovery.** Physical plant may wear out, be made obsolete by new technology, or fail to meet changing requirements. Whatever the reason, it must ultimately be replaced. The capital invested is dissipated by the end of plant life unless it is repaid by some method. Capital recovery is generally accomplished by means of depreciation accounting, a method by which the capital is repaid annually out of current revenues over the life of the plant. Capital expenditures are thus converted to annual costs which repay the initial cost. In a continuing business, the repayment is not actually made to the investor; the money is reinvested in other new plant or assets. The investment is protected by the transfer of capital from old to new plant in installments as the old plant is used up in service.

Depreciation accounting practices must be based upon the service life of the plant to which they are applied and must also be carried out in a manner that satisfies legal requirements. They must also reflect adequately a number of related factors that enter into the costs of the business such as salvage.

*Life of Plant.* In conducting an engineering economy study, it is imperative that the life of the particular plant involved in the study be used rather than some broad average that is applied for the purpose of determining depreciation rates for accounting purposes. Sometimes, plant life may be established by the conditions of the problem. For example, a study may be made of alternate plans for installing additional equipment in a building that is scheduled for retirement in a short period of time, say five years. In such a case, the plans under study must provide for the repayment in five years of all capital expenses involved with suitable adjustments for salvage at the termination of the study period. In other cases, the life of a plant item may depend on the life of other items. Such might be the case if the life of aerial wire or cable were limited by the life of the pole line on which it is placed. The pole line may be near the end of its useful life or it might be terminated by action of public authorities.

If the conditions of the problem do not give an indication of the life of plant, life must be estimated and engineering judgment must be exercised. Even in such cases, estimated life only rarely coincides with the average life used for depreciation accounting purposes.

*Salvage.* A significant capital cost in an engineering economy study is the net salvage value of the plant upon removal. The value may depend on whether salvage is for scrap, trade-in value, or resale. Removal costs (to be subtracted from the gross salvage value) may be quite different depending on whether the salvage is for scrap or reuse. Conservative assumptions should be made as accurately as possible in respect to the possible reuse of plant and removal costs.

*Straight-Line Depreciation.* The accounting method under the Uniform System of Accounts prescribed by the Federal Communications Commission for the Bell System and other common carriers requires the application of straight-line depreciation for financial statements. Because it is used for the book records of the firm, it is also called book depreciation. With this procedure, a capital investment is written off by an equal amount each year during the expected life of the plant; that is, for each year, an amount of revenue (equal to the depreciation rate multiplied by the original first cost amount) is accounted for in the company books as having paid for the depreciation of that item of plant. The rate is determined by the following:

$$\text{Annual depreciation rate} = \frac{100 - \text{percent net salvage}}{\text{plant life (years)}} \%.$$

Recall that life may terminate for a number of reasons (deterioration, obsolescence, rearrangements, etc.) and that for engineering economy studies life must be carefully defined. "Average life" is often unsatisfactory in these studies.

Figure 11-2 illustrates the 100-dollar-per-year, straight-line depreciation of a $1000 investment and also shows a tabulation of annual composite return payments that must be made on the balance (book value) of the investment. Note that the charge is reduced each year since the return on the unrecovered capital is reduced from year to year.

| YEAR END | COMPOSITE RETURN @ 8.5% | UNDEPRECIATED BALANCE BEFORE YEAR-END PAYMENT | YEAR-END PAYMENT | BALANCE AFTER PAYMENT | TOTAL PAID |
|---|---|---|---|---|---|
| 0 | | | | $1000.00 | |
| 1 | $ 85.00 | $1085.00 | $ 185.00 | 900.00 | |
| 2 | 76.50 | 976.50 | 176.50 | 800.00 | |
| 3 | 68.00 | 868.00 | 168.00 | 700.00 | |
| 4 | 59.50 | 759.50 | 159.50 | 600.00 | |
| 5 | 51.00 | 651.00 | 151.00 | 500.00 | |
| 6 | 42.50 | 542.50 | 142.50 | 400.00 | |
| 7 | 34.00 | 434.00 | 134.00 | 300.00 | |
| 8 | 25.50 | 325.50 | 125.50 | 200.00 | |
| 9 | 17.00 | 217.00 | 117.00 | 100.00 | |
| 10 | 8.50 | 108.50 | 108.50 | 0.00 | $1467.50 |

Figure 11-2. Straight-line depreciation and return accounting.

Book depreciation of a capital investment is related to the term, *book value*. After a plant item has been installed and the depreciation of the investment has been started in the accounting procedures, the item is said to have a certain book value. It is computed as the gross plant investment minus the accumulated depreciation. Sometimes the gross plant investment value is called *book cost* and undepreciated plant is called *net plant*.

*Accelerated Depreciation.* In accounting, depreciation is recognized as an expense that may be deducted from revenues before income tax is computed. Tax laws now permit specific forms of accelerated depreciation to be used by public utilities for tax depreciation. Higher rates are applied during the early years of the life of a plant item and lower rates during later years. The rate of depreciation varies somewhat but, in principle, it follows a curve like that of Figure 11-3. The figure compares straight-line and accelerated depreciation of a 1000-dollar plant item having an expected life of 10 years.

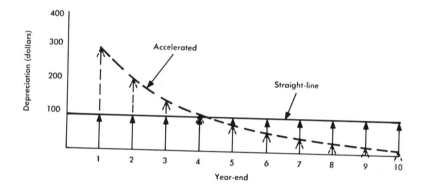

Figure 11-3. Depreciation of a 1000-dollar plant item over a 10-year period.

A number of specific methods of accelerated depreciation accounting may be used [1, 2, 3]. These include the double declining balance (DDB) method, the one-and-one-half declining balance (1.5-DB) method, and the sum-of-years-digits (SOYD) method. In the DDB method, the investment is depreciated at a constant annual rate, $2/n$, of the undepreciated balance where $n$ is the life of the plant item in years. The accrued depreciation cannot exceed the depreciable value of the item, i.e., the initial investment less net salvage. The 1.5-DB method is similar but the depreciation rate is $1.5/n$.

In the SOYD method, the life of plant in years is used to determine the rate of depreciation. For example, suppose a plant item is to be depreciated over a five year period. The sum of years digits is determined by $S_d = 5 + 4 + 3 + 2 + 1 = 15$. The depreciation rate used at the end of each year is remaining life/$S_d$. Thus, in the first year, the depreciation rate is $(5)/15 = 0.333$ and, in the last year, the rate is $(1)/15 = 0.067$.

*Investment Tax Credits.* The investment tax credit is a significant source of capital funds that must be included in many economy studies.

This credit is provided under tax laws to encourage business investment and expansion. It is a direct tax reduction allowed for certain qualifying items of property; buildings and land are usually not included. The tax credit has no effect on the value of the initial investment used to establish the depreciation base.

In many situations, fluctuations in income and other variables are recognized by law and provisions are made to carry gains or losses forward or backward from the year in which they occur. Investment tax credit laws change from time to time. It is essential, when they are to be considered in an economy study, that the law, definitions, and rules current at the time of the analysis be clearly understood.

**Income Tax.** Annual costs associated with any investment must include income tax, a tax on earnings after payment of all expenses. These earnings include dividends paid to stockholders and the amount added to retained earnings. Specific values of income tax to be used in economy studies are not furnished and, as a result, the tax obligation must be determined for each alternative plan. Capital cost tabulations, which reflect accelerated tax depreciation and investment tax credit, are available for operating company use. Thus, annual cost percentages for taxes can be determined on the basis of estimates of life and net salvage for use in economy studies. The burden of income tax, as well as return on the investment, must be borne throughout the life of the investment. Therefore, money invested in new plant costs more than money spent on current expenses such as operations, repairs, and rearrangements.

## Plant Operations Costs

Nearly all costs are either recurring costs associated with the operation of the plant or are capital costs that are dealt with by methods, such as depreciation accounting, that make them equivalent to recurring costs. Plant operations costs are paid out of revenues. They can be regarded as being depreciated immediately, at the time they are incurred. In making engineering economy studies that involve the comparison of alternative plans, many of these costs can be and should be ignored because they are common to the alternative plans. In making such analyses, it is important to choose only those items for extended treatment that differ from plan to plan.

The following list of recurring costs is made up of a number of typical items that should be considered for analysis in any study. This list is not all-inclusive.

(1) *Maintenance Costs:* These are frequently important ingredients of engineering economy analyses. They include the costs of labor and material associated with plant upkeep, the related costs of training, testing of facilities, test equipment, plant rearrangements, and miscellaneous items such as shop repairs, tool expenses, and building maintenance and engineering work.

(2) *Operating Costs:* These include a wide range of costs primarily related to traffic, commercial, marketing, accounting, and administrative work. These costs are usually common to alternate plans; thus, they are seldom involved in engineering economy studies of the type being considered here. They are, of course, important components in overall company economy studies and may enter a detailed engineering study where, for example, network traffic management or the location of operator assignments might be involved in comparing alternative means of providing facilities for traffic management or operator services.

(3) *Rent:* This is a cost that is only occasionally an important element in engineering economy studies.

(4) *Lease:* The leasing of buildings, equipment, and motor vehicles is an increasingly important form of obtaining capital goods. Studies involving leases can be complex since leasing is considered to be an alternate form of debt financing.

(5) *Energy:* The cost of energy must be considered in these analyses where it is not common to alternate plans. Primary power increases with the size and complexity of the plant. The cost of conversion equipment and standby equipment needed to ensure reliability must be included in cost comparisons.

(6) *Miscellaneous Taxes:* Sales, occupation and use, and ad valorem taxes must all be considered where appropriate. These taxes are especially important in considering tax depreciation. Specifically excluded are social security and unemployment taxes, usually treated as loading factors on labor costs.

In general, capital expenditures made in the past cannot be undone or affected by engineering decisions made today. They must be recovered over the anticipated life of the plant through depreciation.

However, when plant is retired, property tax, maintenance, rent, energy, and many operating expenses are no longer incurred.

## Dynamic Effects on Analysis

Since it is necessary to deal with the future, engineering economy studies are bound to be subject to all the uncertainties of prediction. Among these uncertainties are changes in technology, unanticipated demands for new services, and unpredictable variations in the economic climate such as changes of inflationary trends.

**New Technology.** Advances in technology result in improved electronic components, design techniques, and operating efficiencies. Thus, equipment that is less expensive, takes less space, uses less power, and provides more channels or higher speeds of operation becomes available and must be considered as a possible replacement for existing equipment. The partial obsolescence and early retirement of older equipment become subjects of serious engineering economy studies.

**Services and Service Features.** New services, such as DATA-PHONE® digital service, and new service features, such as the custom calling features of stored program electronic switching systems, also have an impact on engineering economy studies and on problems of early equipment retirement due to functional obsolescence. If the new services are to be introduced in an area where existing equipment is incapable of providing them, replacement is mandatory and the economic problems are those of determining the extent, the most efficient means, and the optimum time to carry out the replacement program.

**Inflation.** In recent years, higher returns have been required to protect investments against inflation. As a result, it has been necessary to use higher return rates in cost studies. As inflationary trends have continued, it has become evident that these effects must also be applied to other costs. One straightforward method of accomplishing this end is to estimate future costs explicitly as of the time of occurrence and to use such estimates in the study. For example, if an item currently costing $1,000 is needed now and another is needed one year from now, $1,070 should be used in the study for the second item if a 7 percent change in cost due to inflation is anticipated.

This method of accounting for inflationary effects is effective but tedious to use where many future costs must be considered. In some cost comparison studies, a "convenience" rate may sometimes be used to represent the net impact of inflation on future cost estimates and on their present worth as expressed in Equation (11-4) and (11-7). This convenience rate should be regarded only as an arithmetic short-cut to simplify the analysis. Detailed treatment of inflation and its effects are beyond the scope of this chapter.

*Convenience Rate.* If the cost in each year is greater than that in the previous year by an inflation rate, $h$, and if the cost at the beginning of the first year is $A_0$, the cost at the end of any succeeding year, $n$, may be determined by

$$A_n = A_0(1 + h)^n \quad . \tag{11-10}$$

The present worth of this amount is found by Equation (11-4) as

$$PW(A_n) = A_0(1 + h)^n \left(\frac{1}{1 + i}\right)^n$$

$$= A_0\left(\frac{1 + h}{1 + i}\right)^n \quad . \tag{11-11}$$

The convenience rate, $c$, may be defined as a rate such that the present worth of the initial amount, $A_0$ (uninflated), over $n$ years is equal to the value in Equation (11-11); that is

$$A_0\left(\frac{1}{1 + c}\right)^n = A_0\left(\frac{1 + h}{1 + i}\right)^n \quad .$$

This equation may be manipulated to derive

$$c = \frac{i - h}{1 + h} \approx i - h \quad . \tag{11-12}$$

The convenience rate, $c$, may be used in analyses involving cost items inflating at rate $h$. Since several different convenience rates would be required in a study involving items that are subject to different rates of inflation, this procedure is sometimes difficult to apply. Also, it should be stressed that the present worth of taxes, book depreciation, and the tax factor must be determined by the composite cost of money rate and not by the convenience rate. The convenience rate, $c$, may be substituted for $i$ (the composite cost of money) only in Equations (11-4) and (11-7).

*Economy Study Applications.* While care must be used in applying the convenience rate, it is a valuable concept when properly used.

For example, there are many computer programs, such as the exchange feeder route analysis program (EFRAP), that have wide application in several types of engineering economy studies suitable for computer analysis. Many such programs were written before inflation effects were recognized as important. The convenience rate can be conveniently adapted to these programs which would otherwise have to be completely rewritten to include the effects of inflation.

### Example 11-1:

In this example, only maintenance expenses associated with two alternate plans are computed by two methods to show how results may be distorted if inflation effects are ignored. Plan A involves $500 per year maintenance expense for a single capital expenditure for a plant item having a 10-year life. Plan B involves $600 per year maintenance expense for one capital expenditure (B1) having a 10-year life and $600 per year maintenance expense for a second capital expenditure (B2) having an 8-year life and installed at the beginning of the third year of the plan. The composite cost of money, $i$, is taken as 12 percent and the inflation rate for maintenance expenses is 8 percent per annum. The maintenance expenses of the two plans are to be compared on the basis of present worth of expenditures (PWE), first by ignoring the effect of inflation and second by considering these expenses inflated and by applying the convenience rate. The PWE analysis recognizes cash flows for capital expenditures, net salvage, income taxes, and operations costs when they occur and sums the present worths of these amounts.

For the first analysis, the present worth of maintenance expenses for Plan A may be computed by Equation (11-7) as the present worth of an annuity with $i = 0.12$ and $n = 10$ years. For Plan B, expense B1 is similarly computed but for expense B2, the expense must first be computed as the present worth of an annuity for 8 years ($i = 0.12$) and then by Equation (11-4) to determine the present worth of a future amount for 2 years, the interval between the beginning of the plan and the expenditure of B2. Then, under Plan A, the PWE for maintenance expenses is

$$PWE_A = \frac{(1.12^{10} - 1)\ 500}{0.12 \times 1.12^{10}} = \$2825 \quad .$$

For expense B1,

$$PWE_{B1} = \frac{(1.12^{10} - 1)\ 600}{0.12 \times 1.12^{10}} = \$3390 \quad .$$

For expense B2,

$$PWE_{B2} = \frac{(1.12^8 - 1) \times 600}{0.12 \times 1.12^8} = \$2980$$

at the end of the second year of the study plan. This amount is converted to the beginning of the plan, year 0, by

$$PWE_{B2} = \frac{2980}{(1 + 0.12)^2} = \$2376 \quad .$$

Thus, for Plan B, the total present worth of expenditures for maintenance is

$$PWE_B = PWE_{B1} + PWE_{B2} = 3390 + 2376 = \$5766 \quad .$$

From this analysis, it would be concluded that the present worth of maintenance expenses for Plan B is

$$PWE_B - PWE_A = 5766 - 2825 = \$2941$$

more than for Plan A.

For the second analysis, the effects of inflation are included in computing the $PWE$ for both plans. Equations (11-7) and (11-4) are used as in the earlier analysis but the convenience rate, $c$, is used in place of the composite cost of money, $i$. The convenience rate is determined by Equation (11-12) as

$$c = \frac{1.12 - 1.08}{1.08} = 0.037$$

or 3.7 percent. With this substitution, the present worth of expenditure, $A$, is calculated as

$$PWE_A = \frac{(1.037^{10} - 1)\ 500}{0.037 \times 1.037^{10}} = \$4117 \quad .$$

Expenditure B1 is computed as

$$PWE_{B1} = \frac{(1.037^{10} - 1)\ 600}{0.037 \times 1.037^{10}} = \$4940 \quad .$$

Expenditure B2 is computed in terms of the end of the second year as

$$PWE_{B2} = \frac{(1.037^8 - 1)\ 600}{0.037 \times 1.037^8} = \$4090$$

and brought to year 0, the beginning of the plan, by

$$PWE_{B2} = \frac{4090}{(1 + 0.037)^2} = \$3803 \quad .$$

Thus, for Plan B, the present worth of expenditures is

$$PWE_B = 4940 + 3803 = \$8743 \quad .$$

Now, Plan B costs exceed Plan A costs by $8743 - 4117 = \$4626$, considerably more than the $2941 previously computed.

This example shows the effect of inflation on only one element of a plan comparison economy study. The conclusions illustrate the importance of evaluating the effects of inflation and the manner in which the convenience rate may be used.

## 11-3　ECONOMY STUDY TECHNIQUES

Many approaches and different techniques may be used to achieve the objectives of an engineering economy study, i.e., selecting one alternative course of action in preference to others by comparing their costs. Three of these methods have been found to give equivalent results in that the same alternative is selected. The method used depends on the nature of the available data, the ease of application, and the purposes for which the study is being made. The three methods are called the internal rate of return (IROR), present worth of annual costs (PWAC), and present worth of expenditures (PWE).

In conducting economy studies, certain assumptions must be made and clearly understood in order to be sure that comparisons are based on equivalent conditions. The assumptions must make all alternatives under study equivalent in terms of provision of service, life of plant, and effects of plant retirement. The studies are usually carried out by using only incremental costs, those that are different for the various plans. Common costs are eliminated from consideration.

In transmission engineering studies, it is now found that the PWE method is most easily applied and leads to the most useful results. This method is illustrated by an outline of the entire study process.

## Analytic Alternatives

While most engineering economy studies of the types being considered are based on PWE analyses, some knowledge of the IROR and PWAC methods is desirable. The IROR method is used in some transmission studies though seldom directly in this field.

**Internal Rate of Return.** In the development of IROR analysis, designed to determine the most efficient use of money for each project, the internal rate of return can be defined as the rate that causes the present worth of the net cash flows for the project to be zero. The two main elements of net cash flow are the investment and the recovery; net cash flow thus involves cash flowing into a project (investment) and back (revenues less operations costs and taxes). The equation for IROR is a polynomial that must usually be solved iteratively by trial and error. While this process may be lengthy and complex, it has been programmed for computer solution and can be applied where only a comparison of alternatives is desired. However, roots to the solution may be numerous or there may be no meaningful, finite roots. The IROR method cannot provide an evaluation of the profitability of an alternative.

Another disadvantage to the use of the IROR method of analysis is that as the number of plans is increased, the number of comparisons that must be made increases even faster. For $x$ number of plans, the number of comparisons is $x(x-1)/2$. Where a large number of plans are being considered, the IROR analysis becomes awkward.

Despite these disadvantages, the IROR method of analysis is used in certain situations. Capital funds for the construction program in any one year are finite and it is sometimes difficult to introduce new types of facilities that require high initial capital expenditures. Sometimes, these facility costs appear favorable on the basis of a PWE analysis but are formidible with limited capital funds. The alternatives may then be evaluated on the basis of the IROR. For example, a plan requiring high initial investment may require much lower operating or maintenance funds in comparison with another plan with lower initial costs. Benefits of a higher initial capital outlay can thus be measured by the IROR method to determine the most efficient use of money for each of the projects so analyzed.

**Present Worth of Annual Costs.** This method of analysis and the present worth of expenditures method are essentially alike when both are properly processed. However, in the PWAC approach, certain parameters are often treated in such a manner that the results are invalidated. For example, it is difficult in PWAC studies to account adequately for increasing costs such as those due to inflation and to increased maintenance with equipment aging. These difficulties result from using average cost values for broad categories of equipment; cost changes for individual items can depart significantly from these average values. Thus, the PWAC method is not recommended.

However, this method has been used often because it is possible to group equipment into categories and to assign average values of life, salvage, maintenance costs, operating costs, and ad valorem taxes to each category. From these values, it is a relatively simple procedure to calculate annual cost rates as percentages of installed costs for each category. From these costs, study procedures can be used to derive present worth comparisons rather quickly and simply. The costs may be converted to equivalent present-worth values by considering them as annuities and using Equation (11-7) for conversion. A second reason for working with annual costs is that the treatment of non-coincident equipment placements and retirements is often facilitated.

With noncoincident placements and retirements, two time periods must be defined and treated independently over the period covered by the study. The *planning period* is defined as that between the beginning of the study $(T = 0)$ and the time of the last placement of equipment. The *complementary period* is that time between the last placement to the time of the last retirement of equipment. The planning period covers those years during which additions, removals, and changes are planned in order to meet growth forecasts and other service requirements. The planning period is restricted to the number of years ahead that judgment dictates is reasonable in terms of predictability of needs and availability of resources. The complementary period is the span of years beyond the end of the planning period for which annual costs will continue and will influence present worth evaluations of costs and revenues.

**Present Worth of Expenditures.** This method of analysis may be defined as the summation of the cash flows for capital expenditures, net salvage, income taxes, and operation costs (or savings) for a project after conversion to present worths at the appropriate rate. Equation (11-4) is used for each of the conversions. The method is straightforward, has none of the complications of multiple roots and numerous comparisons found in the IROR analysis, and is the method most often used for engineering economy studies. It is superior to the PWAC method primarily because average costs are not used. Furthermore, since individual costs must be used, it is a simple matter to include in the analysis the effects of variable factors such as inflation.

## Study Assumptions

While the assumptions made for any of the analytic alternatives discussed are similar, those covered here are particularly applicable to a PWE analysis. The important parameters include equivalency of service provided by each of the plans, the life of plant (cotermination or repeated plant), plant retirement effects (sunk costs), and the elimination of common costs (the inclusion of incremental costs only).

**Equivalency of Service.** The alternative plans in an engineering economy study must satisfy service needs equally or allowance must be made for the advantages of one plan relative to another. If the number of new circuits is insufficient to meet the needs, other subsidiary facilities must be provided and allowance must be made in the study for the costs of these additional facilities. Furthermore, the alternative plans should be equivalent in terms of the *quality* of service each provides; the reliability and transmission performance of each must satisfy the overall objectives for the project under study.

One complication arises in respect to the equalization of service capabilities. Modern transmission systems tend to be broadband and capable of providing large numbers of voiceband channels. The growth of demand, on the other hand, tends to be relatively smooth and constant. When growth exceeds capacity, a new system must be installed; thus, the new system provides an excess of capacity until demand again increases to the system limit. Alternative plans usually involve systems of different capacities and costs. These systems fulfill the needs and provide excess capacities in different proportions. The analysis may thus be seriously dependent on short-term versus long-term conditions of meeting service needs and the economic comparison of alternatives must account for the differences.

**Cotermination and Repeated Plant Assumptions.** Either or both of two basic assumptions regarding life of plant may be made in preparing most engineering economy studies. One assumption is for the cotermination of plant and the other is for repeated plant [1]. With cotermination of plant, the retirement dates are identical for all final plant items; this assumption can result in having atypical service life values assigned to various plant items in the study. The cotermination assumption would clearly be valid, for example when various plant items are to be installed in a buiding which is known (or assumed) to have an end-of-life corresponding with the end of the study period.

When repeated plant is assumed, the life of each asset in the study is determined by its physical characteristics. Usually, in studies that involve the repeated plant assumption, the effects of retiring an item of plant at end-of-life must be taken into account by replacing it with one at the same cost that permits the provision of equivalent services over the study period. This replacement must then be evaluated in terms of its effects over the period of the study.

Circumstances and judgment must determine which of the two assumptions is the better in a given study situation. Whichever assumption is made, it is important that the plans be comparable in quantity and quality of service over the same period of time. In addition, it must be recognized that future decisions (and costs) may be affected by present decisions. For example, one of the alternative plans under study might lead to the premature exhaust of building capacity and new building construction might be required. An evaluation of such effects must be undertaken as part of the study.

When the appropriate life-of-plant assumption has been established, the effect of the assumption on costs must also be considered. Average costs are simple to apply but may not be sufficiently accurate. Explicit estimates of costs for maintenance, depreciation, return, and taxes over the expected life span for each item of plant should be included in an economy study. Thus, in most cases, a PWE analysis is required.

In a period of high inflation, the repeated plant assumption of zero inflation is invalid because maintenance and replacement costs increase with time. Thus, the coterminated plant assumption may be more appropriate with all PWE costs properly inflated. An appropriate adjustment in net salvage value is also required. If a PWAC study is being made, the repeated plant assumption may be modified to reflect forecasted price changes. Replacement plant costs may be calculated for a sufficiently long period into the future so that the last plant placements have negligible effects on study results.

**Retirement of Plant.** When the cost of equipment has once been incurred, that cost must be recovered by methods of depreciation accounting whether the item is retired early, at the end of life as originally defined, or at some later time. The cost so incurred is irrelevent to a new engineering economy study involving other alternatives even when one of those alternatives is the early retirement of that item. Such costs are called sunk costs. Sunk costs are irrelevent because they are common to the study alternatives. However, other costs related to that item cannot be ignored. When plant is retired, many other costs are affected; property tax, maintenance, rent, energy, and operating costs are no longer incurred.

**Incremental Costs.** In most engineering economy studies, costs and revenues that are common to alternate plans may be neglected because the comparison of one plan with another involves only the consideration of differences between them. Although the costs (called *incremental costs*) that meet this criterion are usually easy to define, their identification sometimes involves the exercise of engineering judgment.

Revenues can usually be neglected because the selection of plans for comparison is based on equivalent quantity and quality of service and, therefore, of revenue. This equivalency must be considered carefully in each study and, where there are significant differences, incremental revenues must also be included in the analysis.

## Summary of the Comparison Study Process

Since most of the important elements of engineering economy studies have been discussed, the step-by-step procedure used in the conduct of such a study may now be outlined. Such a study starts with the recognition of a need for new facilities and ends with a decision to proceed with the implementation of a specific plan that has been demonstrated as economically superior to alternative plans.

Forecasts of demands for new services and new facilities are made continuously by two kinds of planning groups, one responsible for long-range (fundamental) planning and one for short-range (current) planning. Comparison studies may be made in either type of planning activity. However, comparison studies for current planning activities are of greatest interest here because such studies result in implementation of specific projects in accordance with planning, budgeting, and control procedures required by the construction program. Thus, the initiation of an engineering economy study of alternatives is made in response to a planning-group forecast of needs for new facilities.

After the need for a study has been demonstrated, alternate plans must be proposed for comparison. The number of alternatives to be considered depends on the knowledge and judgment of those involved in the study. Incremental costs (and sometimes revenues) for each of the plans must be determined and documented.

A time-cost diagram is often prepared as an aid to analysis and to presenting an orderly comparison of alternative plans. Such a diagram helps to visualize costs and their times of occurrence; it also provides a mechanism for checking that all important costs are included for analysis and that each of the alternatives provides service over the period of time corresponding with the study period. A time-cost diagram is illustrated in Figure 11-4.

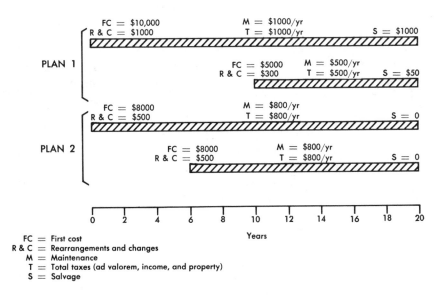

Figure 11-4. Time-cost diagram.

After all pertinent data has been gathered, study assumptions must be established and examined for validity in the specific analysis to be undertaken. Care must be taken to be sure that alternate plans are equivalent in terms of quantity and quality of services provided and that suitable allowances for differences are made where necessary. The applicability of the coterminated or repeated plant assumption must be determined. This assumption is of course closely related to the type of analysis to be carried out (IROR, PWAC, or PWE). Finally, the best possible judgment must be exercised in evaluating intangible aspects of a plan; these might include the esthetic effects of certain designs or the impact of a project on the environment.

Of necessity, economy studies are based on a number of explicit and implicit assumptions. It is often desirable to broaden the scope of the study to determine the sensitivity of the results to variations in the assumptions. For example, the growth rate used in the forecast that initiated the study might be varied or, if it was assumed to be uniform, the effect of a nonuniform rate might be evaluated. Sometimes, project studies are complicated by interactions with other projects. The construction of parallel or crossing routes or succeeding installations may affect initial costs. Estimates must be made of these effects and it may be desirable to determine the sensitivity of the results to variations in the estimates.

When all studies have been completed, the results are compared in respect to economy, uncertainty, and sensitivity to variations. A decision is made in favor of a specific plan and a recommendation is made for management consideration and implementation.

### REFERENCES

1. Engineering Department of The American Telephone and Telegraph Company. *Engineering Economy*, Third Edition (New York: McGraw-Hill Book Company, Inc. 1977).

2. Grant, E. L. and W. G. Ireson. *Principles of Engineering Economy*, Fifth Edition (New York: The Ronald Press Company, 1970).

3. Smith, G. W. *Engineering Economy: Analysis of Capital Expenditures*, Second Edition (Ames, Iowa: Iowa State University Press, 1973).

## Section 3

## Signal Characterization

Telecommunications in the Bell System involve the transmission and, in many cases, the switching of many types of signals which differ materially from one another. To facilitate the evaluation of transmission objectives, the nature and magnitude of various impairments, the performance provided by different systems or facilities, and the manner in which all of these interact, it is necessary that the various types of signals be described in terms that permit the expression of mathematical relationships among all these factors. This section of the book provides such signal characterization for the principal forms of transmitted signals — speech signals, address and supervisory signals, data signals, and video signals. It also covers the characterization of combinations of signals that are found in a frequency division multiplexed load on an analog carrier system.

Chapter 12 covers the characteristics of speech signals typically found in a telephone channel, i.e., a loop or trunk. Bandwidth, amplitude, phase, and frequency variations are described for telephone speech and the characteristics of a multichannel speech signal transmitted on analog carrier systems are described. A brief discussion of radio and television program signals is also given.

Wherever telecommunications signals must be switched, signals must be transmitted for the purpose of directing and controlling the switching apparatus. These signals, called address and supervisory signals, are of many types. The most important are described in Chapter 13. The proliferation of this variety of signals has resulted from the increasing number of switching system types and the increased number of switching features that have been provided. The signal characterization given in this chapter is provided with a minimum of discussion of the equipment or switching features involved.

The material in Chapter 14 represents the characterization of a number of the more important types of data signals found in the Bell System. These signals are, in many cases, digital in format; they involve the provision of channels ranging from bandwidths of tens of hertz to several megahertz. Amplitude, frequency, and phase shift keying techniques are employed in multilevel formats ranging from two to fifteen levels. Some signals are analog in nature and as such, may achieve an infinite number of values over a restricted but continuous range.

The transmission of video signals is among the telecommunication services provided by the Bell System. While the number of video circuits in service is small compared to the number of voice-frequency circuits, the video circuits utilize a substantial portion of the Bell System transmission facility capacity because of the large bandwidth most of them require. Characteristics are described in Chapter 15 for telephoto, video telephone, and black and white and color television signals.

One reason for the extensive and detailed attention given to signal characterization is the fact that signals and transmission systems interact in important ways. It is rare that only one type of signal is to be found in any one transmission system. This is especially true in broadband carrier systems which carry simultaneously a large variety of signals. Some of the effects of such signal combinations are characterized in Chapter 16, where a qualitative discussion of such combinations is presented.

Chapter 12

# Speech Signals

A message channel in the switched message network or in a private line network must carry a wide variety of signals; the most common and, therefore, among the most important is the telephone speech signal. Much research effort has been devoted to an understanding of all the details of the processes of speech and hearing [1,2,3,4]. The concern here, however, is with the electrical signal analog of the acoustic message. This signal and its characterization are related primarily to the processes carried out in the transmitter (microphone) of the telephone station set and the effects on the signal produced by interactions between it and the channels on which it is carried.

The problems of speech signal characterization are made complex by the large number of variables involved and the resulting difficulties of defining and measuring important parameters explicitly. To overcome these difficulties, signal parameters are defined in terms of their statistical properties, such as average values, standard deviations, and activity factors. These parameters are defined first for a hypothetical single continuous talker of constant volume, $V_{0c}$. This value, expressed in vu, is next modified to account for breathing intervals and intersyllabic gaps and to define the single constant-volume talker in terms of power in dBm, $P_{0c}$.

Variables are next introduced to cover the effects of the sex and speaking habits of the talkers, circuit losses, the automatic compensation introduced by talkers to overcome impairments, station set variability, etc. Consideration of these variables introduces the concept of the variable volume talker, one whose average volume is $V_{0c}$ and whose volume has a standard deviation, $\sigma$.

The definition of these parameters is relatively straightforward, but the determination of their values by analytic means is not. Measurements are usually made in working systems to determine the values of average and standard deviations. These measurements must be expressed, of course, in terms of some well-established reference point, such as 0 TLP.

A continuous talker signal is not ordinarily found in a telephone message channel. Activity factors associated with the efficiency of trunk utilization and talk-listen effects must be evaluated. With these factors accounted for, the statistics of talker signals in multiplexed broadband systems can be evaluated and used for the determination of signal-dependent impairments such as intermodulation noise, crosstalk, and overload.

## 12-1  THE SINGLE-CHANNEL SPEECH SIGNAL

Whereas single-frequency signals are easily specified by just a few numbers — one for frequency, one for amplitude, and in some cases, one for phase — in addition to a functional expression such as sine or cosine, a telephone speech signal is not so easily specified or defined. It consists of many frequencies varying in amplitude and relative phases. Its average amplitude fluctuates widely, and even its bandwidth may vary with circumstances. Consider first the speech signal generated at a telephone station set and the way in which it is modified by the transmission elements of the channel between the transmitter and receiver.

### Speech Signal Energy Distribution and Channel Response

The electrical analog of the acoustic speech signal is generated in the station set transmitter. Sound waves from the speaker are impressed on the transmitter of the station set, which typically houses a small container filled with carbon granules. Common battery direct current, supplied from the central office over the loop conductors, passes through these granules. The varying pressure of the speech waves causes the resistance between granules to vary and, in effect, to modulate the direct current passing through them.

Human speech contains significant components extending roughly from 30 to 10,000 Hz. The distribution of the long-term average energy for continuous speech approximates that shown in Figure 12-1. The actual spectral energy density and bandwidth are, of course,

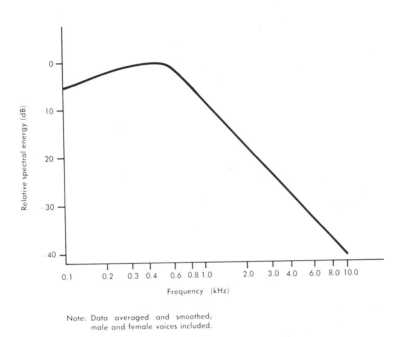

Note: Data averaged and smoothed;
male and female voices included.

Figure 12-1. Approximate long-term average spectral energy density for continuous speech.

highly variable parameters. Nearly 90 percent of the speech energy lies below 1 kHz. This part of the spectrum also contains considerable intelligibility so that speech transmitted through a 1-kHz low-pass filter would be at least partly understandable. However, it would also be quite unnatural and unpleasant. The listener would have to work hard to recover intelligibility, and many of the nuances in speech that permit recognition of the talker would be lost.

In practice, a band extending approximately from about 0.25 to 3.0 kHz has been found to provide commercially acceptable quality for telephone communications. The transmission response at several points in a simple connection is depicted in Figure 12-2. In the Bell System, the transmission band of telephone circuits is defined as that between points that are 10 dB down from the reference frequency, usually taken as 1000 Hz. Figure 12-2 shows that, even for the simple

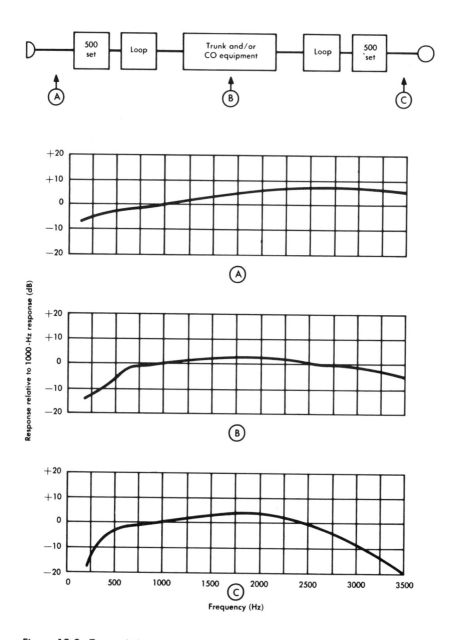

Figure 12-2. Transmission response, normalized at 1000 Hz, at points along a typical connection (trunk and central office equipment assumed distortionless).

connection depicted, the band is already restricted to approximately 0.25 to 3.0 kHz.

## Single Constant-Volume Talker

To develop an understanding of speech signal characterization from the point of view of practical applications of transmission design, layout, and operation, consider first a single *continuous* talker of constant volume, a somewhat hypothetical case. The volume of this talker's telephone speech signal has, by definition, a value of $V_{0c}$ vu.

A continuous talker is not capable of producing truly continuous speech signals. Pauses due to the thought process, to breathing intervals, or to intersyllabic gaps in energy result in an *activity factor*, $\tau_c$, of 0.65 to 0.75.

The value of power in dBm in a speech wave is defined as the value of volume in vu corrected by the activity factor. Thus, the power for a continuous talker may be written

$$P_{0c} = V_{0c} + 10 \log \tau_c \text{ dBm.} \qquad (12\text{-}1a)$$

For example, if $\tau_c = 0.725$, the power in such a signal is

$$P_{0c} = V_{0c} - 1.4 \text{ dBm.} \qquad (12\text{-}1b)$$

This value agrees with an empirically derived relationship between vu and dBm for speech signals which is generally accepted.

## Sources of Volume Variation

Except under specially controlled circumstances, a constant-volume talker is a rarity. Consider some of the important sources of volume variation. First, the telephone speaking habits and sex of the speaker introduce wide variations. He or she may be loud or soft-spoken and may hold the telephone transmitter close or at a distance. In addition, telephone sets have a range of values for the efficiencies with which they transform acoustic waves to electrical waves and vice versa. Further, their efficiencies are, by design, variables which depend on the value of direct current fed to them from the central office. The length of the loop, the wire gauge used, the presence or absence of irregularities such as bridged taps or bridged stations, and the possible use of loading on the loop all contribute to variations from loop to loop. These variations affect the average losses in the loop and the amount of current fed to the transmitter. In addition, these variations affect differently the attenuation at different frequencies. Variations in average loss and in frequency-dependent attenuation are also found in central office wiring and equipment, trunks, and carrier facilities that may be used in a built-up connection. Furthermore, impairments such as sidetone, echo, circuit noise, room noise, and crosstalk have subjective effects on speaking habits, as do distance, trunk loss, and type of call.*

Some of the variable losses involved in a simple interlocal telephone connection are illustrated in Figure 12-3. Station set efficiencies for sound pressure to electrical signal conversion and vice versa are such that, with typical losses in the circuit making up a local connection, a speaker producing at the microphone a sound pressure of 89.5 dBRAP (dB above reference acoustic pressure) would be heard at a sound pressure of 81.5 dBRAP. Reference in this case is an acoustical pressure of 0.0002 dyne per square centimeter. The previously mentioned variables are such that received sound signals have a wide range of values with a standard deviation of nearly $\pm 8$ dB about the average value of 81.5 dBRAP.

In Figure 12-3, the noise impairments shown as introduced in loops, central office equipment, and the trunk might be picked up at any of these points. The figure is illustrative. Room noise at the speaker's end of the connection enters the circuit through the transmitter and appears at the distant receiver along with the speech signal. Room noise at the listener's end affects his hearing directly and, in addition, enters his transmitter and appears in his receiver by transmission through the sidetone path. The decreasing powers in the signal and in each noise component, caused by the increasing circuit loss illustrated at the bottom of the figure, are not assigned values in the figure because they are so highly variable on different connections and under differing circumstances. Even though impairments are not discussed here in detail, the noise and loss impairments are shown qualitatively in Figure 12-3 to illustrate their sources.

---

*It has been observed that volume increases about 1 vu for each 3 dB of trunk loss and about 1 vu for each 1000 miles of distance. Volume on business calls tends to be somewhat higher than on social calls.

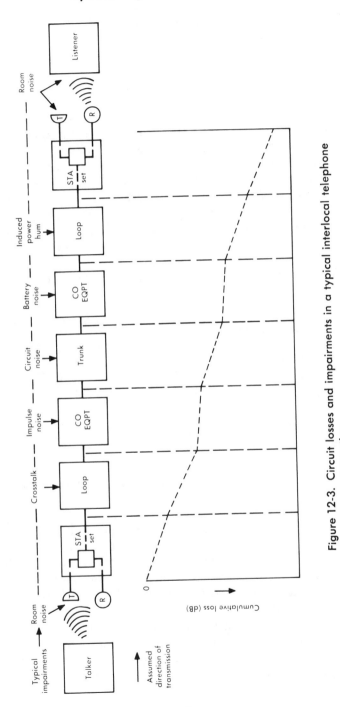

Figure 12-3. Circuit losses and impairments in a typical interlocal telephone connection.

They have an indirect, subjective effect on talker volumes as previously mentioned.

## Single Variable-Volume Talker

As has been pointed out, the single constant-volume talker is, in general, a hypothetical case. The aforementioned variables are so numerous and so difficult to evaluate precisely that it is necessary to rely on measured data in order to characterize the single variable-volume talker. The results of the 1960 survey of speech volume measurements, essentially an evaluation of the variable-volume talker, are summarized in Figure 12-4 [5]. While these are the latest data available, they are somewhat dated, and consideration is being given to conducting a new survey. In such a survey, many variables must be considered, and studies are being made to determine which of these are important in the present day plant [6, 7].

| TYPE OF CONNECTION | SPEECH VOLUMES (VU)* | |
| --- | --- | --- |
| | MEAN | STANDARD DEVIATION |
| Intrabuilding | −24.8 | 7.3 |
| Interbuilding | −23.1 | 7.3 |
| Tandem | −19.6 | 5.9 |
| Toll | −16.8 | 6.4 |

*Measured at transmitting switch, class 5 office.

Figure 12-4. Near-end talker speech volumes, 1960 survey.

A knowledge of the average power per talker of a group of talkers all of whose volumes vary with time is needed for the design of broadband carrier systems. Such designs must be based on total signal power, determined from the mean value and standard deviation of each of the speech signals to be carried. These signals do not combine statistically as normal distributions, even though each is normal in dB. The average power values must be added; this requires conversion from dBm to milliwatts, determination of the average value, addition of the averages in milliwatts, and reconversion of the result to dBm.

Consider a probability density function, normal in dB, having an average value of 0 dBm and standard deviation of 3 dB (these values are illustrative, not typical). Such a function is plotted in Figure 12-5(a). If the dBm values are converted to milliwatts, the density function of Figure 12-5(b) results. Note that this function is skewed and that its mean value is greater than 1 mW. The difference, δ, between the average value in dBm (0 dBm or 1 mW) and the mean value of the distribution increases as $\sigma$ increases. The necessary correction to express the power under a log normal probability density function has been derived elsewhere and is equal to $0.115\sigma^2$ [8]; i.e., to obtain the average power in dBm, $0.115\sigma^2$ must be added to the mean value.

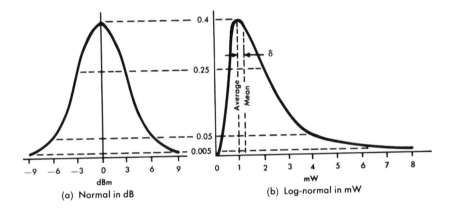

Figure 12-5. Density functions.

Thus, the average power of a variable-volume talker signal having a log normal density function and a standard deviation of $\sigma$ may be expressed by

$$P_{0p} = P_{0c} + 0.115\sigma^2 = V_{0c} - 1.4 + 0.115\sigma^2 \text{ dBm.} \qquad (12\text{-}2)$$

This equation is derived from Equation (12-1b) and from the discussion above which relates the average value of power to the mean value of the density curve, normal in dB.

The mean value of volume for toll calls is given on the last line of Figure 12-4, but further manipulation is necessary to make the data useful for toll system analysis. The first step is to translate the data from the outgoing switch of the class 5, or end office, where the measurements were made, to a comparable point in a toll system. Between the point of measurement and the entrance to the toll portion of the network are, for each connection, a toll connecting trunk and certain items of central office equipment. These have a loss of (VNL + 2.5) dB, which includes 2 dB assigned to the trunk and 0.5-dB allowance for the central office equipment. If 0.5 dB is allowed for the VNL (a typical value), the average −16.8 vu volume for toll calls shown in Figure 12-4 may be translated to toll system values (at the −2 dB TLP) as

$$V_{toll} = -16.8 - (2.0 + 0.5 + 0.5) = -16.8 - 3.0 = -19.8 \text{ vu.}$$

Typically, the losses of toll connecting trunks have a standard deviation of 1 dB. Thus, when combined with the standard deviation of measured toll volumes at the end office, the standard deviation of volume on the toll system is

$$\sigma_{toll} = \sqrt{6.4^2 + 1^2} = 6.47 \text{ vu.}$$

For the values of toll call volumes, the average continuous talker power is

$$P_{0p} = -19.8 - 1.4 + 0.115\sigma^2 = -19.8 - 1.4 + 4.8 \approx -16.5 \text{ dBm.}$$

One further correction is needed. Recall from Chapter 3 that the outgoing switch at which a toll trunk is terminated is defined as a −2 dB TLP. Therefore, the toll average power must be converted to a value at 0 TLP by adding 2 dB; i.e.,

$$P_{0p} = -16.5 + 2 = -14.5 \text{ dBm0.}$$

All of the above discussion relates to volume and power averaged subjectively over an interval of 3 to 10 seconds. In reading the volume indicator, occasional very high and very low readings are ignored. High peaks, however, do occur, and their magnitude is sometimes of considerable interest. The peak factor for a typical continuous talker is approximately 19 dB. For a talker of lower activity, peak magnitudes are not affected, but the average power is reduced relative to the continuous talker.

## 12-2  MULTICHANNEL SPEECH

The need for characterizing the speech signals in multichannel systems arises primarily from the need to control overload performance in analog systems. The characteristics of a multiplexed combination of speech signals are determined by extrapolation of the analysis of single speech signal characteristics.

If there are a number, $N_a$, of independent continuous talker signals of distributed volumes simultaneously present in a broadband system, each signal occupying a different frequency band but at the same TLP, the total power represented by the $N_a$ signals is

$$P_{av} = P_{0p} + 10 \log N_a$$

$$= V_{0c} - 1.4 + 0.115\sigma^2 + 10 \log N_a \qquad \text{dBm.} \qquad (12\text{-}3)$$

In a system containing $N$ channels, the maximum number of simultaneous signals that could be present is $N_a = N$; however, such an event is extremely unlikely, especially when $N$ is large. Thus, it is necessary now to examine the factors that enter into an evaluation of the probable number of simultaneous talkers in such a system.

The speech activity factor for a continuous talker, $\tau_c$, was included in Equation (12-1) for the evaluation of $P_{0c}$. In evaluating $P_{av}$ [Equation (12-3)], other forms of activity must be taken into account. The assumption is made that, on the average, during a conversation the person using a telephone talks half the time and listens half the time. Thus, the value of the talk-listen activity factor, $\tau_s$, may be taken as 0.5. More trunks are provided than are needed, even during the busy hour when traffic is heaviest, because the number of call rejections due to busy circuits must be held to an acceptable minimum. Furthermore, during the time a call is being set up, there is low speech activity on the trunk. These effects may be accounted for by a trunk efficiency factor, $\tau_e$. For domestic circuits, $\tau_e$ is usually taken as 0.7. For overseas calls, this value may be as high as 0.9.

The two activity factors discussed above are usually combined into a single *telephone* load activity factor,

$$\tau_L = \tau_s \tau_e = 0.5 \times 0.7 = 0.35.$$

Other activity considerations not specifically evaluated have led to a commonly accepted value of $\tau_L = 0.25$ for domestic telephone systems. A higher value (usually $\tau_L = 0.35$) is used for transatlantic or transpacific systems. It must be remembered, however, that these are average busy-hour values. The number of speech signals simultaneously present during the busy hour, when such load considerations are important, varies considerably.

For an $N$-channel system having a load activity factor $\tau_L$, the number of independent continuous talker signals, $N_a$, is a variable whose mean value is $N\tau_L$. A system designed to carry just $N\tau_L$ continuous talkers would be overloaded half the time. A system designed to carry $N$ continuous talkers would be impractical because such a signal load would occur only a very small percentage of the time.

It is necessary, therefore, to establish the statistical distribution of channels that would carry continuous talker signal power as a function of time. The variable representing this distribution may be called $N_s$. The probability that the number of channels carrying continuous talker power is $N_s$ may be found from

$$P(N_s) = \frac{N!}{N_s!\,(N-N_s)!}\,\tau_L^{N_s}\,(1-\tau_L)^{N-N_s}.$$

This is a binomial distribution that approaches a normal distribution having a mean value of $N\tau_L$ and a standard deviation $\sqrt{N\tau_L\,(1-\tau_L)}$ if $N\tau_L \geqq 5$.

For design purposes, the number of talkers assumed to generate speech energy simultaneously is the number that may be present one percent of the time. This value, chosen on the basis of experience, shows adequate balance between performance and cost. Thus, $N_a$ is the value of $N_s$ exceeded one percent of the time. From the values of areas under a normal curve (Figure 9-15), this value is $N_a \approx N\tau_L + 2.33\,\sqrt{N\tau_L\,(1-\tau_L)}$.

Examination of this equation shows that the mean, $N\tau_L$, increases more rapidly than the standard deviation, $\sqrt{N\tau_L\,(1-\tau_L)}$, as $N$ becomes larger. Thus, for large values of $N$, $N_a$ approaches $N\tau_L$. Also, it can be seen that the larger the value of $\tau_L$, the smaller $N$ need be for this approximation to be valid. Note that for $\tau_L = 1$, $1 - \tau_L = 0$, and $N_a = N\tau_L$.

Thus, for large values of $N$, Equation (12-3) can be rewritten

$$P_{av} \approx V_{0c} - 1.4 + 0.115\sigma^2 + 10 \log N + 10 \log \tau_L \qquad \text{dBm.}$$

This approximation can be made an equality, even for systems of small $N$, by defining a term which takes into account the deviation of $N_a$ from $N\tau_L$. This term is defined*

$$\Delta_{c1} = 10 \log \frac{N_a}{N\tau_L} . \qquad (12\text{-}4)$$

When terms are rearranged, this may be written

$$10 \log N_a = \Delta_{c1} + 10 \log N + 10 \log \tau_L$$

and substituted in Equation (12-3) to give

$$P_{av} = V_{0c} - 1.4 + 0.115\sigma^2 + 10 \log \tau_L + 10 \log N + \Delta_{c1} \qquad \text{dBm.}$$

$$(12\text{-}5)$$

For the two values of $\tau_L$ given previously (0.25 and 0.35), the relationships among $N_a$, $N\tau_L$, and $\Delta_{c1}$ are shown in Figure 12-6 for systems of various sizes. The value of $\Delta_{c1}$ is shown to become small as $N$ gets larger. It is often ignored in systems in which $N \geqq 2000$ channels.

If $V_{0c}$ is evaluated at 0 TLP, the value of $P_{av}$ in Equation (12-3) is in dBm0. In the total speech load of $N$ signals, $P_{av}$ is the average power at 0 TLP exceeded during one percent of the busy hour when all $N$ channels are busy. (A channel is considered busy when a talking connection is established; speech signals need not be present.)

From Equations (12-2) and (12-5), the long-time average load per channel may be determined (by substituting the previously derived values $P_{0p} = -14.5$ dBm0 and $\tau_L = 0.25$) for broadband toll systems as

$$P_{av}/\text{chan} = P_{0p} + 10 \log \tau_L + \Delta_{c1} \qquad \text{dBm0.}$$

---

*Other near-equivalent definitions of $\Delta_{c1}$ are given in Reference 9, pages 227 and 229. The definition given here, however, is commonly used; its value is conveniently determined and nearly always accurate enough for engineering purposes.

| N | $\tau_L = 0.25$ | | | $\tau_L = 0.35$ | | |
|---|---|---|---|---|---|---|
| | $N_a$ | $N_{T_L}$ | $\Delta_{c_1}$, dB | $N_a$ | $N_{T_L}$ | $\Delta_{c_1}$, dB |
| 6 | 4.84 | 1.5 | 5.1 | 5.60 | 2.1 | 4.3 |
| 12 | 7.37 | 3.0 | 3.9 | 8.78 | 4.2 | 3.2 |
| 24 | 11.80 | 6.0 | 2.9 | 14.59 | 8.4 | 2.4 |
| 36 | 15.88 | 9.0 | 2.5 | 19.94 | 12.6 | 2.0 |
| 48 | 19.84 | 12.0 | 2.2 | 25.19 | 16.8 | 1.8 |
| 96 | 34.74 | 24.0 | 1.6 | 45.18 | 33.6 | 1.3 |
| 300 | 93.32 | 75.0 | 0.9 | 124.92 | 105.0 | 0.7 |
| 600 | 175.55 | 150.0 | 0.7 | 237.89 | 210.0 | 0.5 |
| 2000 | 545.91 | 500.0 | 0.4 | 750.34 | 700.0 | 0.3 |

Figure 12-6. Number of active channels and $\Delta_{c_1}$.

For very broadband systems ($N > 2000$), $\Delta_{c_1}$ approaches zero and the load is

$$P_{av}/\text{chan} = -14.5 - 6 = -20.5 \text{ dBm0}$$

for a telephone signal load of variable volume talkers.*

## 12-3 LOAD CAPACITY OF SYSTEMS

The load capacity of a multichannel telephone transmission system is the peak power generated by the total number of speech signals the system can carry without producing an undue amount of distortion or noise or otherwise affecting system performance or reliability. The maximum signal amplitude impressed on the system depends on the average talker volume, the distribution of volumes, and the talker activity. Overload may be the result of the signal amplitude exceeding the dynamic range of an amplifier or other active device, of frequency deviations exceeding the bandwidth of an angle-modulated system, or of voltages exceeding the quantizing range of a

*None of the material in this chapter considers the effects of address, supervisory, or data signals on average channel loading. These effects are covered in Chapter 16.

digital quantizer. A system is often said to be overloaded when the overload point of the system is exceeded by peaks of the transmitted signal more than 0.001 percent of the time. (It is *not* then said to be overloaded 0.001 percent of the time.)

## Multichannel Speech and Overload

Overload is defined in a number of ways in Chapter 7. These definitions all basically relate to the signal amplitude at which performance is no longer linear enough to satisfy performance objectives. In any of these definitions, it is convenient to use $P_s$ dBm0 to express the average power of a single-frequency signal that causes system overload. The peak instantaneous power of this sinusoid is $(P_s + 3)$ dBm0.

Most systems do not overload on average power but rather when instantaneous peaks exceed some threshold. A multichannel telephone system with $P_{av} = P_s$ overloads severely because the multichannel signal has a peak factor much larger than the 3-dB peak factor for the single frequency, $P_s$. The peak factor for multichannel speech is 13 to 18 dB, depending on the number of channels in the system. It has been found that performance is usually satisfactory if the peak power of the multichannel load exceeded 0.001 percent of the time is set equal to or less than the peak power of the sinusoid, $(P_s + 3)$ dBm0. This may be written

$$P_s + 3 = P_{av} + \Delta_{c2} \qquad \text{dBm0}$$

or

$$P_s = P_{av} + \Delta_{c2} - 3 \qquad \text{dBm0}, \tag{12-6}$$

where $\Delta_{c2}$ is the peak signal amplitude exceeded 0.001 percent of the time. The value of $\Delta_{c2}$ has been determined and is plotted in Figure 12-7. This figure shows that as the number of active channels, $N_a$, increases, the peak factor asymptotically approaches 13 dB.

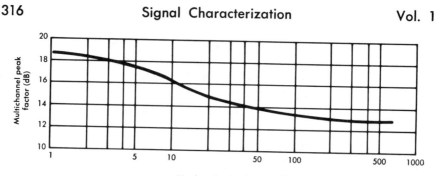

Number of active channels, $N_a$

Figure 12-7. Peak factor, $\Delta_{c2}$, exceeded 0.001% of time for speech channels.

This value corresponds closely to that for random noise.

If Equation (12-5) is substituted in Equation (12-6),

$$P_s = V_{0c} - 1.4 + 0.115\sigma^2 + 10 \log \tau_L + 10 \log N$$

$$+ \Delta_{c1} + \Delta_{c2} - 3 \qquad \text{dBm0} \qquad (12\text{-}7a)$$

or, with $\Delta_c = \Delta_{c1} + \Delta_{c2} - 3$,

$$P_s = V_{0c} - 1.4 + 0.115\sigma^2 + 10 \log \tau_L + 10 \log N$$

$$+ \Delta_c \qquad \text{dBm0.} \qquad (12\text{-}7b)$$

The term $\Delta_c$ is known as the multichannel load factor. It is plotted in Figure 12-8 as a function of $N$ and several values of $\sigma$ for an assumed value of $\tau_L = 0.25$. For other values of $\sigma$ or $\tau_L$, $\Delta_c$ can be found from the empirically derived formula

$$\Delta_c = 10.5 + \frac{40\,\sigma}{N\tau_L + 5\sqrt{2\sigma}} \quad \text{dB.} \qquad (12\text{-}8)$$

Single-frequency signals are used in the analysis of system load capacity, but they are seldom used in load testing. A band of Gaussian noise is frequently used in system testing to simulate a multichannel signal.

## Effect of Shaped TLP Characteristics

The discussion of the multichannel speech signal load and its relation to overload phenomena has been carried out in terms of 0 TLP characterizations of speech signals. Implied in the discussion is the

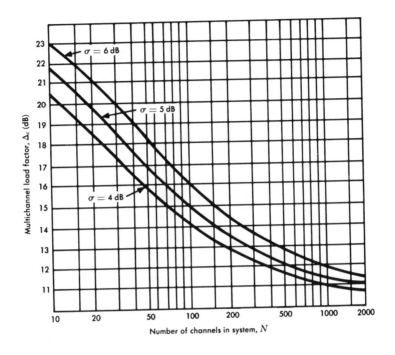

Figure 12-8. The multichannel load factor, $\Delta_c$, for $\tau_L = 0.25$.

assumption that the transmission from 0 TLP to the point of interest where overload may occur (for example, at the output of a line repeater) is flat with frequency and thus the same for all channels in the broadband system. This is not necessarily so. Noise advantage in the system can frequently be obtained by shaping the transmission between the two points. In this case, the TLP at the point of interest is not flat with frequency; as a result, the volume distribution at the point of interest is modified according to the line frequencies of the individual channels and the transmission characteristic between the two points.

The average power of such a shaped signal load may be determined [9] at the point of interest by

$$P'_{av} = P_{av} + 10 \log \int_{f_B}^{f_T} \frac{10^{C(f)/10} df}{f_T - f_B} \quad \text{dBm.} \tag{12-9}$$

Here, $f_T$ and $f_B$ are the top and bottom frequencies, respectively, of the signal spectrum at the point of interest, and $C(f)$ is the gain shape in dB between 0 TLP and the point of interest.

With shaping between 0 TLP and the overload point, the peak factor, and hence $\Delta_c$, are more complex. The effects of signal shaping on overload have been studied, using a computer, for normally distributed talker volumes having various gain characteristics over the multiplexed band. It has thus been found empirically that, for the same overload condition (0.001 percent), the value that should be used for $\Delta_c$ is very well approximated if the system is assumed to have $\eta$ channels instead of $N$ channels. The value for $\eta$ is taken as that number of channels whose TLPs are within 6 dB of the channel having the highest TLP at the point where overload occurs.

## 12-4  PROGRAM SIGNALS

Program transmission is a nationwide service provided by the Bell System to transmit the audio programs of radio and television broadcasters between points of program origination and one or more transmitting stations. In addition, "wired music" material is also transmitted for distribution to customers subscribing to such services. Other program services include conference calls and calls connected to public address systems for a large audience. While such signals are audio signals, regular telephone circuits cannot be used for program transmission because of the more stringent objectives generally applicable to program service. The more stringent objectives arise from the necessity of transmitting music and from the need for higher fidelity speech when the receiver is not a telephone set receiver.

At the present time, the majority of program circuits used in toll transmission systems employs a band of frequencies from about 50 to 5000 Hz. For special broadcasts in which the program is speech alone, such as newscasts, the broadcaster may use specially conditioned message circuits that transmit a band of frequencies from 200 to 3500 Hz. Other program services, less frequently used, cover frequency ranges of 50 to 8000 Hz and 35 to 15,000 Hz. The latter two are used primarily to transmit high quality music for FM and FM-stereo broadcasts in local areas and to satisfy the needs of educational television services.

The bandwidth is specified differently for program facilities than for telephone speech. The bandwidth of program circuits is defined as that between the frequencies at which the response is 1 dB below the 1-kHz response, as contrasted with 10-dB response points for message circuits. Program circuit filters must roll off more gently than message circuit filters because program signals are more susceptible to delay distortion impairments than are ordinary message signals. Program channel equipment is often provided with a modest amount of delay distortion equalization.

The energy distribution in program signals is difficult to specify because of the wide range of program material transmitted — speech, drama with sound effects, music of different varieties, etc. No generally accepted program spectrum has been established.

The average volume and the dynamic range of program signals are somewhat higher than for telephone speech. There are relatively few program channels, however, and contributions to system load effects are generally small enough to be ignored. A possible exception is the coverage often given to special events such as a presidential speech or a political convention. All program facilities leaving one location may be carrying the identical program. Careful study is necessary to guard against overload of systems in these circumstances.

#### REFERENCES

1. Dudley, H. "The Carrier Nature of Speech," *Bell System Tech. J.*, Vol. 19 (Oct. 1940), pp. 495-515.

2. French, N. R. and J. C. Steinberg. "Factors Governing the Intelligibility of Speech Sounds," *The Journal of the Acoustical Society of America* (Jan. 1947).

3. Potter, R. K. and J. C. Steinberg. "Towards the Specification of Speech," *The Journal of the Acoustical Society of America* (Nov. 1950).

4. Sullivan, J. L. "A Laboratory System for Measuring Loudness Loss of Telephone Connections," *Bell System Tech. J.*, Vol. 50 (Oct. 1971), pp. 2663-2739.

5. McAdoo, Kathryn L. "Speech Volumes on Bell System Message Circuits — 1960 Survey," *Bell System Tech. J.*, Vol. 42 (Sept. 1963), pp. 1999-2012.

6. Sen, T. K. "Subjective Effects of Noise and Loss in Telephone Transmission," *Conference Record*, IEEE International Conference on Communications, 1971.

7. Brady, P. T. "Equivalent Peak Level: A Threshold-Independent Speech Level Measure," *The Journal of the Acoustical Society of America*, Vol. 44 (Sept. 1968), pp. 695-699.

8. Bennett, W. R. "Cross-Modulation Requirements on Multichannel Amplifiers Below Overload," *Bell System Tech. J.*, Vol. 19 (Oct. 1940), pp. 587-610.

9. Technical Staff of Bell Telephone Laboratories. *Transmission Systems for Communications*, Fourth Edition (Winston-Salem, N. C.: Western Electric Company, Inc., 1970), Chapter 9.

10. Holbrook, B. D. and J. T. Dixon. "Load Rating Theory for Multichannel Amplifiers," *Bell System Tech. J.*, Vol. 18 (Oct. 1939), pp. 624-644.

Chapter 13

# Signalling

Signalling involves the generation, transmission, reception, and application of a class of signals needed for directing and controlling automatic switching machines and conveying to telephone system users information needed for using the network. Such signals may be functionally categorized as follows:

(1) Address signals

(2) Supervisory signals*

(3) Alerting signals

(4) Information signals

(5) Test signals.

Address signals are used to set up connections (i.e., to route calls) by controlling the operation of automatic switching machines. Such signals may be generated at station sets, switchboards, or switching machines. Many types of address signals are used on both loops and trunks.

Supervisory signals are used to convey, to a switching machine or to an operator, information regarding the status of a loop or trunk. The four service conditions that supervisory signals convey are as follows:

(1) *Idle circuit,* which is indicated by the combination of an on-hook signal and the absence of any connection in the central office between the loop and another loop or trunk.

---

*Although address and supervisory signals are both used to control switching machines, they are considered separately in this chapter.

(2) *Busy circuit,* which is indicated by an off-hook signal and a connection to a trunk or another loop.

(3) *Seizure,* or call for service, which is indicated by an off-hook signal and the absence of any connection to another loop or trunk.

(4) *Disconnect,* which is indicated by an on-hook signal and a connection to a trunk or another loop.

The terms *on-hook* and *off-hook* are derived from supervisory conditions that exist on a loop. If the station set is on-hook, it is idle; if it is off-hook, it is busy. The terms are so descriptive that they are commonly applied to trunks as well as to loops. Supervisory signals must be extended over a connection to the point at which billing information can be used by a message accounting machine or by an operator. Details of how such signals are used are beyond the scope of this chapter.

Alerting signals are those whose primary function is to alert an operator or a customer to some need. Included in this group are such signals as flashing, ringing, rering, recall, and receiver-off-hook signals.

Information signals include machine announcements, audible ring, busy tone, and dial tone. While many of these signals are normally transmitted at low enough amplitudes or are used infrequently enough that they have little impact on transmission, the reverse is not true. For example, machine announcement arrangements, such as the Automatic Intercept System, have been carefully engineered so that the customer hears the announcement at about the same amplitude as he would hear an operator. This avoids contrast and ensures a good overall grade of service. Also, in order to be compatible with acceptable transmission standards, the design of tone generators for dial tone, audible ringing, busy tone, etc., is controlled by a precise tone plan which specifies the frequencies and amplitudes of all such tones.

Test signals are of many types. They are not covered in detail in this chapter, but discussions of several types of test signals are found elsewhere in the text.

The characterization of signals covered in this chapter is important from a transmission point of view for a number of reasons. There is a great variety of such signals and some are used frequently, in large numbers, and for long periods of time. It is important to know their characteristics if they are likely to affect the transmission performance of other signals sharing the same facility or transmission medium. Furthermore, such signals sometimes have transmission requirements that are more stringent than other "pay-load" types of signals and, as a result, may be a controlling factor in establishing overall design limits for transmission facilities. In addition, the signalling circuits interconnect with transmission circuits and may contribute to transmission loss and distortion. Finally, on loops, the signalling circuits affect the amount of current that is delivered to the station set transmitter.

The incompatibilities between signalling and transmission circuits could cause distortion of the address signals. Pulse splitting, a serious form of mutilation that can make a single pulse look like two, is an example. It can occur in four-wire terminating sets as a result of spurious low-frequency oscillation caused by parallel resonance between a transmission capacitor and the inductance of a signalling relay. This type of problem must be avoided in the design of signalling-transmission interface circuits. Typically, a nonlinear device, e.g., a diode, may be connected in series with the oscillatory elements to break up the low-frequency oscillations.

## 13-1  SIGNALLING ON LOOPS

Three aspects of signalling on loops are important from a transmission standpoint. These are supervision, addressing, and customer alerting. All of these aspects of signalling on loops are related to what is known as common battery operation.

### Common Battery Operation

Most of the equipment associated with an individual telephone central office is operated from a single large centralized battery.* Current supplied from such a battery to the loops connected to the

---

*Some local battery operation and manual switchboards may still be found in rural areas. This type of operation and the signalling arrangements required are rapidly becoming obsolete and are not covered here.

central office is modulated by speech in the transmitter to form the speech signal. The same battery current is used to implement signalling functions that must be provided from the station set toward the central office equipment.

One type of connection of loops to the common battery supply is illustrated in Figure 13-1. Three loops and station sets are shown

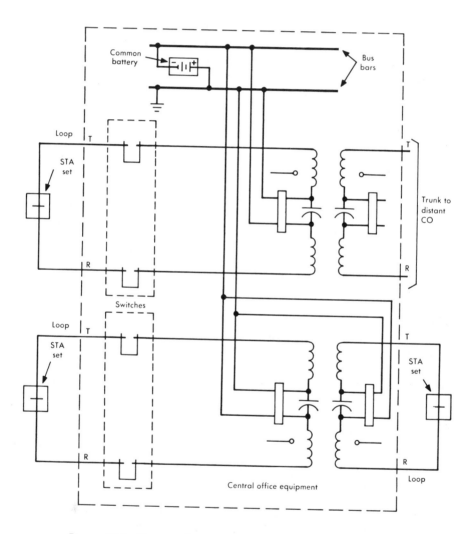

Figure 13-1. Common battery connections — repeat coil circuits.

with the *tip* loop conductors, T, connected to the grounded positive-side bus bar of the battery. The *ring* conductors, R, are connected to the ungrounded negative side of the battery. The repeat coils (or transformers) and capacitors in each of the battery feed circuits to which the loops are connected couple the transmission from the loops into the switches to complete connections to trunks or to other loops. Another circuit configuration commonly used as a battery feed circuit is known as a bridged-impedance-type circuit. This circuit, shown in Figure 13-2, couples the loop to the switches by capacitors rather than repeat coils. Both types of battery feed circuits are designed to minimize the transmission of speech or noise signals from the loops into the common battery. These are oversimplified schematics that do not show details of the signalling functions.

## Supervision on Loops and PBX-CO Trunks

During various stages of a call (call for service, dial tone, dialing, connecting, ringing, talking, etc.), battery and ground are supplied to the loop or Private Branch Exchange (PBX)-central office (CO)

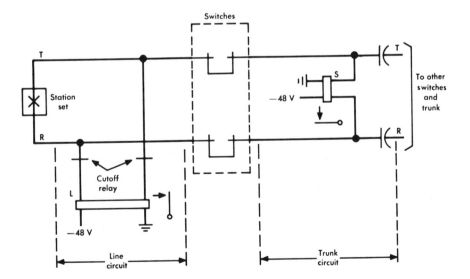

Figure 13-2. Transfer of loop supervision — bridged impedance battery feed.

trunk* by a circuit somewhat like those illustrated in Figure 13-1. The battery supply may be a different circuit, however, for each stage of the call and may be different for either an incoming call or an outgoing call. Furthermore, while idle loops always have negative battery on the ring conductor, the battery-ground connections to a calling party may be reversed during the progress of setting up a call. In the talking condition, either calling or called party loops may have the polarity reversed, particularly when served by a step-by-step switching machine. Each battery supply circuit must include a relay or other device which can respond to changes in the signalling or supervisory condition on the loop and, in responding, extend the information regarding the changed conditions to other circuits.

Figure 13-2 illustrates the process for an outgoing call. When the station set is on-hook, battery and ground are connected to the loop conductors through the windings of the L relay in the loop circuit and the closed cutoff relay contacts. Operation of the L relay, caused by the flow of current through its windings when the station set changes to an off-hook condition (call for service), results in switching system operations which disconnect the L relay from the loop (by operating the cutoff relay) and which connect the loop to a trunk circuit. Thus, during the first part of the call sequence, supervision of the loop is provided by the flow of current through the L relay; during the second part of this sequence, supervision is provided by the S relay in the trunk circuit through whose windings current is supplied to the loop.

It should be stressed that the circuits of Figures 13-1 and 13-2 and the sequence of operation just described are illustrative only. Although many variations exist in different types of switching systems, the basic function of loop supervision is performed in all systems by circuits very similar to those described.

The process just described is known as the loop-start process. Another process used to initiate a call is known as ground-start. In some cases, for example on certain dial-selected PBX trunks, the calling sequence is started by applying a ground to the ring side of the line. In such cases, the line relay is wired to accept only this call-for-service signal and responds accordingly. It is used in this application in order to minimize the probability of simultaneous

*In many respects PBX-CO trunks are functionally similar to loops.

seizure of a trunk from both ends for an incoming and an outgoing call. The simultaneous seizure of a trunk from both ends, a condition called *glare*, would be a serious problem on dial-selected PBX trunks because there can be an interval of 4 seconds after an incoming call is connected to the trunk before it is rung. Additional time may pass until the incoming call is answered and the trunk is made busy at the PBX. With ground-start operation, the trunk, while in the idle state, has no ground on the tip conductor. Upon seizure by the central office equipment, ground is applied to the tip conductor, a condition used immediately to make the trunk busy at the PBX; removal of the tip ground is recognized by the PBX as a disconnect signal. When a call is originated at the PBX, ground is placed on the ring conductor. When a central office connection is established, the normal battery and ground connections to ring and tip are made. Either state (ground on ring or loop closure) is recognized immediately by the central office equipment as a trunk seizure. The central office equipment later recognizes the opening of the loop as the disconnect signal.

The parameters that enter into the calculation of loop supervision relationships include the resistance of the station set, the resistance of the loop conductors, the resistance of the central office equipment and wiring, the resistance of the battery supply circuit (nominally 400 ohms in most central offices), the sensitivity of the relay or other device that must respond to changes in loop status, and the battery voltage itself. These parameters all vary within their respective ranges. The station set resistance has manufacturing variations and, in addition, is designed to be a function of the loop current. The resistance of the loop conductors is dependent on the distance of the station set from the central office and the gauge of wire employed. The resistance of the central office wiring is also dependent on length and wire gauge. In addition, the resistance of the paths through the CO equipment is different according to the circuit type and type of switching machine and must be accounted for along with manufacturing tolerances. Allowance must also be made for loop conductor leakage currents.

The battery voltage has, in most central offices, a nominal value of −48 volts; it varies approximately ±4 volts about the nominal. Provision is sometimes made to increase the supply voltage to 72 volts for groups of long loops designed for operation on a single relatively small gauge of cable (Unigauge design) or when dial long line equipment is used to extend loop length.

The large number of variables involved in supervisory signal computations makes it necessary to apply a set of rules that can be used universally to determine if signalling or some other function limits loop performance. One such rule for laying out loop plant is that the conductor loop resistance must be equal to or less than the signalling limit or 1300 ohms, whichever is lower (in most cases, loop resistance may exceed 1300 ohms for signalling). The 1300-ohm limit has been established to assure adequate transmission. Other rules apply to loading, allowable number of bridged taps, etc. Signalling limits must be determined for each case.

## Address Signalling on Loops

Two modes of generating address signals are used at common battery telephone station sets operating in a machine switching environment. These are dial pulsing and TOUCH-TONE signalling. They are described in some detail because different transmission problems are related to each.

**Dial Pulsing.** Address signalling occurs when a rotary dial is moved to its off-normal position and then released. The signals consist of pulses which result from interruption of the loop current by the pulsing contacts of the dial. The number of pulses corresponds to the digit dialed. The central office equipment responds to the dialed digits to establish the desired connection.

Timing relationships are important in this process in a number of ways. Note first that the dial pulse signals differ from supervisory signals on a loop only in respect to timing. On-hook and off-hook supervisory signals are of long duration while dial pulsing signals are measured in small fractions of a second. The process of transferring address information from the station set dial to the central office equipment is dependent on these timing relationships and on the designs of the dials, the central office equipment, and the loops.

Some basic time relationships are shown in Figure 13-3, where the digits *2* and *3* are assumed to have been dialed sequentially. The first of these time relationships is illustrated by the first pulse in Figure 13-3. The complete pulse cycle is made up of a *break* interval during which the pulse contacts of the dial are open, and a *make* interval during which the pulse contacts are closed. The two intervals are related by the expression

$$\% \text{ Break} = \frac{\text{Break interval} \times 100}{\text{Break} + \text{make intervals}} \quad .$$

The percent break used in Bell System dials is 58 to 64 percent.

The second time relationship of importance is the pulse repetition rate or number of pulse cycles per second that can be successfully transmitted. Most dials used on station sets are designed to operate at 10 pulses per second (pps). While many parameters influence the maximum, the pulse rate of these dials is primarily set by the operating speed capabilities of step-by-step switching equipment. The dials used at PBX or manual central office switchboard positions are often of a 20-pps design. The higher pulse rate is used to achieve higher operating efficiency. The higher speed dials can only be used where tie trunks or foreign exchange trunks on the PBX do not involve DX or SF signalling arrangements. These signalling systems, described later, are not capable of operating at the higher speeds. In addition, the switching system involved must be capable of responding to the higher pulse rate; step-by-step systems are generally not capable of such operation. The use of the higher speed dials is facilitated by the lack of bridged ringers on PBX trunks and the short trunk length usually associated with PBX operation.

The two timing relationships given so far, percent break and pulse repetition rate, are governed largely by the operate and release char-

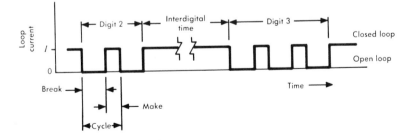

Figure 13-3. Dial pulsing of digits 2 and 3.

acteristics of central office equipment as they are affected by the loop characteristics. The pulse waveforms of Figure 13-3 are highly idealized. As illustrated in Figure 13-4, impedance characteristics of the loop, station set, and ringer circuits cause distortions of the pulses that must be taken into account when station set signalling problems are being considered. The dashed-line pulses in Figure 13-4 are again highly idealized; the solid-line pulses show how one form of distortion (caused by ringer and cable capacitance charge and discharge) causes changes in the percent break of the repeated pulses. Margins for such distortion must be provided in the design of central office control circuits.

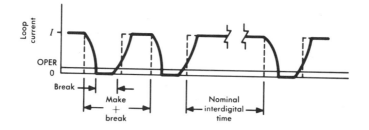

Figure 13-4. Effect of pulse distortion on dial pulse time relationships.

The third timing relationship in dial pulsing is shown in Figures 13-3 and 13-4 as the interdigital time. This is the time that the loop is closed after a digit has been dialed until the first pulse of the next digit. It includes the time required by the customer or operator to search for the next digit, to pull the dial around to its stop, and to release it to start pulsing the next digit. The central office equipment must contain timing circuits to recognize this interval with allowances (or margins) for pulse distortion caused by the loop and other equipment.

**TOUCH-TONE Signalling.** A second form of address signalling used on station sets is implemented by a set of pushbuttons rather than by a rotary dial. This form of signalling, called TOUCH-TONE, is usually superior to conventional dial pulsing because it is more accurate, more convenient, and faster. (It is also somewhat more costly.) Operation of any pushbutton results in the generation of two single-frequency tones which are transmitted as long as the button is depressed. Oscillators, activated by pushbutton operation, are powered by the line current furnished from the central office. While a button is depressed, the telephone transmitter circuit is opened, and a resistor is inserted in series with the receiver so that the tones are heard in the receiver at a comfortable sound amplitude.

The layout of the standard 12-button TOUCH-TONE matrix pad and the frequencies generated by each button are depicted in Figure 13-5. If the number 7 pushbutton is operated, for example,

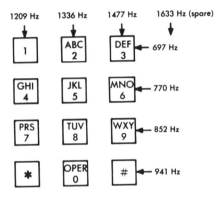

Figure 13-5. Pushbutton layout on TOUCH-TONE station set pad showing signalling frequencies.

the 1209-Hz and 852-Hz frequencies are generated. Central office equipment, different from that used to receive dial pulse signals, recognizes these tones as representing the numeral 7. This equipment, called TOUCH-TONE converters, translates the oscillator signals to digital signals similar to dial pulse signals for machine switching recognition and operation. The pushbuttons marked * and # are used for certain special signalling. Some 10-button sets, lacking the * and # pushbuttons, are still in service. A 16-button set (4-by-4 matrix) is also available for use in private line network service provided to the U.S. government.

The signals in the low-frequency group, 697 to 941 Hz, are transmitted nominally at −6 dBm; those in the high-frequency group are transmitted nominally at −4 dBm. The actual amplitudes are dependent on the amount of loop current. These high signal amplitudes, and the fact that this type of signalling is not as susceptible to distortion caused by the medium as are dial pulse signals, make the design of pulse receiving equipment at the central office quite straightforward. Although these amplitudes are higher than those of many other signals transmitted in the voiceband, they are considered acceptable because they have a low duty cycle; i.e., they are transmitted only occasionally and they are of short duration. Nevertheless, these amplitudes are being reviewed for a possible downward adjustment which may result in somewhat more stringent sensitivity requirements for the pulse receivers. Since TOUCH-TONE signals fall in the voiceband, they may be transmitted through the switched message network. Thus, they may be used as a form of data communication.

## Alerting Signals on Loops

There are two types of alerting signals transmitted towards the station set that are considered here, namely, ringing signals and the receiver-off-hook signal used to alert a customer that his receiver has been left off-hook.

**Ringing.** Conceptually, the alerting signal used to ring the station set bell is simple.* However, details of signal generation, coding for party-line operation, variables that may affect the ringing process, and instrumentalities used to achieve ringing objectives make a conceptually simple process rather complex in practice.

The ringing signal is used mainly on loops, although some 20-Hz signalling is used on ring-down (manual) trunks, and as a ring-back and ring-forward signal on other types of trunks. On loops, it is usually applied at the central office** as a composite ac and dc signal. The forward-ringing ac component has a frequency of 20 Hz. In some types of switching machines, an ac component of about 420 Hz is superimposed; this component is fed back to the calling party to serve as an audible ring, giving assurance that the called number is being rung. The dc component of the ringing signal may be of either polarity with respect to ground.

---

*In addition to the complexities discussed here, the alerting of a customer to an incoming call is sometimes accomplished by in-band, coded tone signals. Such signals are used primarily in special service arrangements.

**One exception, for example, is in the Subscriber Loop Multiplex System in which the ringing signal is applied to the loop at a terminal remote from the central office.

It would, of course, be possible to ring the station bell continuously until the station set is answered. Early tests, however, indicated that continuous ringing would be undesirable and irritating. The standard central office ringing cycle has been set as a 2-second ringing interval followed by a 4-second silent interval. This cycle is sometimes modified to provide coded ringing to alert the desired one of several party-line stations on a single line. The standard ringing cycle used at PBXs is a 1-second ringing interval followed by a 3-second silent interval.

It is desirable to set the magnitude of ringing signals as high as possible in order to maximize the length of loop over which station sets operate satisfactorily. However, since the telephone plant is designed generally to operate at low currents and voltages, the maximum ringing-signal voltages are limited to values that do not operate protective devices, cause dielectric failure or overheating of equipment, or present a hazard to operating personnel.

The station set ringer may be connected to the loop in a number of ways, depending on the type of service. On individual lines, the ringer is normally connected across the line in series with a capacitor, as illustrated in Figure 13-6(a). With the types of high-impedance ringers presently used, a total of five ringers can be connected in parallel as illustrated by Figure 13-6(a). The number is limited by ringing and dial pulsing requirements. Ringing ranges vary with the number of ringers used and with the characteristics of the switching system involved; they are less than dialing and supervisory ranges when the number of ringers is a maximum.

For party-line service, other types of ringer connections are required. One is illustrated in Figure 13-6(b). The types of service include 2-party, 4-party, and 8-party service in many suburban and rural areas. In more remote rural areas, 10- and 20-party service is sometimes provided. Full selective ringing (only the called party hears the ring) can be provided on 2-party and 4-party lines. Semiselective ringing (where only a limited number of parties hear each ring) is provided on some 4-party lines and all 8-party lines. Nonselective or semiselective code ringing is provided on the rural lines with large numbers of parties.

As shown in Figure 13-6(b), party-line ringing often involves ringer connections between one side of the line and ground. Due to

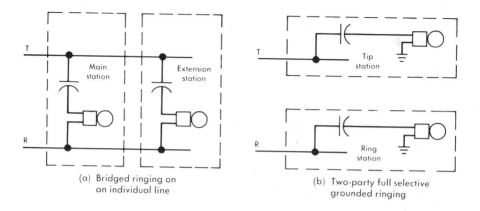

(a) Bridged ringing on
an individual line

(b) Two-party full selective
grounded ringing

Figure 13-6. Two common types of station set ringer connections.

unbalanced conditions that might exist on such lines, caused by different numbers of ringers on each side or very different loop lengths to each, such lines may be quite noisy and may cause crosstalk due to interference currents. In these cases it may be necessary to use gas tubes or solid-state ringer isolator circuits which balance the lines so that induced currents are not converted to excessively large interferences. Care must be used in the application of such circuits so that additional noise impairments caused by gas-tube breakdown are not introduced.

**Receiver-Off-Hook Signal.** When a station set is left in the off-hook condition, a tone may be applied to the loop to attract the attention of someone at the station to this condition. The tone used was at one time known as a howler. The howler, a very high-amplitude signal in the voiceband, proved to be unsatisfactory for use with the 500-type station set because of the clipping action of the equalizer in the set. Furthermore, when the howler signal was transmitted over telephone lines using carrier facilities, there was danger of seriously overloading some transmission paths.

The howler signal has been almost universally replaced by a signal called the receiver-off-hook (ROH) tone. The signal is made up of a combination of 1400, 2060, 2450, and 2600 Hz. When applied automatically, the ROH signal appears on the loop for about 50 seconds. It is interrupted at a rate of five times per second. It can also be applied manually from the local test desk as a continuous or interrupted signal.

## 13-2  SIGNALLING ON TRUNKS

While there are significant differences in detail, most of the same general functions of signalling must be accomplished on interoffice trunks as on loops. These functions include addressing, supervision, alerting, transfer of information, and testing. As may be expected, many of the characteristics of signals used on trunks are similar or identical to those used on loops.

Signals that relate directly to station operation, such as ringing and ROH signals, are not generally used on interoffice trunks.* On the other hand, the types of switching systems that must be controlled and the functional characteristics of the trunks themselves are so diverse that the variety of signals used on trunks is considerably greater than on loops. Two general types of signals are described; they are classified as dc or ac signals. Under each type there are many variations.

The address information required to route a call must be forwarded from the originating central office through various toll offices to the terminating central office. In general, dc signals are used within the switching machines. Such signals are often unsuitable for transmission over trunks, and it is necessary to transform the signals at one end of a trunk to a form more suitable for transmission and then back to the original form at the other end of the trunk. If the trunk length exceeds the range limits of the dc systems or if the trunk cannot pass dc, ac or derived dc techniques must be used. These conversions require equipment which is described elsewhere.

One form of signalling interface is used frequently for both dc and ac signalling. The name, E and M lead signalling, is derived from lead designations historically used on applicable circuit drawings. The E and M lead interface between a signalling path on a transmission facility and a switching system trunk circuit is shown in Figure 13-7. The circuit conditions on the E and M leads are standard in all systems employing this method of connection. (Some later systems utilize paired leads rather than single-wire leads to reduce interferences; these applications involve some departures from the simple E and M lead circuit conditions.) The manner in which signals are

*Ring-forward and ring-back signals are transmitted over operators' trunks to the local office where they are then applied in appropriate form to the loop.

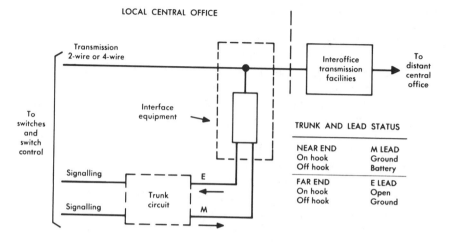

Figure 13-7. E and M leads.

converted to dc or ac types for transmission over the trunk, the characteristics of the transmitted signals, and the method of combining the signalling path with the transmission path vary widely. Systems that employ E and M leads have the advantage that signals can be transmitted independently in both directions on a trunk.

## DC Loop Signalling on Trunks

Since the transfer of address and supervisory signalling information is most economically accomplished by dc signalling, such methods are used whenever technically feasible. There are two forms of dc signalling. The first is called loop signalling, a name which is derived from the fact that a dc circuit, or loop, is available between the two ends of a trunk. (It is not related to the loop that connects a station set to the central office.) The second form of dc signalling, called derived dc signalling, is discussed subsequently. One or the other of these dc signalling arrangements is used extensively for all inter-local, toll connecting, or toll trunks that operate at voice frequency and that are short enough to permit their application.

The dc loop signalling systems operate generally by altering the direct current flow in the trunk conductors. At one end of a trunk, the current may be changed between high and low values, it may be interrupted, or its polarity (direction of flow) may be reversed. These changes are detected by suitable relays or other types of apparatus at the other end of the trunk. The signalling systems are known as reverse-battery, battery and ground, high-low, and wet-dry.

Signals cannot be transmitted in both directions independently in dc loop systems. Thus, such systems are used on one-way trunks, primarily on local and on toll connecting trunks.

**Reverse Battery Signalling.** Because of its economy and reliability, this is the most widely used dc loop signalling method on local trunks. Battery and ground for signalling purposes are furnished through the windings of the A relay at the terminating end of the trunk as shown on Figure 13-8. Supervision is provided at the originating end of the trunk, usually by opening (on-hook) or closing (off-hook) contacts in the trunk transmission path under the control of the originating station set through relay S1. At the terminating end of the trunk, supervision is provided by the station set and relay S2. Normal battery and ground are connected to the trunk conductors for the on-hook signal and are reversed by operating the T relay to represent the off-hook condition. Address signals may be under the control of the calling station set or under the control of dial pulsing equipment in the originating central office.

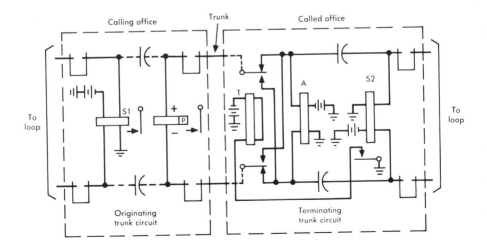

Figure 13-8. Reverse battery signalling.

**Battery and Ground Signalling.** This mode of signalling is used to extend the range of loop signalling. It is accomplished by connecting battery and ground at both ends of the loop in a series aiding configuration. This type of connection, illustrated in Figure 13-9, is usually provided only during the period that addressing information is being transmitted.

The current available for signalling is nearly doubled as compared with the ordinary dc loop connection with battery and ground at one end only. For supervision, the battery and ground at the originating office is usually removed, and a dry polar bridge is substituted to function with the reverse-battery supervision signal from the terminating end. However, reverse-battery supervision can also be provided in the battery and ground arrangement of Figure 13-9; the battery and ground must be reversed at both ends of the trunk.

**Miscellaneous DC Loop Arrangements.** A number of other dc loop signalling arrangements are used to provide address or supervisory signalling information on voice-frequency trunks. Most are being replaced by the reverse-battery or battery and ground systems previously discussed or have so little impact on transmission problems that detailed discussion here is not justified. These include wet-dry signalling and high-low signalling. Both of these signalling methods utilize dial pulsing or ac signalling for the transmission of address information and may thus be considered primarily as supervisory systems. Wet-dry signalling provides dc loop supervision in the form of presence or absence of battery and ground. The trunk is *wet* (battery and ground connected) for one set of supervisory states

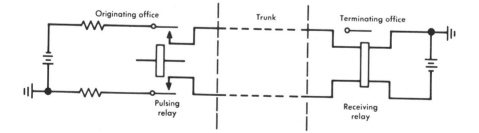

Figure 13-9. Battery and ground pulsing.

and *dry* (battery and ground disconnected) for the opposite. High-low signalling refers to the impedance bridged across the trunk, high impedance being used for one set of supervisory conditions, low impedance for the other. This method is still used occasionally to provide supervision at the originating ends of reverse-battery signalling systems.

### Panel Call Indicator (PCI) System

This system utilizes a 4-bit code to transmit address information. Originally, the system was designed to transmit address information from a panel-type switching machine to an operator position at a manual B-type switchboard. The use has been extended to signalling between panel-type switching machines and manual switchboards, between crossbar switching machines and manual switchboards, between panel or crossbar switching machines and other panel or crossbar machines, and in specialized applications within panel or crossbar machines.

At the receiving end of a PCI system, the called number may be displayed on a lamp field before an operator, or the transmitted address may directly drive switching system registers. In either case, supervisory information must be transmitted by another means.

The 4-bit code in PCI signalling is designed so that the first and third bits of each digit code are defined by open-circuit or light positive pulses, and the second and fourth bits by light or heavy negative pulses. The negative pulses in the second and fourth time slots are used to synchronize the receiving with the sending end of the trunk and to advance a register to successive digits. Four decimal digits, sent consecutively with no pause in between, require a total transmission time of about 1 second. A heavy positive pulse is transmitted to indicate end of pulsing. Thus, the complete system consists of a five-state signalling system. Figure 13-10 gives the various permissible signal conditions; Figure 13-11 gives the complete PCI code; Figure 13-12 shows a portion of a typical transmitted signal.

### Revertive Pulsing

As in the case of PCI, revertive pulsing was developed to satisfy the signalling needs of panel-type switching systems. It was later adopted for use with certain crossbar systems because it is capable

| SYMBOLS | | | | LOOP CONDITION | | | |
|---|---|---|---|---|---|---|---|
| — | | | | Open, zero current | | | |
| p | | | | Light positive current | | | |
| n | | | | Light negative current | | | |
| N | | | | Heavy negative current | | | |
| P | | | | Heavy positive current | | | |
| (Polarities are ring relative to tip.) | | | | | | | |
| BASIC PSI CODE CYCLE | | | | | | | |
| Time slot (interval) | | | A | B | C | D | |
| Normal | | | — | n | — | n | |
| Permissible | | | p | N | p | N | |

Figure 13-10. PCI signal conditions.

| DIGIT | HUNDREDS TENS AND UNITS | | | | THOUSANDS | | | |
|---|---|---|---|---|---|---|---|
| | A | B | C | D | A | B | C | D |
| 0 | — | n | — | n | — | n | — | n |
| 1 | p | n | — | n | — | n | — | N |
| 2 | — | N | — | n | p | n | — | n |
| 3 | p | N | — | n | p | n | — | N |
| 4 | — | n | p | n | — | N | — | n |
| 5 | — | n | — | N | — | N | — | N |
| 6 | p | n | — | N | p | N | — | n |
| 7 | — | N | — | N | p | N | — | N |
| 8 | p | N | — | N | — | n | p | n |
| 9 | — | n | p | N | — | n | p | N |

Figure 13-11. PCI codes.

of operating somewhat faster (up to 22 pps in crossbar and 32 pps in panel) than more conventional dial pulse systems and because the crossbar systems, designed to replace the panel, had to interconnect with existing panel systems. This mode of signalling is no longer recommended for new installations, although many revertive pulsing systems are still in operation.

The mode of operation is quite different from other systems; address signals with different functions are transmitted in both directions, as shown in Figure 13-13. The terminating office, which may

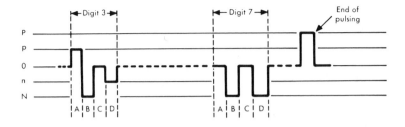

Figure 13-12. Typical PCI pulse signals — digits 3 and 7 in hundreds, tens, or units position.

be of the panel, crossbar, or ESS type, receives a start signal from the originating office. Equipment at the terminating office generates pulses in accordance with its operation. These pulses are sent back to the originating office, and when the number of pulses received at the originating office corresponds to the digit being transmitted, a stop signal is sent to the terminating office to end that phase of the operation. After the appropriate number of digits has been recorded in the terminating office equipment, an incoming advance pulse is returned to the originating office; the trunk is then ready to be connected through to the talking paths at each end. As can be seen in Figure 13-13, revertive pulsing requires three signalling states for its operation, two to convey the pulsing count and one for the incoming advance signal.

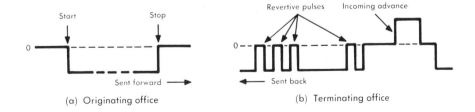

Figure 13-13. Revertive pulsing signals.

## Derived DC Signalling on Trunks

Derived dc signalling paths are used for many long local trunks and short-haul toll connecting and intertoll trunks where a complete dc loop is not available or where extended ranges are desired for dc signalling. In these cases, dc signalling paths are sometimes derived from the transmission path, which, of course, must be a physical facility. Derived systems utilize E and M lead connections and, as a result, may be used on one-way or two-way trunks. The types of derived paths now in use include simplex (SX), composite (CX), and duplex (DX) circuits, all of which may be used to transmit both supervisory and address signals.

**Simplex Signalling.** The method of connecting a simplex signalling circuit to a voice-frequency trunk is illustrated in Figure 13-14. By feeding the signalling currents through the center taps of line transformers, signalling current flux is cancelled in the transformers and the signals are not transmitted beyond the transformers in either direction. However, the trunk resistance is halved by paralleling the two conductors, thus extending the range compared to loop signalling. Simplex signalling has largely been superseded by DX signalling.

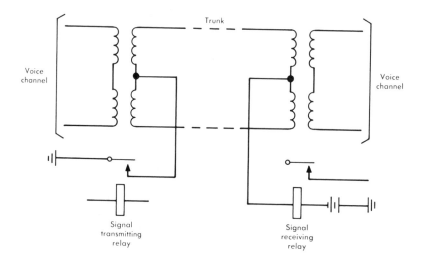

Figure 13-14. Simplex signalling connections.

**Composite Signalling.** This method of signalling consists essentially of combining a voice transmission path with dc signalling paths by means of a high-pass, low-pass filter arrangement as illustrated in Figure 13-15. The dc address and supervisory signals are transmitted between central offices over one wire of the transmission circuit with ground return. Where necessary, the second conductor of the transmission path can be used to compensate for differences in earth potential between the two offices.

The crossover frequency of the filter characteristics is approximately 100 Hz. Thus, interference from signalling currents is blocked from the voice-frequency band.

One arrangement of such a composite signalling system is shown in Figure 13-16. The connection through the P windings of the CX relays is used for earth potential compensation. The arrangement may be extended to several other signalling circuits, each using the same trunk conductor, by wiring the P windings of the CX relays in series with the ones shown in the figure. Thus, the signalling and transmission are not necessarily associated; signalling on a given trunk transmission path may be associated with a different trunk.

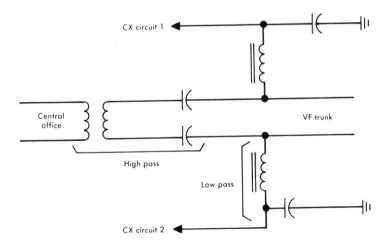

**Figure 13-15. Composite signalling circuit for one end of a trunk pair.**

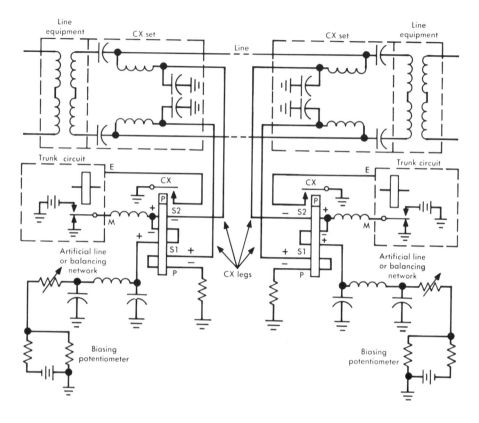

Figure 13-16. Composite signalling circuit — one voice channel.

**Duplex Signalling.** Duplex signalling, illustrated in Figure 13-17, is based on the use of a symmetrical and balanced circuit that is identical at both ends of the trunk. The circuit and its mode of operation are patterned after those used in CX signalling, but a composite set is not required. Signalling and transmission are on the same transmission path and hence do not occur simultaneously. One wire of the trunk conductor pair is used for signalling and the other for ground potential compensation. Its chief advantage is that it can operate on circuits having loop resistance higher than can be tolerated by other systems, i.e., up to 5000 ohms.

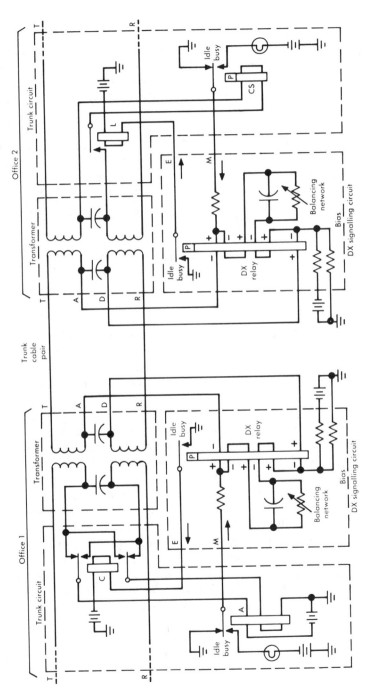

Figure 13-17. Duplex signalling system.

## AC Signalling on Trunks

Use of ac signalling may be dictated by the limitations of distance on dc systems or by the inability to transmit dc signals over commonly used carrier systems. Thus, even though conversion equipment and ac generators are required, such systems are a necessity in the telephone plant.

It is theoretically possible to transmit ac signals for address and supervisory information at any frequency in the voiceband, defined for these purposes as approximately 200 Hz to 3500 Hz. Carrier transmission systems usually provide 4000 Hz spacing between channel carriers. Present voice-frequency signalling systems operate in the range of 500 to 2600 Hz.

A number of ac signalling systems have been designed to operate by using inband frequencies. Two are commonly used at present, the 2600-Hz single-frequency. system and the multifrequency pulsing system.

**Inband 2600-Hz Signalling.** This system is commonly referred to as single-frequency (SF) signalling. With certain adaptations that involve other single-frequency signals, the system may be used to transmit address and supervisory signals in both directions on most types of trunks.

One of the design considerations in voice-frequency signalling is the prevention of mutual interference between transmission and signalling systems. Voice-frequency signals are audible; consequently, signalling must not take place during conversation. In most applications of this type of system, the presence of a 2600-Hz signal corresponds to the on-hook condition and the absence of 2600 Hz corresponds to the off-hook condition. Thus, there is no 2600-Hz tone normally present on the line during conversation. Signal receiving equipment, however, must remain connected during conversation in anticipation of incoming signals and may be subject to false operation due to speech signal components that resemble the tones used for signalling. Several methods are used to protect against false operation of signalling circuits:

(1) Where possible, signal tones of a character not likely to occur in normal speech are chosen.

(2) Time delay is used in the signalling and trunk circuits so that normal speech currents are ignored.

(3) Speech signal energy, when detected at frequencies other than the signalling frequency, is used to inhibit operation of the circuits in the signalling receiver.

This system may be used to signal independently in both directions on four-wire facilities. When a 2600-Hz SF signal is being transmitted to reflect the on-hook, steady-state supervisory condition, it is applied to the trunk at an amplitude of —20 dBm0. Thus, the steady-state load effect of such a signal is somewhat below the long term average speech power in a telephone channel. When being used to transmit address information, however, the 2600-Hz signal is increased in amplitude by 12 dB to —8 dBm0. Such a high signal amplitude is permissible because of the short duration of the address signal pulses and because of the low probability of large numbers of such high-amplitude signals being simultaneously present in a transmission system.

There are a number of SF signalling characteristics that interact importantly with transmission systems. These have particularly serious implications in their interactions with carrier systems. Problems arise as a result of two conditions: (1) a majority of trunks utilizing a carrier system may be equipped with SF signalling, and (2) most of these trunks may originate at the same office. In the latter case, there may be high coherence in the relative phases of the many 2600-Hz signals. As a result, the way in which these signals combine may cause overload or excessive peaks of intermodulation noise at certain frequencies and at unpredictable times. In such situations, action must be taken to break up phase coherence among 2600-Hz signals in different channels.

Furthermore, where large numbers of trunks in a carrier system employ SF signalling and terminate in the same office, serious disruption of the switching system operations can occur as a result of carrier system failure. If most of the trunks are in the idle condition, the carrier system failure causes the sudden interruption of all of the 2600-Hz signals. This is interpreted by the switching machine as simultaneous calls for service from many trunks. As a result, the switching machine is momentarily overloaded until it can dispose of the disabled trunks. Some carrier systems employ trunk conditioning circuits which cause all affected trunks to appear busy so they will not be seized until after repairs have been made; the conditioning circuits then remove the busy condition and restore the trunks to service.

Another adaption of SF signalling (really a two-frequency system) is used for selective signalling in a multistation four-wire private-line network such as might be used as an order-wire facility for carrier systems or for private customer communications networks interconnecting separate locations. Dial pulses are used to signal selectively any one of a maximum of 81 stations. The dc dial pulses are converted, at the customer's premises or at the central office, to a frequency shift format utilizing 2600 Hz and 2400 Hz.

**Multifrequency Pulsing.** Multifrequency (MF) pulsing signals are used to transmit address information on trunks. Signalling is accomplished by the transmission of combinations of two, and only two, of six frequencies in the voiceband. The principal advantages of this system are speed, accuracy, and range. However, this system is not capable of transmitting supervisory signals and, as a result, supervision must be provided by another system such as DX, loop, or SF.

## 13-3 OUT-OF-BAND SIGNALLING

Any signalling arrangement that utilizes frequencies out of the voiceband of the trunk over which signalling is taking place may be considered as an out-of-band signalling system. By such a broad definition, dc systems and a common channel interoffice signalling (CCIS) system would be considered out-of-band systems. However, it has been convenient to discuss dc systems as a separate class of systems. In the CCIS system, signalling information is transmitted on a voice-frequency data channel independent of the trunks involved in the connection.

Out-of-band systems that are in current use include early N1, O, and ON carrier systems, which use a single-frequency signal at 3700 Hz in the voice channel, and the digital signalling arrangements used in the time division multiplex T-type systems. In addition, systems like the 43A1 Carrier Telegraph System are sometimes used (as in submarine cable operation) to signal over a channel separate from the voice channel.

## Out-of-Band SF Signalling

The out-of-band signal at 3700 Hz falls in the passband of a voice-frequency channel but at a high enough frequency that it is above the cutoff of the channel filters, hence above the band occupied by speech energy. This mode of operation has the advantages that no provision need be made for protection against inadvertent voice operation of the signalling circuits; in addition, signalling can take place during the talking interval if required.

During the trunk idle condition, the 3700-Hz signals are present in both directions of transmission; trunk control, supervisory, and address signals are transmitted by interrupting the 3700-Hz signal in a fashion similar to that described for 2600-Hz inband signalling. Interconnection between the transmission system signalling and other transmission circuits is made by E and M lead facilities.

The 3700-Hz signals are applied to the high-frequency line of the carrier system at the transmitting end of the carrier system after the compressor portion of the compandor. Thus, compandor action has no effect on the 3700-Hz signals.

## Out-of-Band Digital Systems

In the coding of PCM signals for transmission over T-type carrier lines, address and supervisory signals are assigned specified bits in the carrier pulse stream. In some cases, this assignment of bits is permanent; as a result, a significant portion of the system's theoretical channel capacity is assigned to signalling. In other cases, the address and supervisory bits are assigned on a borrowed basis in such a way that speech transmission is of higher quality than would otherwise be possible. The signalling bits are used for speech coding when not required for signalling.

## 13-4  SPECIAL SERVICES SIGNALLING

Most of the discussion of signalling and the characterization of alerting, address, and supervisory signals that have been given in this chapter apply equally to signalling in the switched message network and to special service arrangements. However, there are some significant differences; some are due to the nature of special service circuits themselves, and some are due to the manner in which the special services are administered.

## Tandem Signalling Links

In the switched message network, the tandem connection of a number of trunks usually involves the regeneration of address signals at the point of interconnection. Thus, in signalling over long distances through a combination of trunk types and transmission system types, the signalling equipment and signal transmission impairments need only be considered on a trunk-by-trunk basis.

One type of special service trunk (by definition) is the PBX-central office trunk. A PBX station, connected through the PBX trunk to the central office must signal by dial pulsing through the PBX station line and PBX trunk without regeneration. This situation may be further compounded by the need to signal through an intermediate PBX tie trunk. These conditions often result in marginal signalling conditions in PBX station signalling over tie trunks and PBX-central office trunks.

## Service Demands and Plant Complexities

Many special service circuits must traverse parts of the plant in which a mix of trunk plant and loop plant occurs. One example is a foreign exchange (FX) line whose station set is located in one central office area but whose home central office may be many miles away. The final loop connection is from the local central office, but the loop must then be extended through cables normally used for interoffice trunks to the distant serving office. Another example is an off-premise extension from a PBX that may require a transmission path involving connections through both loop and trunk cables. Inward and outward WATS (wide area telephone service) lines provide additional examples.

#### REFERENCES

1. Breen, C. and C. A. Dahlbom. "Signalling Systems for Control of Telephone Switching," *Bell System Tech. J.*, Vol. 39 (Nov. 1960), pp. 1381-1444.

2. Weaver, A. and N. A. Newell. "Inband Single-Frequency Signalling," *Bell System Tech. J.*, Vol. 33 (Nov. 1954), pp. 1309-1330.

Chapter 14

# Data Signals

The transmission of data signals in the Bell System involves the transmission of coded information between machines or between man and a machine. In some cases the transmitted information is coded into some digital form that is convenient to the operation of a machine, such as a computer, and also convenient to the necessary interpretation by man at the input and output of the machine. In other cases, the transmitted signal is more conveniently coded as a direct electrical analog of the information and digital encoding is not utilized. Thus, there are two important forms of data signals, digital and analog. Digital signals are those that can assume only discrete values of the parameter that is varied to convey information; analog signals can assume a continuum of values between given maxima and minima. A common application of digital transmission techniques requires *digital data* signal characterization for transmission over analog systems. Other digital signals and and certain forms of analog data signals must also be characterized.

Digital data signals are transmitted at signalling rates that range from a few bits per second to millions of bits per second. The most commonly used rates, several thousand bits per second, are those compatible with voiceband circuits. In many private line applications and in the switched public network, transmission circuits that are normally used for voice communications are alternatively used for digital or analog data signal transmission.

In many cases of interest here, data signals are those used by computers; they are usually binary signals transmitted serially on a pair of conductors, but they are seldom in a form convenient for transmission over Bell System facilities. Processing to transform a signal to a suitable format often takes place at two locations — first at the station set to make the signal suitable for transmission on telephone loops, and then at carrier terminals to prepare the signal for transmission over a carrier system in a form suitable for modulating and multiplexing with other types of signals. The processing may involve special coding for error detection and correction. Each of the processes must be reversed so that the signal delivered at the receiving end of the circuit is a faithful replica of the signal accepted from customer equipment. These processes are similar to those described in Chapter 8.

Since many existing transmission systems were designed as analog facilities for the transmission of analog speech signals, the processing of a data signal for transmission over these systems must be such as to make the signal compatible with the transmission system. This compatibility involves loading effects, channel characterization, and intermodulation and signal-to-noise performance. Thus, the nature of the processes must be described here in some detail.

Processing of data signals for transmission over digital transmission systems is not covered here because the coding is unique to each digital system and the operation of the digital system is not materially affected by the characteristics of the signals. On the other hand, the line signal of a digital transmission system (suitably processed) is sometimes transmitted over an analog transmission system. The analytic treatment of digital data signals and digital line signals is identical when they are transmitted over analog systems, and both must be characterized. Hereafter, they are generally referred to simply as digital signals.

## 14-1   DIGITAL SIGNAL TRANMISSION CONSIDERATIONS

A number of considerations related to digital signal transmission on analog facilities have had important effects on the design of signal formats. These include restrictions on signal amplitudes, signal-to-noise ratios and error rates, and the relationships between signal and channel characteristics.

### Signal Amplitudes

A number of criteria must be considered in setting the amplitude of a digital signal using the switched public network or sharing facilities with the network. One such consideration is that the power in the signal should not cause excessive intermodulation or overload in transmission systems, especially in analog carrier systems where service to many other customers might be jeopardized. The established requirement is that signals operating in the voiceband are to be limited to −13 dBm0*, defined as the maximum allowable power averaged over a 3-second interval. When the activity factors and other statistics of data transmission are accounted for (e.g., the number of operating half-duplex versus full-duplex channels), this value is equivalent to a long-term average power of −16 dBm0 per 4-kHz channel.

The signal amplitude requirement for narrowband data signals, several of which may be multiplexed in a single voice channel, is also −16 dBm0, or a 3-second maximum of −13 dBm0, for the composite signal. The power of each individual signal must be sufficiently lower so that the total power in the channel does not exceed the objective.

For a wideband digital signal, one occupying more than a 4-kHz channel, the amplitude criterion is sometimes expressed somewhat differently, namely, that the signal power may not exceed the total power of the displaced channels.

The gain of some analog transmission system repeaters is regulated by the power in the transmitted signal. When wideband digital signals are processed for transmission over this type of system, the power in the transmitted carrier and its sidebands must be essentially constant at a value equal to that of the displaced message channel carriers and their sidebands.

Irrespective of its form or the bandwidth it occupies, one more constraint is imposed on a wideband digital signal. No single-frequency component may exceed an average power of −14 dBm0 [1]. This limit, established to avoid the generation of intelligible crosstalk intermodulation products in analog carrier systems, may sometimes be exceeded on the basis of low probability of occurrence or because of the short duration involved. Where danger of intelligible crosstalk exists, a scrambler or other means of reducing single-frequency components must be used [2].

Signal components at frequencies above the nominal band must be limited in amplitude to low values that can not interfere with adjacent channels in a carrier system, or interfere through any

---

*This value is equivalent to −12 dBm as measured at the serving central office.

crosstalk path with some other wider band signal or a cable carrier system that might share the same facility. For example, these unwanted signal component amplitudes are specified for voiceband signals at frequencies of 3995 Hz and higher [3]. In addition, it is required that the power in the band between 2450 and 2750 Hz not exceed that in the band between 800 and 2450 Hz in order to minimize interference with single-frequency signalling systems.

## Error Rate and Signal-to-Noise Ratio

Unlike speech or video signals, which must be evaluated on a subjective basis because of human responses to various types of signal impairment, digital signals are evaluated objectively. The evaluation of digital signal transmission is often expressed in terms of error rate, i.e., the number of errors in a given number of transmitted bits (e.g., one error in $10^6$ bits or an error rate of $10^{-6}$). It is sometimes convenient, however, to evaluate performance in terms of the signal-to-noise ratio because signal and noise amplitudes are easy to measure. When this is done, the noise characteristics must be specified; usually the Gaussian distribution (see Chapter 17) is used because there is a definite and demonstrable relationship between Gaussian noise and error rate.

While the effects of impairments other than noise (such as gain distortion or delay distortion) may also be expressed in terms of error rate, transmission studies are often facilitated by converting the impairment into an equivalent signal-to-noise ratio. This is done by evaluating the error rate for the impairment being studied and, from that error rate, determining the reduction in signal-to-noise ratio that would produce the same error rate in an unimpaired channel compared with the impaired channel. The reduction in signal-to-noise ratio is called *noise impairment*. While noise impairments cannot be added directly to give an overall impairment or error rate, the technique provides a convenient method of comparing the merits of one mode of transmission with another over a real channel.

Consideration is being given to the possibility of using a figure of merit other than signal-to-noise ratio or error rate. In practice, errors often occur in bursts that produce a high error density for only a small portion of the time involved in transmitting a digital signal; the remaining time may be error-free. Such a burst may cause a high apparent error rate, yet have little effect on the efficiency of transmission. One possibility being considered as a figure-of-merit involves calculating or counting the percentage of data blocks (time intervals) in which error-free transmission occurs. Another involves the number of error-free seconds per minute or per hour.

## Channel Characteristics

The format into which a digital signal is to be processed for transmission on a analog channel must represent a compromise between maximizing the rate of information transmitted (bits per second per hertz of bandwidth) and minimizing the impairments due to extraneous noise or intersymbol interference. The transmission characteristics of the channel bear an important relationship to the design compromises that are made, as does the cost of the terminal equipment.

For a transmitted pulse to retain a rectangular shape, the bandwidth of the transmission channel would have to be very great (theoretically infinite). Bandwidth is expensive and, furthermore, the wide band would admit interference from noise or other perturbations appearing at frequencies outside the band which contains the major portion of the signal energy. It is desirable, therefore, to curtail the signal spectrum as much as possible without undue impairment of the signal. Nyquist's criteria (1) and (2), defined in Chapter 10, give important leads to how the band may be limited and pulses shaped to minimize errors at the receiver. There are several satisfactory ways of shaping the pulses by appropriate design of the channel characteristic [4]. One is the *raised cosine* characteristic; it has the virtues of meeting simultaneously Nyquist's criteria (1) and (2) in response to an applied impulse and does so without undue penalty in added bandwidth. It tends to produce less noise impairment than other channel characteirstics and also has the virtue of being physically realizable to a close approximation by straightforward design techniques. The raised cosine channel characteristic, near optimum for transmission of an impulse, requires some modification to accommodate commonly transmitted rectangular pulses.

Figure 14-1(a) shows an idealized channel characteristic, curve $p$. Curves $r$ and $s$ are modifications that follow cosine-shaped roll-off characteristics at the high end of the band. They are symmetrical

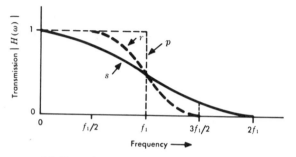

(a) Channel characteristics, raised cosine roll-offs

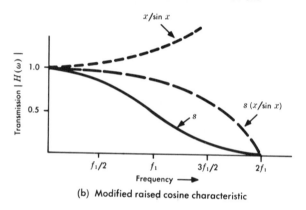

(b) Modified raised cosine characteristic

Figure 14-1. Channel shaping.

about the frequency $f_1$ where their values are 0.5 relative to the values at zero frequency. These raised cosine channels are said to have Nyquist shaping. If the characteristic yields zero transmission at frequency $3f_1/2$ (curve $r$ approximately), it has a 50 percent roll-off. If the characteristic yields zero transmission at $2f$ (curve $s$ approximately), it has a 100 percent roll-off. Roll-off is thus defined as excess bandwidth expressed as a percentage of the theoretically minimum requirement.

If the bandwidth is extended beyond $f_1$ and the roll-off characteristic has Nyquist shaping, a linear phase/frequency characteristic can be closely approximated in practice. When these characteristics (cosine roll-off and linear phase) are provided, the zero amplitude crossings of output pulses resulting from applied impulses still occur at times corresponding to $\pm n/(2f_1)$. If the roll-off extends to $2f_1$, the raised cosine pulses at the output have additional zero crossings at odd multiples of one half the intervals that occur in the idealized channel transmission.

Rectangular pulses, transmitted through a channel having any of the characteristics in Figure 14-1(a), cannot be readily detected because excessive intersymbol distortion occurs. As discussed in Chapter 6, the $(\sin x)/x$ channel response results from the application of an impulse, a signal which has a flat energy distribution. To make the channel respond in the desired fashion to rectangular pulses—that is, so that output pulse waveforms have the desired $(\sin x)/x$ format—it is necessary that the spectrum of the applied rectangular pulses be modified to approach the flat spectrum of an impulse. The normal spectrum of the applied rectangular pulse has a $(\sin x)/x$ spectrum. To make the spectrum appear flat, the pulse must be multiplied by the inverse function, $x/\sin x$.

The desired modification of the signal can be accomplished by modifying the raised cosine channel by an $x/\sin x$ function. In practice, only the first lobe of the $x/\sin x$ function need be considered. Figure 14-1(b) illustrates. Since the $(\sin x)/x$ function becomes zero at $2f_1$, $x/\sin x$ theoretically becomes infinity at this frequency. It can be shown, however, that the product of the $x/\sin x$ function and the cosine function representing curve $s$ also becomes zero at $2f_1$.

A stream of rectangular pulses having a 100 percent duty cycle $(\tau = T)$ transmitted over a channel having a characteristic, $s(x/\sin x)$, like that of Figure 14-1(b), appears at the channel output as shown in Figure 14-2(b) where $T = 1/(2f_1)$. The time delay between the input, Figure 14-2(a), and the output, Figure 14-2(b), is ignored. Note that the $(\sin x)/x$ form of each output pulse is such that the zero crossings correspond to the sampling points of successive pulse intervals.

The interval, $1/(2f_1)$, is known as the *Nyquist interval*. In channels having sharp cutoffs, the signalling interval, $T$, must closely approximate this interval in order to minimize intersymbol interference.

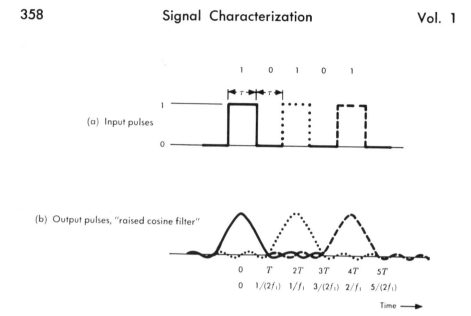

Figure 14-2. Effects of a cosine channel on pulse shaping.

With a 100 percent roll-off characteristic, larger departures from $T = 1/(2f_1)$ can be tolerated than in a channel having a sharp cutoff.

The shaping of the channel characteristic may be placed at any point in the channel. If the characteristic of the medium is predictable, its characteristic can be incorporated in the overall channel characteristic. Since there are a number of places where shaping may be used, the detailed effect on characterization of the transmitted signal cannot be generalized.

## 14-2   DIGITAL SIGNAL CHARACTERISTICS

A large number of different digital signal formats are possible and have been used in the Bell System. Many formats have been tried and found unsatisfactory because of low efficiency or susceptibility to various forms of impairment and are now considered obsolete. An important stimulus to continued development, in addition to the burgeoning demands of the business machine and computer industries, is the desire to make signal transmission more economical by increasing efficiency, i.e., by increasing the number of transmitted bits per second per hertz of available bandwidth at less cost per unit of information transmitted.

## Amplitude Shift Keyed Signals

Initially, the generation of digital information was by amplitude shift keying (ASK) techniques in a binary baseband mode. This mode, basically that used to operate computers, is still used in many telephone network signalling systems. Because of its simplicity, the ASK binary baseband mode is used for transmission of digital data signals over Bell System facilities, but only for relatively short distances. The binary baseband signal format is neither as efficient for a given bandwidth as other formats nor is it suitable for transmission on facilities which provide no dc continuity or are subject to quadrature distortion, low-frequency cutoff, or significant envelope delay distortion. As a result, where these restrictions are important and signal-to-noise performance is adequate, equipment is installed to process the binary baseband signals into forms more suitable to the environment.

The nature of ASK signals is such that when they are used as baseband signals in the form of simple on-off pulses with average amplitude of zero, low- and zero-frequency components are important to their characterization and to their recovery by detection circuits in the receivers. Because of their nature, such signals may be regarded as formed by a process of modulation of a direct current. The difficulties associated with transmission of zero- and very-low-frequency components through transmission facilities and networks are among the important reasons for the infrequent use of ASK modes of signal transmission without additional processing for transmission over Bell System facilities. When an ASK mode is used, special provision must be made to eliminate the low- and zero-frequency components of the signal at the transmitter and at the outputs of regenerative repeaters and to restore these components at the repeater inputs and at the receiver.

Terminal equipment in the form of station sets, sometimes called data sets, has been developed by the Bell System and many other manufacturers to process data signals in a variety of ways. Some of this equipment was initially arranged to have the binary data signal amplitude-modulate a carrier in a 4-kHz voiceband channel.

A signalling rate of 750 bits per second was achieved by using double sideband modulation with transmitted carrier, and a rate of 1600 bits per second was achieved by using vestigial sideband techniques. These AM techniques provide frequency translation to eliminate the dc continuity and low-frequency cutoff problems. The transmitted carrier is recovered for demodulation at the receiver in proper frequency and phase to eliminate quadrature distortion impairment from the received signal.

**ASK Signal Waveforms.** Digital symbols may be represented by any of a large variety of electrical signal formats. As previously mentioned, the control of computers usually involves the use of binary ASK signals. Logically, the operation of computers relates to the *0* and *1* representation of binary numbers which in turn correspond to the two states of a binary signal. Some alternate ways of representing these two states are illustrated by the formats shown in Figure 14-3.

The waveforms of Figure 14-3 have several features in common. First, all of the waveforms represent the same sequence of digits, namely, *0110001101*. Each of the waves depicted represents a synchronous system in which the receiving equipment is timed by some mechanism so that the incoming signal is sampled at the instants indicated. The sampling is required in most cases in order to determine if the signal amplitude at the sampling instant is above or below one or more of the decision thresholds indicated. In Figure 14-3(f), the sampling would take the form of a zero-crossing detector since, in that case, *0*s are represented by transitions in the signal and *1*s by no transitions. Finally, the peak-to-peak amplitudes of the signals are all shown as equal to two units of voltage, *V*.

Figure 14-3(a) illustrates the simplest of these signal formats. A *1* is represented by the presence of a voltage, and a *0* by the absence of voltage. If the wave is between half and full amplitude at a sampling instant, it represents a *1*; if it is between zero and half amplitude, it represents a *0*. Thus, the half-amplitude value is the decision threshold.

Figure 14-3(b) is similar in all respects to Figure 14-3(a) except that opposite polarities of voltage are used to represent *0* and *1* instead of voltage and no-voltage. The decision threshold in this case is zero volts. The polar form illustrated here is sometimes adapted to the transmission of nonsynchronous digital signals.

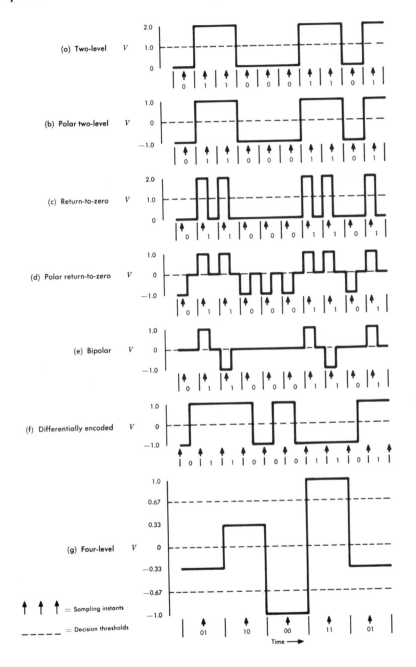

Figure 14-3. Basic ASK waveforms.

Figure 14-3(c), called *return-to-zero*, differs from the first two in that pulse length for a symbol is less than the time allotted to the symbol interval. The extent to which the pulse length differs from the symbol interval determines the *duty cycle*, defined as $100\tau/T$ percent where $\tau$ is the pulse length and $T$ is the symbol interval. As in Figure 14-3(a), voltage is used to represent a *1* and no-voltage to represent a *0* for the return-to-zero signal.

Figure 14-3(d) is the return-to-zero counterpart of the polar two-level signal of Figure 14-3(b). Note that the polar return-to-zero signal is really a three-level signal having less than a 100 percent duty cycle.* However, since only two of the values, plus-voltage and minus-voltage, are used to represent the digital information in the signal, it may be regarded as a binary signal. Note that each symbol, whether a *1* or a *0*, is associated with the presence of a pulse. For this reason, the synchronization of the receiving equipment may be accomplished by using the information in the signal, thus making the receiver self-clocking. This feature allows this signal format to be used for the transmission of nonsynchronous data.

The bipolar signal of Figure 14-3(e) has valuable properties that have caused it to be the format chosen for the line signal for the T1 Carrier System. The symbol *0* is represented by zero voltage and the symbol *1* is represented by the presence of voltage. However, the polarity of voltage for successive *1* symbols is alternated. Two important results are achieved. First, the dc component of the signal is virtually eliminated. This permits transformer coupling of the repeater to the line, facilitates the separation of the signal from dc power, and makes decision threshold circuits more practical by effectively eliminating the phenomenon known as baseline wander caused by a varying dc signal component. Second, the concentration of energy in the signal is shifted from the frequency corresponding to the baud rate to one-half the baud rate. This reduces near-end crosstalk coupling, reduces the required bandwidth to about one-half of that needed for a polar signal of the same duty cycle and repetition rate, and makes the design of timing recovery circuits more practical.

In Figure 14-3(f), the information is coded in terms of transitions that occur in the transmitted signal. Successive pulse intervals are

---

*The polar return-to-zero and bipolar signals are sometimes called pseudo three-level signals.

compared. If they are identical, a *1* was transmitted in the original signal; if successive intervals show a transition, a *0* was transmitted.

The signals of Figures 14-3(a) through 14-3(f) may all be considered binary, either in the number of values of voltage transmitted or in the significant number of values used to represent binary information. Figure 14-3(g) is not binary; it is illustrative of a class of signals which can be used to transmit data quite efficiently when the signal-to-noise ratio that can be realized is high enough to permit signal detection at a number of different decision threshold values that generally are smaller than those for the binary signals previously discussed.

The signals of Figure 14-3 are used in many ways. In some cases, they are the signals delivered to the station set by the customer and in other cases they represent the signals transmitted over Bell System facilities. The several forms commonly used may be characterized somewhat more fully to illustrate their use.

**Wideband Binary ASK Signals.** A limited number of applications of this signal format are used for digital data transmission in the Bell System. Data station and carrier terminal facilities are available to permit the transmission of a polar form of signal, somewhat like that of 14-3(b), at synchronous rates of 19.2, 50.0, or 230.4 kb/s; for nonsynchronous service, the signal elements must have corresponding minimum durations of 52.0, 20.0, and 4.0 $\mu$s. The three arrangements have been developed to permit wideband data transmission in 24-kHz, 48-kHz, and 240-kHz bands found in commonly used FDM equipment. The 50-kb/s arrangement is the one most frequently used; its operation is typical of these arrangements.

As mentioned, the signal is transmitted in a polar form, called *restored polar*, different from the format of Figure 14-3(b) in that the dc component and some of the low-frequency components are filtered out at the transmitting data station and restored at the receiver. As a result of the filtering, the transmitted signal is sharply skewed, as shown in Figure 14-4(b). This mode of transmission obviates the need for high fidelity transmission at zero and very low frequencies.

The power spectral densities for synchronous and nonsynchronous polar signals and restored polar signals are shown in Figure 14-5. For the synchronous signal, the spectra are those of a signal having a rate of 50 kb/s ($T = 20$ $\mu$s) and random bits having an equal probability of being *1* or *0* ($p = 0.5$). The nonsynchronous signal spectra represent a two-valued facsimile signal in which the average rate of black-white transitions is 4000 per second, and pages are 10 percent black and 90 percent white (the probability of a *1*, p = 0.1). The low-frequency power density spectra are very similar for the synchronous and nonsynchronous signals.

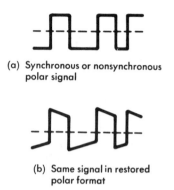

(a) Synchronous or nonsynchronous polar signal

(b) Same signal in restored polar format

Figure 14-4. Restored polar signal.

Thus, Figures 14-4(b) and 14-5 illustrate, in the time and frequency domains, respectively, the characteristics of the processed 50-kb/s signal as it is transmitted on loops. If a carrier system is used, further processing is necessary. Two cases are of interest, i.e., processing for transmission in the 48-kHz group band of the L-type multiplex and processing for transmission in the 96-kHz N-type carrier band. Transmission in both cases is by amplitude modulation with vestigial sideband (VSB).

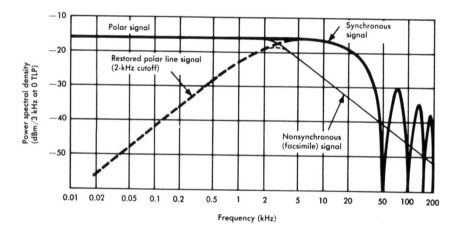

Figure 14-5. Power density spectra of polar and restored polar signals.

The frequency allocation, channel characteristic, and resulting signal spectrum for transmission in L-type multiplex are shown in Figure 14-6. The channel transmission characteristic and frequency allocation are shown in Figure 14-6(a). In the 60- to 108-kHz spectrum, a speech channel for coordination of operations may be provided in addition to the data channel for the VSB data signal. The baseband signal spectrum, shown in Figure 14-6(b), is the same as that of Figure 14-5 but modified by the group frequency channel characteristic; the modification is most notable at high frequencies where channel characteristics limit the baseband top frequency to 37 kHz. The modulation process results in a VSB signal with the carrier suppressed. However, the fact that signal components at

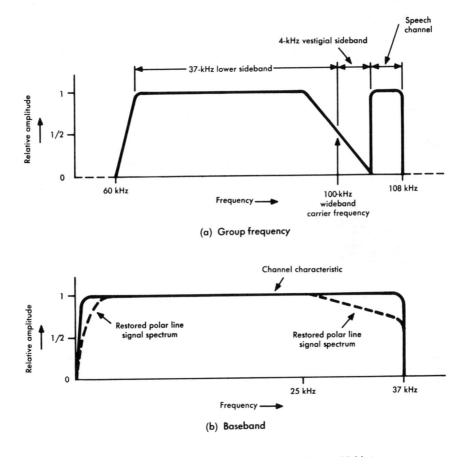

(a) Group frequency

(b) Baseband

Figure 14-6. Data signals in L-type multiplex at 50 kb/s.

zero and low baseband frequencies have been removed at the station set permits the reinsertion of a low-amplitude carrier component. This component is recovered at the receiving terminal equipment to control the phase and frequency of the carrier used in the demodulation process [5].

When the 50-kb/s signal is processed for transmission over an N-type carrier system, it is modulated into the high group of the N-type system as shown in Figure 14-7. The modulation process is VSB with carrier transmitted. Since the nominal bandwidth in N-type carrier is 96 kHz, the signal is not as severely band limited as in the L-type multiplex. Two voice channel carriers are transmitted, with or without voiceband modulating signals. These carriers and the data carrier are not quite sufficient to supply the signal power for N-carrier line regulation. Therefore, a single-frequency signal is added at 176 kHz at an amplitude sufficient to make the total power of the transmitted composite signal equivalent to that of the normal N-carrier line signal.

As mentioned previously, similar signal formats are provided for transmission at 19.2 kb/s (one-half group band) and at 230.4 kb/s (supergroup band). The transmission arrangements are quite similar except that in the case of 19.2-kb/s transmission the vestigial shaping is accomplished at the data station instead of the carrier terminal.

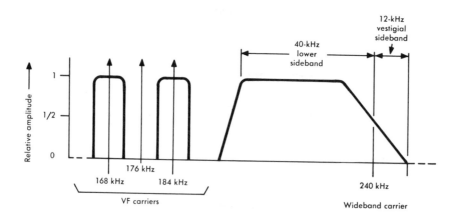

Figure 14-7. Data signals in N-type carrier at 50 kb/s.

Thus, the 19.2-kb/s signal is transmitted at carrier frequency over the data loops. The carrier is at 29.6 kHz [1].

**Bipolar Line Signals.** Bipolar signals find their greatest use in the Bell System as line signals in T-type carrier systems. The line signal in the T1 Carrier System is bipolar, like that shown in Figure 14-3(e), in all respects. The line signal in the T2 Carrier System is similar, but with one important exception; in T2, the line signal is prevented from containing more than five successive *0*s by a method that modifies the bipolar signal format. This is accomplished by logic circuits in the transmitting terminal which examine the line signal before it is applied to the line. If the signal contains six consecutive *0*s and if the last *1* was a +, a *0+−0−+* signal is substituted for the six *0*s; if the last *1* was a −, a *0−+0+−* signal is substituted for the six *0*s. The resulting violation of the bipolar rule (alternate *1*s must be of alternate polarity) is a means for recognizing the need for six *0*s which must be reinserted in the pulse stream at the receiving terminal. The substitution is made in order to guarantee a minimum density of *1*s in the line signal.

This code substitution eases the design and increases the accuracy of repeater timing circuits. The price paid is the additional logic circuits that must be used to accomplish the substitution and the additional complication of ignoring the substituted codes when bipolar violations are used as a measure of system performance.

The timing problem in the T1 Carrier System is also solved by limiting the maximum number of successive *0*s in the line signal but in a manner different from that used in the T2 system. In the T1 line signal, the number of consecutive *0*s that can be transmitted is limited to 15. For example, if encoded speech signals are being transmitted, this limitation is imposed by preventing any 8-bit word containing all *0*s from being transmitted to the line. If such a word is generated, it is modified in the terminal equipment by inserting a *1* in the seventh digit of the coded word. This is the least significant digit of the code representing the amplitude sample. The eighth digit is used for signalling. The code substitution permits the true bipolar feature to be maintained in the line signal. Its cost is a slight increase in channel coding noise.

The power spectral densities of the T1 and T2 signals are, of course, functions of the statistical makeup of transmitted signals and of the signalling rate employed. The power spectral densities of the two signals are conveniently represented in terms of the probability, $p$, of a $1$ in the signal sequence. The bipolar signal used in T1 is represented in Figure 14-8 for a range of $p$ from 0.4 to 0.6. The T2 signal with code substitution for six successive $0$s is shown in Figure 14-9 for the same range of values of $p$, 0.4 to 0.6. In both figures the abscissas are normalized to unity, $fT = 1$, where $f$ is in hertz and $T$ is the signalling rate.

In the T1 system, the signal is transmitted at $1.544 \times 10^6$ bits per second with a 50 percent duty cycle. The minimum pulse width is then $\tau = 1/(2 \times 1.544 \times 10^6)$, or about 0.324 $\mu$s. In the T2 system, the signal is transmitted at $6.312 \times 10^6$ bits per second, also with a 50 percent duty cycle. For T2, the minimum pulse width is $\tau = 1/(2 \times 6.312 \times 10^6)$, or about 0.079 $\mu$s.

The bipolar nature of the two signals and the coding sequence employed result in negligibly small discrete frequency components

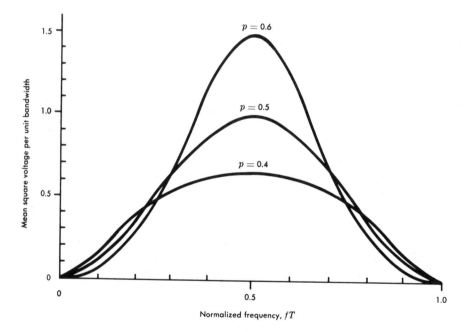

Figure 14-8. Power spectra for T1 bipolar signals.

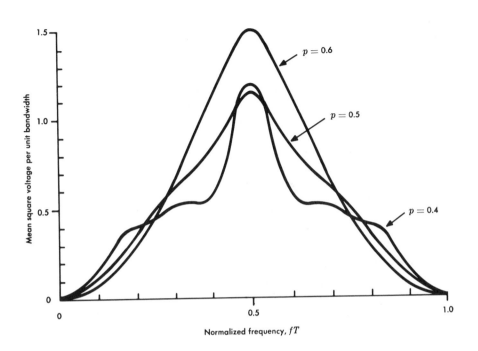

Figure 14-9. Power spectra for T2 bipolar signals coded for restricted number of sequential zeros.

and nulls in the spectrum at zero frequency and at integral multiples of frequencies corresponding to the signalling rate. For each system, the losses above the frequency corresponding to the signalling rate are high; as a result, the spectra of Figures 14-8 and 14-9 may be ignored above the signalling frequency. In T1 carrier, the top transmitted frequency then may be regarded as $f = 1/(2\tau) = 1/(2 \times 0.324) = 1.544$ MHz; in T2 carrier, the top transmitted frequency is $f = 1/(2\tau) = 1/(2 \times 0.079) = 6.312$ MHz.

Signal amplitudes vary widely through the system in both T1 and T2. One reference point often used is the output of a line (regenerative) repeater. Even here, the amplitude varies somewhat. The nominal value in T1 is 3 volts, peak; in T2, 4.2 volts, peak.

**Multilevel ASK Signals.** The ASK signals described thus far have been two-level or, at most, three-level (T1 and T2 line signals).

Where channels and transmission facilities exhibit a high signal-to-noise ratio and have well controlled transmission characteristics (gain/frequency and phase/frequency), the efficiency of transmitting information can be significantly improved by coding the digital signal into a multilevel ASK format such as the four-level signal illustrated in Figure 14-3 (g). Such multilevel signals are used in the voiceband [6] and in the mastergroup band of broadband carrier systems [1]. In addition, a system is being introduced that permits the coding of a 1.544-megabit-per-second line signal, such as that used in T1 carrier, into a multilevel signal that can be transmitted over microwave radio systems at frequencies below the normal frequencies allocated to message channels [7]. The multilevel ASK signals presently transmitted in the Bell System are listed in Figure 14-10.

The first four groups of signals, those transmitted at 1.8, 2.4, 3.2, and 3.6 kilobauds, are transmitted in the voiceband. The signalling

| BAUD RATE (kilobauds) | BIT RATE (kb/s) | NO. OF LEVELS | BANDWIDTH |
|---|---|---|---|
| 1.8 | 1.8 | 2* | Voice |
|  | 3.6 | 4 | Voice |
|  | 5.4 | 8 | Voice |
| 2.4 | 2.4 | 2* | Voice |
|  | 4.8 | 4 | Voice |
|  | 7.2 | 8 | Voice |
| 3.2 | 3.2 | 2* | Voice |
|  | 6.4 | 4 | Voice |
|  | 9.6 | 8 | Voice |
| 3.6 | 3.6 | 2* | Voice |
|  | 7.2 | 4 | Voice |
|  | 10.8 | 8 | Voice |
| 772 | 1,544 | 7 | 440 kHz |

*Optional binary mode.

Figure 14-10. Multilevel ASK signals in the Bell System.

rates are optional and are provided to the customer in accordance with appropriate tariffs which govern the extent to which transmission facilities are equalized to provide adequate transmission quality. Note that for these signals a general rule may be applied; i.e., if $n$-valued coding is used at a rate of $x$ bauds, the transmitted information rate may be expressed as $x \log_2 n$ bits per second.

The signal transmitted at the 772-kilobaud rate is the digital line signal used for transmission over microwave radio systems. The rate at which information is transmitted for this signal cannot be computed by the expression previously given for voiceband rates because the method of coding is different. The mode involves what is known as class IV coding for partial response transmission [8, 9], for which the transmission rate is $x \log_2 (n + 1)/2$ bits per second, where $x$ again represents the baud rate and $n$ represents the number of values.

**Signal Spectra.** Binary baseband signals are coded into the multilevel format in a data set or in terminal equipment associated with a carrier system. All of the multilevel signals listed in Figure 14-10 are transmitted in a partial response format and have an energy distribution as shown in Figure 14-11. Energy above frequency $f_1$, the Nyquist frequency, is removed by an appropriate cutoff filter which leaves only a vestige of the second lobe, as illustrated in Figure 14-12. A timing signal, $P_t$, is added to facilitate recovery of the synchronizing, or timing, information at the receiver.

When multilevel baseband signal transmission is appropriate, the signal of Figure 14-12 may be transmitted directly. Note that there is no energy at zero frequency and that low-frequency components are of low amplitude. Only slight signal impairment is suffered when such a signal is transmitted over baseband facilities which cannot pass direct current and low-frequency components. This mode of transmission (baseband) is used for transmit-

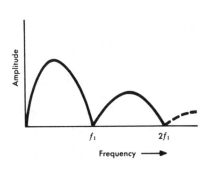

Figure 14-11. Spectral energy distribution, multilevel partial response signal.

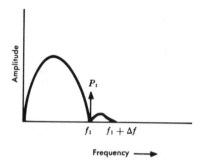

Figure 14-12. Partial response signal — high-frequency lobes removed — timing signal added at $f_1$.

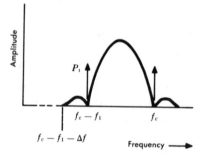

Figure 14-13. Partial response signal at carrier frequency — vestigial upper sideband.

ting the 772-kilobaud signal of Figure 14-10 on microwave radio systems. It is frequency modulated in the FM transmitter along with whatever speech channels are carried by the system.

All signals in Figure 14-10 other than the 772-kilobaud signal are transmitted by VSB techniques. These may be upper or lower sideband; a lower sideband version with a vestigial upper sideband is illustrated in Figure 14-13.

### Phase Shift Keyed Signals

Signal transmission by phase shift keying (PSK) techniques is accomplished in the voiceband by the use of data sets which code incoming binary signals into multilevel phase shifts of appropriate carrier signals. In PSK transmission, sideband frequencies are generated by the modulating signal so that the bandwidth required is equal to twice the highest frequency component in the modulating signal. The distribution of energy in the modulated carrier wave follows a pattern like that illustrated in Figure 14-14.

Phase shift keying has advantage over other modes in certain situations. In a band-limited channel, PSK signals are relatively immune to amplitude changes. The signal is transmitted at essentially constant power. The mode lends itself well to the recovery of a clock signal at the receiver and offers speed advantages by multilevel coding techniques. This mode has the disadvantage of being sensitive to phase distortions, phase jitter, and impulse noise. These impairments may affect the zero crossings of the signal.

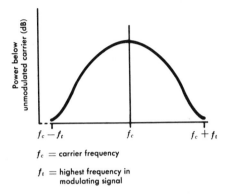

Figure 14-14. Power spectral density of a PSK data signal.

A number of successful PSK data sets and types of terminal equipment have been developed. One such arrangement is used to transmit a number of telegraph signals simultaneously in a single voiceband. Each signal is modulated to a separate part of the voiceband, and the signals are combined by FDM methods. Some four-phase voiceband systems provide for the transmission of data at rates up to 2400 bits per second. The baseband signal phase-modulates a carrier by ±45 degrees and ±135 degrees relative to its nominal zero-phase condition. (Such a system, no longer used extensively, was also developed to transmit digital data at 40.8 kb/s.) A newer voiceband system using eight-phase modulation transmits data at 4800 and 9600 bits per second [11, 12].


### Frequency Shift Keyed Signals

Data sets have been developed to exploit the use of frequency shift keying (FSK) techniques. In one such system 1200 Hz and 2200 Hz were used to represent the "space" (no pulse) and "mark" (pulse) signals, respectively. (Space and mark are terms handed down from telegraph usage.) This system operated at about 1200 bits per second over the switched public network and up to 1800 bits per second on conditioned private lines. Later, more sophisticated terminal equipment utilized FSK techniques to achieve 2400 bits per second.

These arrangements display most of the advantages of the PSK mode of transmission and have the additional advantage of being relatively simple to implement; circuits are readily designed. However, FSK signals have the usual feature of achieving high immunity to noise at the cost of wider required bandwidths. As a result, the FSK systems are commonly used in voicebands or less at low signalling rates relative to those achievable with multivalued PSK and ASK arrangements.

Many private line and switched network telegraph and teletypewriter services are provided by FSK techniques. In addition, some moderate-speed voiceband data is also transmitted FSK. TOUCH-TONE signalling as a means of data transmission is, in effect, also a form of FSK.

**Telegraph Signals.** Telegraph signals are commonly transmitted at various rates, including 75 and 150 bauds, by FSK techniques in narrow channels which are combined into a voiceband by frequency division multiplex techniques. Seventeen separate channels can be provided with frequencies as given in the top portion of Figure 14-15 to transmit 75-baud telegraph signals. For 150-baud signals, the channel assignments are shown at the bottom of the figure.

The binary (*1* and *0*) input signals are translated into FSK signals at the terminal equipment. A *1*, or mark, signal corresponds to a frequency in the passband of the channel 35 Hz (for 75-baud signals) or 70 Hz (for 150-baud signals) above the channel center frequency. The *0*, or space, signal is represented by a frequency 35 (or 70) Hz below the channel center frequency. The center frequency is not transmitted as a discrete signal.

**Low-Speed Voiceband Data.** Low-speed teletypewriter signals are provided in the voiceband by FSK techniques using 1070 and 1270 Hz as the mark and space signals, respectively, for transmitting in one direction, and 2025 and 2225 Hz for signalling in the opposite direction. This arrangement, in effect, provides equivalent four-wire transmission and allows full duplex operation. Data rates vary from about 100 bauds to 300 bauds.

**Medium-Speed Data.** Data at rates up to 1800 bits per second can be transmitted in a voiceband by an arrangement that uses 1200 Hz and 2200 Hz as the mark and space frequencies, respectively.

| SINGLE BANDWIDTH | | | |
|---|---|---|---|
| Channel number | Space frequency | Center frequency | Mark frequency |
| 1 | 390 | 425 | 460 |
| 2 | 560 | 595 | 630 |
| 3 | 730 | 765 | 800 |
| 4 | 900 | 935 | 970 |
| 5 | 1070 | 1105 | 1140 |
| 6 | 1240 | 1275 | 1310 |
| 7 | 1410 | 1445 | 1480 |
| 8 | 1580 | 1615 | 1650 |
| 9 | 1750 | 1785 | 1820 |
| 10 | 1920 | 1955 | 1990 |
| 11 | 2090 | 2125 | 2160 |
| 12 | 2260 | 2295 | 2330 |
| 13 | 2430 | 2465 | 2500 |
| 14 | 2600 | 2635 | 2670 |
| 15 | 2770 | 2805 | 2840 |
| 16 | 2940 | 2975 | 3010 |
| 17 | 3110 | 3145 | 3180 |
| DOUBLE BANDWIDTH | | | |
| 21 | 610 | 680 | 750 |
| 22 | 950 | 1020 | 1090 |
| 23 | 1290 | 1360 | 1430 |
| 24 | 1630 | 1700 | 1770 |
| 25 | 1970 | 2040 | 2110 |
| 26 | 2310 | 2380 | 2450 |
| 27 | 2650 | 2720 | 2790 |
| 28 | 2990 | 3060 | 3130 |

Figure 14-15. Voice-frequency carrier data channel assignments.

**TOUCH-TONE Signalling.** TOUCH-TONE signalling, described briefly in Chapter 13, was introduced initially as a means for transmitting address signals from telephone station sets to the central office switching machine. These signals may be impressed on the telephone line while a connection is established. Since the TOUCH-TONE signals fall in the voice-frequency band, they offer the possibility of being used to transmit data [13]. Special receivers are provided for data transmitted by TOUCH-TONE signalling.

## 14-3   ANALOG DATA SIGNALS

A number of voiceband analog data signals are transmitted by FM techniques. Three of these are found in the plant in sufficient quantity to warrant individual description. In each case, signal power is limited to −16 dBm0, long-term average. Many other signals are transmitted in quantities too small to warrant individual characterization. Among these are several types of telemetry and telewriter signals.

## Medium-Speed Voiceband Data

One type of analog data signal transmission involves the translation of a 0 to +7 volt continuous signal from the customer to an FM signal which is transmitted over telephone facilities in the voiceband. A zero-volt input signal is transmitted as a 1500-Hz signal on the line, and a +7 volt signal is transmitted as a 2450-Hz signal on the line; intermediate input voltages and line frequencies are linearly related. The baseband signal may contain components from zero frequency up to about 1000 Hz.

In the direction of transmission opposite to that used for data transmission, a 60 ± 1 Hz signal is sent for synchronization. This signal is translated to 600 Hz for transmission over the line. The powers contained in the data and synchronizing signals are equal.

The data signal can be used for facsimile transmission with resolution equivalent to 100 lines per inch and at speeds of up to 180 lines per minute for copy reproduced on 8-1/2 by 11 inch paper.

## Low-Speed Medical Data

Medical data such as electrocardiagrams and electroencephalograms can be transmitted over telephone facilities by FM techniques. Input signals are accepted with components from zero to about 100 Hz and with amplitudes of −2 to +2 volts. Such signals frequency-modulate a carrier at 1988 Hz; the carrier frequency varies linearly with the input voltage at frequencies between −262 and +262 Hz relative to the carrier. A signal at 387 Hz is transmitted in the opposite direction to permit signalling from the receiver to the transmitter.

One feature of this type of transmission that is different from others is that the signal generated in medical electronic equipment may be coupled to the telephone line electronically or acoustically. The latter method has the virtue of allowing for portable equipment.

The signal can be coupled to the line at any telephone connected to the network.

## Low-Speed Analog Data

Multichannel analog data transmission is provided in the voiceband at low speed by an FM arrangement utilizing carriers at 1075, 1935, and 2365 Hz. Each of these carriers may be frequency-modulated by baseband signals having components from zero to 105 Hz and amplitudes between $-2.5$ and $+2.5$ volts. A reverse direction signal is used to permit the receiver to communicate with the transmitter during data transmission. This arrangement is used to transmit medical or other types of analog data.

### REFERENCES

1. Mahoney, J. J., Jr. "Transmission Plan for General Purpose Wideband Services," *IEEE Transactions on Communications Technology*, Vol. COM-14, No. 5 (Oct. 1966), pp. 641-648.

2. Fracassi, R. D. and F. E. Froehlich. "A Wideband Data Station," *IEEE Transactions on Communications Technology*, Vol. COM-14, No. 5 (Oct. 1966), pp. 648-654.

3. Bell System Technical Reference PUB41005, *Data Communications Using the Switched Telecommunications Network* (American Telephone and Telegraph Company, May 1971).

4. Bennett, W. R. and J. R. Davey. *Data Transmission* (New York: McGraw-Hill Book Company, Inc., 1965).

5. Ronne, J. S. "Transmission Facilities for General Purpose Wideband Services on Analog Carrier Systems," *IEEE Transactions on Communications Technology*, Vol. COM-14, No. 5 (Oct. 1966), pp. 655-659.

6. Bell System Technical Reference PUB21204, *Data Set 203-Type* (American Telephone and Telegraph Company, June 1970).

7. Seastrand, K. L. and L. L. Sheets. "Digital Transmission over Analog Microwave Radio Systems," IEEE International Conference on Communications, June 19-21, 1972.

8. Gerrish, A. M. and R. D. Howson. "Multilevel Partial-Response Signaling," *IEEE Conference on Communications Digest*, 1967.

9. Kretzmer, E. R. "Generalization of a Technique for Binary Data Communication," *IEEE Transactions on Communications Technology*, Vol. COM-14, No. 1 (Feb. 1966), pp. 67-68.

10. Bell System Technical Reference PUB41301, *Data Set 301B — Interface Specification* (American Telephone and Telegraph Company, Mar. 1967).

11. Baker, P. A. "Phase-Modulation Data Sets for Serial Transmission at 2,000 and 2,400 Bits per Second," *AIEE Transactions*, Part 1, *Communications and Electronics*, No. 61 (July 1962), pp. 166-171.

12. Schilling, D. L. and H. Taub. *Principles of Communications Systems* (New York: McGraw-Hill Book Company, Inc., 1971).

13. Bell System Technical Reference PUB41802A, *Data Couplers CBS and CBT for Automatic Terminals* (American Telephone and Telegraph Company, Mar. 1971).

Chapter 15

# Video Signals

A wide variety of video services is provided over the transmission facilities of the Bell System. These vary from narrowband telephotograph service, operating in the voiceband, to multimegahertz bandwidth television services provided for the television broadcast industry and for educational, industrial, and private distribution systems.

As in the transmission of other types of information, the transmitting and receiving equipment at both ends of a video transmission path must include transducers capable of translating one form of energy to another. In this case, different values of luminance (light intensity), together with color information in the transmission of color television signals, are converted to electrical signals at the transmitter; at the receiver, the transducer must translate the electrical signal back to light signals so that the transmitted image can be viewed or recorded on film or paper.

Many natural characteristics of the human recipients of video information have influenced the design of transmitting and receiving equipment as well as the design of the transmitted video signals. Among these are the persistence of vision, the preferred viewing distance of visual images, the resolution capability of the eye, the human tolerance to departures from accurate color rendition, and the effects on viewing preferences of ambient conditions such as lighting [1]. These human factors have influenced the rate at which picture images are transmitted; the format of color television signals; the resolution and, therefore, the bandwidth of signals; and many other aspects of video signal transmission.

## 15-1 TELEVISION SIGNALS

In 1970, the Bell System operated well over 100,000 route miles of part-time and full-time television circuits. The majority of these

379

were provided for network television broadcast signals* [2]. As a convenience, the black and white (monochrome) signal is used here as the basis of television signal description even though most television signals now transmitted in the United States are color. The chrominance information in a color signal is regarded as being superimposed on the monochrome signal and is so described.

### Standard Monochrome Baseband Signals

Intra-urban transmission needs are nearly always provided in the Bell System by baseband facilities. Interurban needs are usually provided by long-haul or short-haul microwave relay facilities; these are usually fed by baseband facilities which interconnect the broadcaster's equipment and the terminals of the microwave radio system.† The baseband signal received at the microwave system terminal frequency-modulates the microwave carrier. Only the baseband signal is characterized in detail here, although some attention is subsequently given to the signal format used in commercial television broadcasting.

The conversion of light signals to electrical signals and the reconversion to light signals at the receiver involves a scanning operation which differs in details for different systems. However, all systems must provide, in the scanning operation, for the synchronization of the receiver with the transmitter (a coding process); they must also provide for the conversion from luminance variations to electrical signal variations (a modulation process).

**Scanning and Synchronization.** Figure 15-1 illustrates the scanning pattern used for broadcast television signals in the United States. The scanning mechanism causes the exploring element and the reproducing spot to move in synchronism across the image field and the receiving field (picture tube) in nearly horizontal lines from left to right. The scanning lines are started at the top; successive

---

*The signals described here have been standardized in the United States. Other standards have been estabilshed elsewhere; e.g., 625-line, 50 frame-per-second signals are used in Europe. As a result of satellite transmission, Bell System facilities are being adapted for transmission of such signals. Conversion of the signal to the USA standard is presently the responsibility of the broadcaster.

†Coaxial cable systems were once used for television signal transmission, but they are no longer used for this purpose in the Bell System.

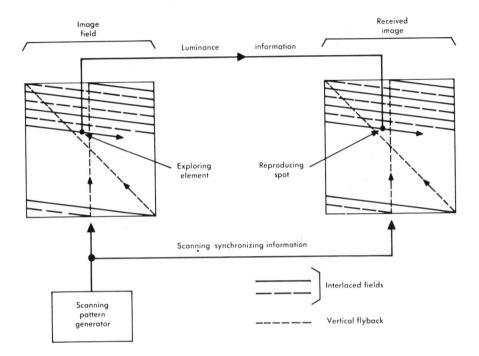

Figure 15-1. Broadcast television scanning process.

scans are made at successively lower parts of the field until the bottom is reached. The spots are then returned to the top to begin a second field. Succeeding scans in alternate fields are interlaced. The interlaced scans of two successive fields make up a frame.

The exploring element at the transmitter is caused to scan the image field by the scanning pattern generator. The scanning pattern must cover every part of the image in a systematic and specified manner. Information regarding the location of the exploring spot and the direction in which it is being moved are coded at the transmitter and sent to the receiver so that the reproducing spot can be located in the received image field at a position corresponding to that of the exploring element in the transmitter. This process of synchronizing the receiver to the transmitter is also illustrated functionally in Figure 15-1.

While Figure 15-1 shows the transmission of scanning/synchronizing and luminance information over separate channels, the information is in reality combined into one composite signal for transmission or broadcast. The two kinds of information, separated in polarity and time, are illustrated in Figures 15-2, 15-3, and 15-4.

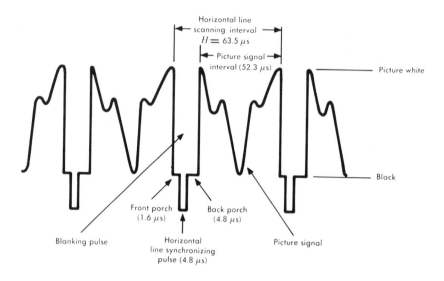

Figure 15-2. Monochrome television horizontal line scanning and synchronization.

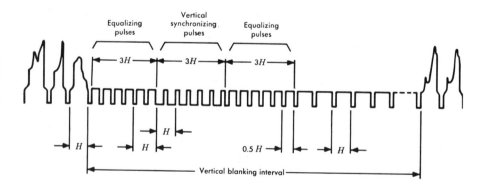

Figure 15-3. Monochrome television vertical synchronization.

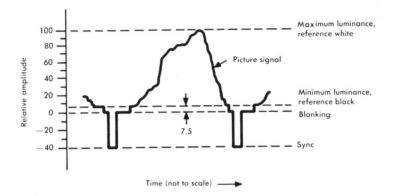

Figure 15-4. Monochrome television signal, relative amplitudes.

Consider first the horizontal line scanning function illustrated in Figure 15-2. The total line scanning interval, $H$, is 63.5 $\mu$s, equivalent to about 15,750 line scans per second. Of this interval, 11.2 $\mu$s are assigned to a blanking pulse. The blanking pulse interval is divided into a 1.6-$\mu$s front porch interval, a 4.8-$\mu$s back porch interval, and a 4.8-$\mu$s horizontal synchronizing pulse interval. The porches isolate the synchronizing pulse from transients or overshoots of the picture signal. The synchronizing pulse, recognized at the receiver by its polarity and duration, triggers circuits that drive the reproducing spot at the receiver to the left side of the image field (flyback). During the blanking pulse interval, the receiving tube is normally blanked out; the luminance of its reproducing spot is driven into the ultrablack region so that the synchronizing pulse and the flyback of the spot are not visible. After each blanking pulse interval, the scanning circuits drive the exploring and reproducing spots in synchronism across the image from left to right.

When the last line of a field scan is completed, the scanning pattern enters a vertical blanking interval as shown in Figure 15-3. This interval is about 1200 $\mu$s long, equivalent to the duration of about 20 horizontal scan intervals. During a part of this interval, a number of vertical synchronizing pulses, each about 25 $\mu$s long, drive the spot to the top of the screen (vertical flyback) to begin a new field scan. A series of horizontal pulses (called equalizing pulses), transmitted at twice the normal line scan rate, precede and follow the vertical synchronizing pulses. Following the second burst of these equalizing pulses is a series of horizontal synchronizing pulses transmitted at the normal scan rate as a part of the vertical blanking interval. The equalizing and normal horizontal synchronizing pulses transmitted during the vertical blanking interval are provided to condition the synchronization circuits of the transmitter and receiver so that the two are indeed in synchronism and to guarantee that the interlace pattern is properly implemented in successive field scans.

The timing of the vertical blanking intervals is such that a field is produced every 1/60 second. The interlacing of the next field with the first makes up a frame, one of which is produced each 1/30 second. This 30-per-second frame rate, combined with the 15,750 horizontal line rate, results in a picture nominally formed of 525 lines per frame.

**Luminance Signal.** As shown in Figure 15-1, an exploring element measures the intensity of the light at a given spot on the image to be transmitted. The light intensity is converted to an electrical signal whose amplitude is a specified function of the measured intensity. This electrical analog of the light intensity is transmitted to the receiver where the inverse process, the conversion of the electrical signal to appropriate light intensity values, takes place. The reproducing spot at the receiver then illuminates the receiving mechanism to the proper intensity.

If the image being scanned at the transmitter is one having no motion, the light intensity at a given point and its electrical signal counterpart depend only on the position of the exploring element. Thus, the signal can be regarded as a function of two variables which describe the two-dimensional image field. If the image at any spot involves time variations of light intensity, say due to motion, the luminance signal is a function of time also, and the corresponding electrical signal is a function of three independent variables.

The electrical signal amplitude is defined by a scale that was originally standardized by the IRE (now IEEE). The scale*, used as a convenience in examining television waveforms on an oscilloscope, is illustrated in Figure 15-4.

*This scale can be derived from Figure 7, Reference 2.

**Bandwidth and Resolution.** During the development of television, subjective viewing tests were used to determine that a bandwidth of about 4.2 MHz results in a satisfactory received television image. This conclusion involved the combined evaluation of many parameters such as acceptable vertical and horizontal resolution, frame and field rates, equipment costs, etc. Bandwidth in excess of 4.2 MHz does provide somewhat better performance, but the improvement is considered uneconomical.

Low-frequency transmission requirements are set primarily by the low rate of 60 fields per second. Vertical blanking pulses appear in the complex signal waveform at that rate, and due to the complexity of the signal, there are also sideband components around that frequency. As a result, good transmission response must be provided to nearly zero frequency in order to maintain good phase response at and near 60 Hz.

As mentioned, the required bandwidth was determined by subjective tests. These were, in turn, conducted on the basis of previous judgments regarding the desirable vertical and horizontal resolution that was to be provided. These parameters, bandwidth and resolution, are importantly and intimately related.

*Vertical Resolution.* As previously described, the scanning process produces a standard pattern of 525 horizontal lines per frame. The picture width is 4/3 its height. This ratio is defined as the aspect ratio. Horizontal lines are lost during the vertical blanking period, reducing the effective (visible) number of lines to about 93 percent of the total. Further loss of resolution, inherent in the scanning process, is due to the finite width of the scanning line and to the shape of the scanning spot. The relative position of horizontal image lines and *scanning lines* affects reproduction. In the extreme, if the image has alternate black and white lines of the same width as the scanning lines and coincident with them, a faithful reproduction results; however, if the same scanned lines were centered on the boundary between scanning lines, they would produce a flat gray picture. On the average, this effect decreases vertical resolution to about 70 percent. The net effect, then, is that the number of vertical elements which can be resolved is

$$n_v = 525 \times 0.93 \times 0.7 \approx 342.$$

*Horizontal Resolution.* Horizontal resolution is determined by the highest frequency component that can be resolved along a line. Assume that a simple sinusoid generates a series of black and white dots along the line. If spot size is not limiting, the finest detail that can be resolved is determined by the highest frequency that can be transmitted. If the horizontal resolution is to be about equal to the vertical resolution, the number of picture elements per line scan should be $n_h = 342 \times 4/3 = 456$ (the multiplier of 4/3 is used to account for the aspect ratio).

A sinusoid that would generate 456 alternate black and white dots would go through 228 cycles along a line. As shown in Figure 15-2, the duration of the visible portion of a line scan is about 52.3 $\mu$s. Thus, to satisfy the criterion that horizontal resolution should be about equal to vertical resolution, the top transmitted frequency should be $f_t = 228/52.3 \approx 4.3$ MHz, a value close to that mentioned earlier, 4.2 MHz, obtained from subjective tests.

**Spectrum.** The line scanning rate of a monochrome television signal determines to a great extent the distribution of energy in the signal spectrum. Thus, strong signal components are found in the signal at 15,750 Hz; since the signal waveform is complex, many harmonics of this fundamental are also produced. The spectral distribution is illustrated in Figure 15-5. Each component varies with time by approximately ±3 dB according to picture content and motion; average values are illustrated. No voltage is shown at zero frequency because the amplitude of that component is under design control and is related to the design of the transmission circuits used. However, it is customary to clamp the signal so that the dc component is relatively constant in order to avoid excessive base-line wander [3]. Note that the envelope of the distribution decreases at a rate of 6 dB per octave; i.e., for each doubling of the frequency, the line scan component is about 6 dB lower (one-half the voltage).

The 60-per-second field frequency generated by the vertical blanking pulses also influences signal energy distribution. These blanking pulses produce upper and lower sidebands of 60 Hz and 60-Hz harmonics about each multiple of the line scanning frequency.

**Signal Amplitudes.** In television signal transmission, amplitudes are limited by signal-to-noise and system overload considerations just as in any other form of signal transmission. These limitations are ex-

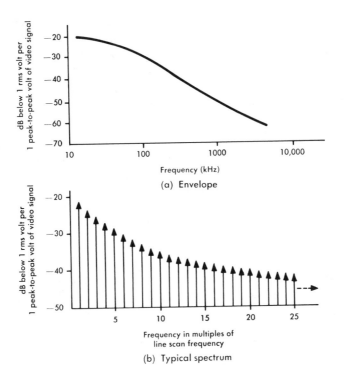

Figure 15-5. Monochrome television signal voltage spectrum.

pressed in terms of average, peak, or single-frequency power at specific TLPs in the telephone system if the signal parameter can be translated to the voiceband. However, TLP relationships cannot usually be directly applied if the signal is wideband (as in television), i.e., if it occupies more than a telephone channel bandwidth.

Generally, television signal amplitude measurements are more conveniently expressed in voltage than in power. It is important, therefore, to recognize the voltage-impedance-power relationships that exist in television circuits and to recognize the complications inherent in properly translating the voltage expressions and their points of application to the TLPs used in telephone system operation.

Television signal transmission in the Bell System is controlled from a television operating center (TOC), where signals are received at baseband from the broadcasters. It is from such a center that signals may be switched to other baseband circuits for local distribution or pick-up and/or to terminal locations for transmission over the microwave radio channels used in the makeup of the television network in the United States. The relation of the TOC to the network is illustrated in Figure 15-6.

In order that the baseband and microwave radio transmission system designs may properly take into account the amplitude-bandwidth-spectral energy distribution relationships, television signal amplitude is maintained at one volt peak-to-peak into 124 ohms at the TOC. While it is never referred to as such, this point is somewhat analogous to the 0-dB TLP in the telephone network. It is sometimes called the 0-dBV point.

### Baseband Color Signals

The National Television System Committee (NTSC) was formed by the television industry during the early 1950s [4] for the purpose

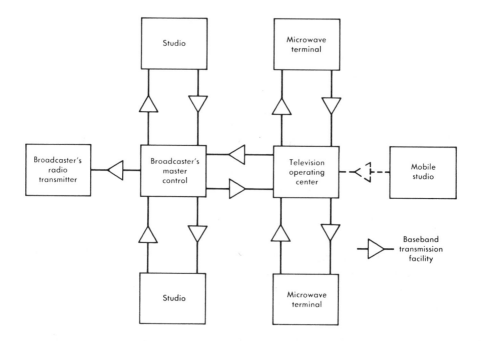

Figure 15-6. Typical intracity television network layout.

of developing a set of color television standards that would be compatible with existing monochrome standards. The signal format that finally evolved is one that superimposes information regarding the image color content on the monochrome luminance signal.

While luminance is transmitted as previously described for a monochrome picture signal, chrominance (hue and saturation) must be coded for transmission and is accomplished by modulating a carrier signal at 3.579545 MHz. The amplitudes and relative phases of this carrier and its sideband components are carefully controlled and together carry the necessary color information.

**Scanning and Synchronization.** The scanning process is identical to that used for monochrome signal transmission. The color carrier at the receiver is synchronized to that of the transmitter by means of a burst of the color carrier frequency superimposed at a reference phase on the back porch of each horizontal synchronizing pulse. The burst signal contains about nine full cycles of carrier frequency (a minimum of eight) as shown in Figure 15-7. The phase relationship of the color carrier and its sidebands with respect to these color bursts determines the hue of the color. A scanned line is also illustrated in Figure 15-7 to show how the color information modifies the normal monochrome luminance signal.

**Spectrum.** Concentrations of energy are found at line scan frequency multiples above and below the color carrier. The color signal carrier, 3.579545 MHz, was chosen as an odd multiple of one-half the monochrome line rate*, so that the color signal components fall in the spectral spaces between components of the luminance signal. The power in chrominance signal components is 10 to 15 dB lower than in the corresponding components of the luminance signal. The spectrum of a color signal is illustrated in Figure 15-8.

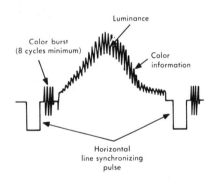

Figure 15-7. Color television signal waveform.

*For color signals, the line rate is 15.734264 kHz. This frequency is used to avoid a high-frequency multiple of 60 Hz.

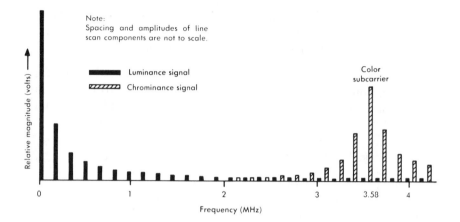

Figure 15-8. NTSC color TV spectrum illustrating interleaving of monochrome and color line scan components.

## Broadcast Signals

Baseband monochrome or color television signals are processed for broadcasting in accordance with standards specified by the FCC [2]. The baseband audio signal, which frequency-modulates a carrier at 5.75 MHz, is added to the baseband video signal; the composite video/audio signal is then used to amplitude-modulate a carrier of assigned frequency in the radio-frequency spectrum. The channel assignments in the RF spectrum, each 6 MHz wide, are shown in Figure 15-9. The assigned carrier frequency for each channel is located 1.25 MHz above the bottom frequency assigned to the channel.

The transmitted signal is a truncated double-sideband, amplitude-modulated signal with transmitted carrier as illustrated in Figure 15-10. The overall transmission plan specified is one involving vestigial sideband transmission. As shown in Figure 15-10, there is no vestigial roll-off shaping at the transmitter. Thus, vestigial shaping must be provided in each television receiver. This mode of transmission tends to optimize overall signal-to-noise performance in the channel.

At present, it is general practice in the Bell System to transmit the audio signal over facilities separate from those used for the video signal. Several methods of combining the signals have been tried, in-

| CHANNEL NO. | FREQUENCY BAND (MHz) | CHANNEL NO. | FREQUENCY BAND (MHz) |
|---|---|---|---|
| 2 | 54-60 | 35 | 596-602 |
| 3 | 60-66 | 36 | 602-608 |
| 4 | 66-72 | 37 | 608-614 |
| 5 | 76-82 | 38 | 614-620 |
| 6 | 82-88 | 39 | 620-626 |
| 7 | 174-180 | 40 | 626-632 |
| 8 | 180-186 | 41 | 632-638 |
| 9 | 186-192 | 42 | 638-644 |
| 10 | 192-198 | 43 | 644-650 |
| 11 | 198-204 | 44 | 650-656 |
| 12 | 204-210 | 45 | 656-662 |
| 13 | 210-216 | 46 | 662-668 |
| 14 | 470-476 | 47 | 668-674 |
| 15 | 476-482 | 48 | 674-680 |
| 16 | 482-488 | 49 | 680-686 |
| 17 | 488-494 | 50 | 686-692 |
| 18 | 494-500 | 51 | 692-698 |
| 19 | 500-506 | 52 | 698-704 |
| 20 | 506-512 | 53 | 704-710 |
| 21 | 512-518 | 54 | 710-716 |
| 22 | 518-524 | 55 | 716-722 |
| 23 | 524-530 | 56 | 722-728 |
| 24 | 530-536 | 57 | 728-734 |
| 25 | 536-542 | 58 | 734-740 |
| 26 | 542-548 | 59 | 740-746 |
| 27 | 548-554 | 60 | 746-752 |
| 28 | 554-560 | 61 | 752-758 |
| 29 | 560-566 | 62 | 758-764 |
| 30 | 566-572 | 63 | 764-770 |
| 31 | 572-578 | 64 | 770-776 |
| 32 | 578-584 | 65 | 776-782 |
| 33 | 584-590 | 66 | 782-788 |
| 34 | 590-596 | 67 | 788-794 |
| | | 68 | 794-800 |

Figure 15-9. FCC radio spectrum broadcast television channel assignments.

cluding one that involves the pulse amplitude modulation of the audio signal and its super-position on the front porch of the video signal. None of these has proved to be satisfactory. Practical means are still being explored so that the two signals may be transmitted on the same facility without undue penalty in cost, performance, or bandwidth.

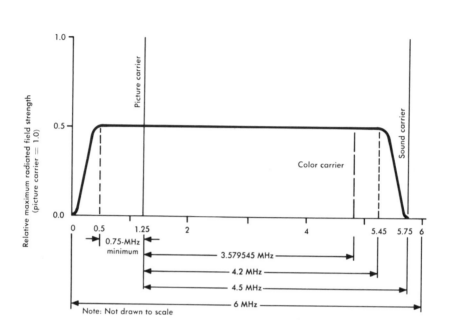

Figure 15-10. Idealized television channel amplitude characteristic — transmitter.

## Closed Circuit Signals

The Bell System transmits a variety of closed circuit signals such as industrial television (ITV) and educational television (ETV) as well as broadcast-type signals transmitted over community antenna television (CATV) systems. Because these are closed circuit arrangements, FCC standards need not always be met in all respects; however, the signal format is sometimes designed to meet FCC standards so that standard television receivers can be used for viewing.

**Industrial Television.** Industrial television systems are always operated on a closed circuit basis; that is, ITV signals are not transmitted over normal broadcast facilities. Because of this, the scanning and synchronizing mechanisms for ITV transmitting and receiving equipment are often less sophisticated than those required for broadcast television, and less operating margin may be provided. Interlaced scanning is often not used because transmission objectives are less stringent than in other types of services. However, the bandwidth provided is usually about equal to that used for broadcast quality service. Amplitude control is provided at a TOC or equivalent when such signals are transmitted over Bell System facilities.

The overall effect is that for transmission analysis, an ITV signal may safely be assumed to be equivalent to a broadcast quality television signal. Audio signals are transmitted only as required and usually not in accordance with FCC standards for broadcast TV signals.

**Educational Television.** While educational television network arrangements may be quite different from standard broadcast TV network arrangements, the receiving equipment is often a standard television receiver; therefore, the signal format is usually identical to the standard signal previously described. Sometimes, ETV signals are transmitted over cable systems which provide six channels. At the viewing locations, carrier-to-baseband converters are used with baseband viewing sets.

**Community Antenna Television.** In CATV systems, broadcast signals are received at a common point and distributed by cable distribution arrangements to CATV subscribers. In such systems the signal format is again generally constrained to the standard broadcast format by the use of standard television viewing sets. Two exceptions are found. First, the distribution system need not have the same total band as is assigned in the radio spectrum although the FCC does prescribe a minimum of 20 channels. Thus, a limited number of channels might be distributed to the CATV subscribers. Second, to ease the design of amplifiers in the distribution system, the sound signal is usually transmitted at a lower amplitude relative to the video signal than in normal broadcast practice. This is accomplished at the antenna location ("head end") of the CATV system by separating the two signals, demodulating the audio signal, and then remodulating and recombining at the new relative amplitudes.

## 15-2 PICTUREPHONE SIGNALS

PICTUREPHONE service is now being introduced in the Bell System. A signal format has evolved that provides satisfactory results, but bandwidth requirements are high. Evaluations of all aspects of this service are continuing with the objective of providing more economical modes of transmission. A brief description of the PICTUREPHONE signal as it is presently constituted is given here [5]. Significant changes in the signal format are likely in the future.

The basic methods of transducing a picture to an electrical signal and then back to picture information are very similar in PICTUREPHONE service and in television. The differences are in the details. The standards that have been established to date provide full motion capability (adequate, for example, for lip reading); resolution is sufficient for a life-like image of the face.

Figures 15-11 and 15-12 depict the baseband PICTUREPHONE signal. One significant difference in the treatment of this signal is that high-frequency energy in the video signal (not in the synchronizing pulses) is pre-emphasized in the transmitting station set and de-emphasized in the receiving station set. As a result, the received signal-to-noise ratio is significantly improved, but there are overshoots in the signal which must be considered when PICTUREPHONE signal transmission analyses are undertaken. Margin must be provided so that these overshoots do not overload carrier systems, and clipping levels must be carefully established so that the picture quality is not excessively degraded.

## Scanning

The scanning process for PICTUREPHONE signals follows that used for television in that the image is scanned from left to right and lines are formed from the top of the image to the bottom. The detailed dimensions of the scanning pattern are given in Figure 15-11. There are nominally 60 fields per second with alternate fields B and A interlaced, thus providing 30 frames per second. Note that there are 125.5 active lines per field and the equivalent of eight horizontal lines per vertical synchronizing pulse interval. The total number of horizontal scans, then, is 267 per frame.

## Modulation

Picture signal modulation of a scanned line is illustrated in Figure 15-12. The relative amplitudes of synchronizing and picture signals are shown on a relative amplitude scale similar to that used for television (see Figure 15-4). However, the peaking effect of

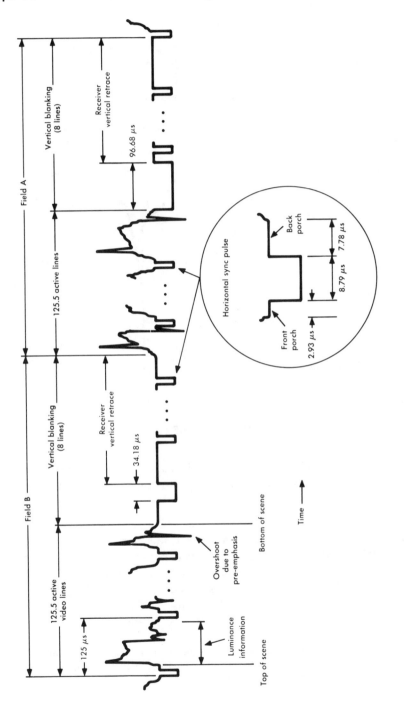

Figure 15-11. Composite PICTUREPHONE video signal showing scanning intervals.

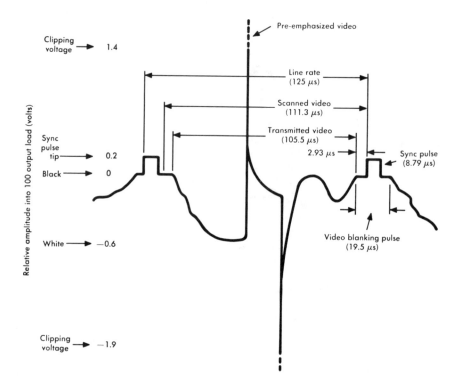

Figure 15-12. PICTUREPHONE signal line format.

pre-emphasis, previously mentioned, is seen to cause signal excursions well in excess of the normal sync-tip-to-reference-white voltage.

## Amplitude

The PICTUREPHONE signal amplitude is controlled at the 0-dB PICTUREPHONE transmission level point (0 PTLP). For this service, the 0 PTLP is defined as the output of the central office loop equalizer in the direction of transmission from the station set to the central office. At the 0 PTLP, the signal is maintained at a nominal value of 0.8 volts peak-to-peak across 100 ohms.*

*This value is subject to interpretation due to the fact that the signal is pre-emphasized at the station set. Amplitudes, such as that given, are based on the equivalent, de-emphasized signal format.

## 15-3   TELEPHOTOGRAPH SIGNALS

In spite of the fact that telephotograph is one of the oldest of the video services in the Bell System, dating back to the early 1930s, and in spite of the fact that little new development work has been done in recent years, many thousands of miles of telephotograph circuits are still in operation. These are used primarily to satisfy the needs of the news photo services [6].

Several different equipment types are currently in use, but the signal format is sufficiently similar that one general description should suffice. The systems of interest operate on private line voice-band circuits that are equalized to meet the necessary transmission requirements for satisfactory picture transmission and reception.

### Scanning

In telephotograph systems, the light beam used to scan the image is held in a constant position. The picture being scanned is wrapped around and fastened to a cylinder which is rotated and advanced axially by a synchronous motor.

The scanning light beam, modulated by the various shades in the picture, is reflected from the image surface into a photoelectric cell. The scanning density and rate achieved by the optical mechanism are 100 lines per inch and 20 inches per second (circumferentially), respectively. One vertical inch of picture, up to 11 inches wide, is scanned in one minute.

Synchronism is achieved by absolute control of the motor speed at the transmitter and at the many receivers which can be operated simultaneously. The control is maintained by a 300-Hz tuning fork which is housed in a temperature-controlled enclosure. The accuracy of synchronization is maintained to within a few parts per million.

The holding bar used to clamp the image picture in place on its cylinder is made of highly polished metal. The high-amplitude pulse resulting from the reflected light from this bar is used to make the initial adjustment of the receiver to start the receiver motor in step with the motor at the transmitter.

## Bandwidth

The scanning and modulation processes just described are accomplished by the intensity modulation of a light beam which is itself varied by a sine wave function at a rate of 2000 Hz in some systems and 2400 Hz in others. The resulting modulated electrical signal at the output of the photoelectric cell is then transmitted as a double-sideband signal (2000-Hz carrier) or a vestigial-sideband signal (2400-Hz carrier). Double-sideband signals are transmitted at a somewhat lower rate than that given previously which applied to the VSB mode of transmission. The band must be gain and delay equalized between 1000 Hz and 2800 Hz. This useful band, 1800 Hz wide, is capable of producing the resolution demanded by the scanning rates given.

### REFERENCES

1. Glasford, G. M. *Fundamentals of Television Engineering* (New York: McGraw-Hill Book Company, Inc., 1955).

2. Federal Communications Commission. *Rules and Regulations*, Vol. III (March 1968), pp. 234 B, C, and D, and 259-267.

3. Doba, S., Jr. and J. W. Rieke. "Clampers in Video Transmission," *Trans. AIEE*, Part I, Vol. 69 (1950), pp. 477-487.

4. Fink, D. G., Editor. "Color Television Standards," *Selected Papers and Records of the National Television System Committee* (New York: McGraw-Hill Book Company, Inc., 1955).

5. Baird, J. A. et al. "The PICTUREPHONE System," *Bell System Tech. J.*, Vol. 50 (Feb. 1971).

6. Reynolds, F. W. "A New Telephotograph System," *Bell System Tech. J.*, Vol. 15 (Oct. 1936), pp. 549-573.

Chapter 16

# Mixed Signal Loading

The signals described in Chapters 12 through 15 are found in various facilities of the Bell System, but seldom is any one type of signal the only one ever found at any given location or facility. In some instances, a variety of signal types are simultaneously transmitted on the same facility; in other instances, the type of signal transmitted on the facility changes with time. Broadband carrier systems, for example, may simultaneously carry speech, narrowband and wideband data, facsimile, PICTUREPHONE, and address and supervisory signals. A single trunk, on the other hand, may carry speech, data, facsimile, and address and supervisory signals at different times.

Important combinations or mixtures of signals must be characterized primarily so that the composite signal may be properly related to overload phenomena in carrier transmission systems. In some cases, overload effects are relatively minor, causing only partial deterioration of performance. However, the effects are accentuated with increased signal amplitudes so that transmission may be seriously impaired and, ultimately, the entire system may fail. Any study of mixed signal loading must then be concerned with the characterization of signals known to be transmitted simultaneously in today's environment. In addition, the characterization must be in terms that permit continuous re-evaluation of signals to account for the effects of new technology, the introduction of new instrumentalities or new services, or the implementation of new policies such as interconnection with customer-provided equipment.

Since a very large number of combinations of signal loads may occur in broadband systems, it is extremely difficult to characterize mixed signal loading effects explicitly. Intermodulation phenomena, whose effects are accentuated as signal amplitudes increase, can produce noise, crosstalk, or other distortions such as signal compression. These are due to the nonlinear input/output characteristics of active circuits and devices, such as the amplifiers used to compensate for media losses in transmission systems. All transmitted signal components interact to form new, unwanted signals. In some cases, these combine to form a noise-like impairment. In other cases, components of the interference signal fall directly upon and in phase with the components of a particular wanted signal so that its internal magnitude relationships become distorted due to the fact that some components are more distorted (compressed) than others. Sometimes, signal components combine with pilot or control frequencies to fall into other channels as intelligible crosstalk.

## 16-1 MIXED SIGNALS AND OVERLOAD

As previously mentioned, the effects of intermodulation are sometimes relatively minor and result in only partial deterioration of performance; but as amplitudes increase, transmission may be seriously impaired. To avoid this, the signal load on a system must be carefully controlled and limited to well-defined maximum values. When distortion or noise results from excess amplitudes and performance is seriously impaired, a transmission system is said to be *overloaded*. If the effects are so serious that communication is impossible, the phenomenon is sometimes called *hard overload*. Often, the effects of overload must be evaluated statistically. Signal and system characteristics interact in ways that are strongly dependent on how long and by how much a signal exceeds its nominal value, how the system responds to the signal, and how quickly the system recovers from the overload condition.

Signal load criteria have been established to guard against overload. In general, the simplest statement of these criteria for signals transmitted in the Bell System is that the long-term average power in any 4-kHz band shall not exceed −16 dBm0. The statistical properties of individual types of signals are applied to specific situations to allow higher amplitudes for short time intervals. The most important of these statistical properties are the variations of signal amplitude with time and the activity factors that may properly be applied to various modes of transmission for each signal type — simplex, half duplex, or duplex.

The same broad criterion is applied to signals requiring more than a 4-kHz frequency allocation, but it is expressed somewhat differently.

In this case, the total power in the signal may be no greater than the long-term average power resulting from a signal of −16 dBm0 in each of the displaced 4-kHz channels. Again, the random nature of the broadband signal amplitude variations sometimes permits excess amplitudes for short periods of time or in restricted portions of the band.

Since many of the more serious problems relating to system loading are experienced in wideband systems capable of transmitting signals in 600-channel blocks (a mastergroup) or more, the following considerations of signal loading are mastergroup-oriented. For comparison purposes, the mastergroup *speech* signal load is first analyzed, and mixed signal loads are then compared to this analysis. For systems wider or narrower than one mastergroup, the same approach may be used by extrapolation and with appropriate care in the treatment of variables that are functions of system capacity.

## 16-2   MASTERGROUP SPEECH SIGNAL LOAD

Consider a 600-channel mastergroup loaded with speech signals only. To determine overload relationships, it is necessary to know the average power and the peak power in the composite signal. The average power at 0 TLP (that exceeded no more than 1 percent of the time during the busy hour) may be computed by using Equation (12-5):

$$P_{av} = V_{0c} - 1.4 + 0.115\sigma^2 + 10 \log \tau_L + 10 \log N + \Delta_{c1} \quad \text{dBm0.}$$

From Figure 12-4, the average value of $V_{0c}$ for toll calls is −16.8 vu. Since this value must be converted to its equivalent value at 0 TLP, allowances of −3 dB for toll connecting trunk loss (VNL + 2.5 dB) and + 2 dB for conversion from the outgoing toll switch (−2 dB TLP) are added. Thus, $V_{0c} = -17.8$ vu and, from Figure 12-4, it has a standard deviation of 6.4 vu. A standard deviation of 1 dB is also allowed for toll connecting trunk loss variations. Thus, $\sigma = \sqrt{6.4^2 + 1^2} = 6.47$ vu, and $0.115\,\sigma^2 = 4.8$ vu. The activity, $\tau_L$, is taken as 25 percent; thus $10 \log \tau_L = -6$ dB. The number of channels, $N$, is 600; $10 \log 600 = 27.8$ dB. The value of $\Delta_{c1}$, which accounts for the maximum number of active channels, is found in Figure 12-6 to be +0.7 dB for a 600-channel mastergroup. If these values are summed, the average power in a mastergroup of 600 telephone channels *carrying speech only* is found to be

$$P_{av} = -17.8 - 1.4 + 4.8 - 6 + 27.8 + 0.7 = +8.1 \quad \text{dBm0.}$$

The peak power can be determined from the average power by adding a correction factor, $\Delta_{c2}$, which represents the rms signal amplitude exceeded 0.001 percent of the time. Thus,

$$P_{max} = P_{av} + \Delta_{c2} \qquad \text{dBm0.} \qquad (16\text{-}1)$$

For $N = 600$ channels and $\tau_L = 25$ percent, the number of active channels exceeded no more than 0.001 percent of the time is found from Figure 12-6 to be $N_a = 175.55$. From Figure 12-7 the corresponding value of $\Delta_{c2}$ is about 13 dB. The 0.001 percent peak, then, is found from Equation (16-1) as

$$P_{max} = P_{av} + \Delta_{c2} \approx +8.1 + 13 \approx +21.1 \qquad \text{dBm0.}$$

Note in Figure 12-7 that for large numbers of active channels $(N_a \geqq 100)$, $\Delta_{c2}$ approaches a constant value of about 13 dB. This relationship permits the use of Gaussian noise, which has the same peak factor, as an excellent simulation for a busy hour, multichannel telephone load for large systems.

The peak power capacity required of a transmission system $(P_s = P_{max} - 3 \text{ dB})$ is usually expressed in terms of the peak power of a single-frequency sinusoid applied at 0 TLP [see Equations (12-6) and (12-7)].

## 16-3 MASTERGROUP MIXED SIGNAL LOAD

In the present plant, no mastergroup can be expected to carry speech signals only. Thus, it is necessary to consider the effects of mixing various other types of signals with speech signals.

### Speech and Idle Channel Signals

The signalling system most commonly used on trunks employing broadband carrier facilities is the 2600-Hz SF system described in Chapter 13. When a trunk that is so equipped is idle, 2600 Hz is transmitted continuously as a supervisory signal at an amplitude of $-20$ dBm0. Thus, theoretically, a mastergroup may carry 600 such randomly phased idle channel signals (translated to carrier frequencies). In this case, the total average power in the mastergroup would be equal to $- 20 + 10 \log 600 = +7.8$ dBm0, about the same as the previously determined average power in a mastergroup carrying only speech signals. However, activity factors affect the mastergroup signal in a complicated manner when the combined speech and idle channel signal load is considered.

**Average Mastergroup Power.** It was shown previously that the busy-hour speech (only) load is $+8.1$ dBm0 if the speech activity factor, $\tau_L$, is 0.25. That value of $\tau_L$ is, in turn, based on a trunk efficiency factor, $\tau_e$, of 0.7. Thus, 180 trunks [$600\ (1 - 0.7)$] may carry idle channel supervisory signals. The power in these signals totals $-20 + 10 \log 180 = +2.6$ dBm0. When this power is combined with the speech power by power addition (Figure 3-5), the total busy-hour load may be found as 2.6 "$+$" $8.1 \approx 9.0$ dBm0.

While the ratio of speech and idle channel signals in a mastergroup varies from the busy hour to the nonbusy hour, the total power in a mastergroup remains relatively constant (from a maximum of 9.0 dBm0 to a minimum of 7.8 dBm0).

To illustrate the way in which the 600-channel mastergroup loading varies with time, consider the load when the busy-hour effect has been reduced so that the equivalent speech load is that of a 300-channel system. The speech load for $N = 300$ channels may be computed by the same method as that previously used; its value is found to be $+5.3$ dBm0. It is assumed that the other 300 channels carry idle channel signals; the power in these signals is $-20 + 10 \log 300 = +4.8$ dBm0. The remaining 300 channels are subject to the trunk efficiency factor, $\tau_e = 0.7$. Thus an additional 300 $(1 - 0.7) = 90$ channels carry idle channel signals. The power in these additional signals is $-20 + 10 \log 90 = -0.5$ dBm0. The total power is thus 5.3 "$+$" 4.8 "$+$" $(-0.5) = 8.7$ dBm0.

**Average Channel Power.** In Chapter 12, it was shown that the long-time average (or rms) load per channel in a broadband toll system is about $-20.5$ dBm0 when speech signals only are considered. The mastergroup power for speech was found to be about $+8$ dBm0 during the busy hour. When single-frequency signal loading is included with the speech load, the average mastergroup busy-hour load is 1 dB higher, $+9$ dBm0. Therefore, it may be concluded that the long-time rms channel load is also 1 dB higher, or about $-19.5$ dBm0. This value is the long-term average channel power based on the average speech volume of $-16.8$ vu for toll calls measured in the 1960 survey, which also indicated a tendency for volumes to increase slightly on longer toll calls [1]. Thus, the speech load on long-haul systems could increase by about 1 dB. Even with the 1-dB allowance there appears to be some margin between the speech load with present station sets and the design objective of −16 dBm0. In the future this margin may be used to permit somewhat greater transmitter efficiency in new telephone sets, particularly on longer loops. Therefore, long range planning should be based on the assumption that both speech and data signals eventually may have a long-term average of approximately −16 dBm0.

**Maximum Mastergroup Power.** In a mastergroup made up only of speech signals, the value of power that is exceeded 0.001 percent of the time may be found by Equation (16-1),

$$P_{max} = P_{av} + \Delta_{c2} \qquad \text{dBm0.}$$

It is now desirable to determine this maximum power value for a mastergroup having a speech and idle channel signal load. It has been shown that the average value of a composite busy-hour mastergroup load is +9 dBm0. A value of $\Delta_{c2}$ for the composite signal remains to be determined.

As shown in Figure 12-7, $\Delta_{c2}$ is equal to about 13 dB for speech signal loads in excess of about 75 active channels. It can be shown [2] that the instantaneous value of the sum of $n$ sine waves of equal amplitude, different frequencies, and random phase relationships also exceeds the rms value of the $n$ signals by about 13 dB 0.001 percent of the time for values of $n$ in excess of 100. Thus, in a combined signal, one part consisting of 300 channels containing speech signals and one part consisting of 300 single-frequency supervisory signals, each having approximately the same peak factor, it can be safely assumed that the total also has a peak factor of 13 dB. Thus, the maximum power in a mastergroup carrying 300 speech signals and 300 idle channel signals is

$$P_{max} = 9 + 13 = +22 \qquad \text{dBm0.} \qquad (16\text{-}2)$$

This result depends on the assumption that the 2600-Hz signals are randomly related to one another in phase. Suppose, for example, that the phases of the 300 single-frequency signals assumed in developing Equation (16-2) were coherent so that their peak amplitudes coincided 0.001 percent of the time (an unlikely event). The peak voltage of the 300 signals then would be 300 times the peak voltage of one, and the peak power would be $300^2 = 90,000$ times the power of one such signal. Thus, the peak power would be

$$-20 + 10 \log 90,000 = -20 + 49.5 = +29.5 \qquad \text{dBm0.}$$

This is a peak value 7.5 dB higher than that of Equation (16-2), a value that would surely be expected to cause overload.

Phase coherence would not be expected under past field operating conditions and design practices. A multiplicity of 2600-Hz oscillators and a random physical association of signalling equipment and transmission multiplexing equipment caused a dispersion of frequency and phase relationships that prevented any significant effects due to coherence. More recent trends in the layout of office equipment and in equipment design practices provide fixed wiring patterns between a single supervisory signal generator and many carrier channels. Special wiring patterns with controlled phase reversals must be designed to guarantee partial cancellation of composite signal peaks.

### Speech and Address Signals

Address signals are transmitted at amplitudes higher than the −16 dBm0 long-term average power objective for a channel, but the statistics of these signals do not cause the average power objective to be exceeded. High-amplitude speech signal bursts of short duration also occur; these are limited to a maximum of +3 to +10 dBm0, depending on the system, by channel terminal equipment. The limiting has negligible effect on the individual speech signal; the distortion is masked by other distortions such as that in the carbon transmitter of the station set.

Under normal operating conditions, these high-amplitude address and speech signals do not seriously affect carrier system operation. Their amplitudes are low compared to the total signal power in a mastergroup (+9 dBm0), and/or the frequency of occurrence is so low that such signals do not usually cause trouble. However, abnormal operating conditions or system designs which change the statistical relationships can lead to serious overload troubles related to the transmission of these high-amplitude signals. The maladjustment of channel equipment, operating errors (for example, the improper application of a test tone to a circuit), or the improper maintenance of a carrier system (which might result in some frequencies being transmitted at much higher amplitude than the designed values) are all trouble conditions to guard against. A system design which requires the use of shaped TLP characteristics, such as the pre-emphasis used in microwave radio systems and the signal shaping used in coaxial systems, results in peak factors that are equivalent to those in systems of fewer channels than are provided in the design (see Chapter 12-3). Thus, peak factors are higher, the effective signal band is smaller, and even a single channel carrying an inordinately high signal can cause system overload.

## Speech and Data Signals

Carrier system speech channels are frequently used for the transmission of data signals. These signals are sometimes tranmitted over trunks which are parts of the switched public network or switched private line networks and sometimes over dedicated, point-to-point private line circuits. Any of these circuits may involve interconnection with customer-provided equipment through a Bell System connecting arrangement. The control of signal amplitudes tends to vary somewhat depending on the source of the signal.

As discussed in Chapter 14, the maximum amplitudes of voiceband digital data signals are specified not to exceed —13 dBm0 when averaged over a 3-second interval. Allowing for channel activity factors and a mix of duplex and half-duplex operation, the resulting long-term average should not exceed —16 dBm0. Modern systems are designed to operate satisfactorily over a range of data and speech signal combinations that meet the —16 dBm0 objective. However, a heavy concentration of private line duplex data channels applied to a given system could cause this average to be exceeded and should be avoided by dispersion of these channels over several systems.

Data signals, as transmitted over carrier systems, may be regarded simply as single-frequency signals insofar as their overloading effect is concerned. Therefore, a peak factor of 13 dB for multichannel systems (mastergroup or higher) may be safely assumed, since the peak factors for all contributors — speech, data, and supervisory signals — are equal.

## Speech and Video Signals

While several broadband carrier systems were initially designed to carry a combined signal consisting of voice and broadcast television signals, none of them are so used today primarily because of difficulties encountered in controlling intermodulation products between the two types of signal. Therefore, the characterization of such a combined signal is of no consequence and is not discussed.

PICTUREPHONE service has not yet been provided in any significant amount on systems that transmit combinations of speech and analog PICTUREPHONE signals. Studies are now under way to determine how such signals interact and how the combined signals may affect the systems over which they are transmitted.


## 16-4  SYSTEM-SIGNAL INTERACTIONS

The characterization of signals in this chapter has been presented to relate the average and peak powers of the signals to carrier system overload. In some cases, system type, design, operation, or maintenance interacts with the transmitted signal to change its characteristics so that overload effects may be accentuated or mitigated.


### System Misalignment

Multirepeatered broadband analog systems are designed so that the gain of a repeater compensates for the loss of the preceding section of the transmission medium. Because the compensation is not perfect, the signal amplitudes depart from their nominal values. These departures are called misalignment; they may be positive at some frequencies, causing the signals to be higher than nominal, and negative at other frequencies, causing the corresponding signal components to be lower than nominal. The effects on signal characteristics can be negligible for small misalignment, or they may be quite significant. The analysis of such effects is similar to that relating to the shaped TLP concept of Chapter 12.


### Carrier and Pilot Signals

The discussion of phase coherence among single-frequency supervisory signals and the importance of guaranteeing random phase relationships among such signals applies also to carrier signals (such as may be transmitted in a DSBTC system) or to single-frequency pilot signals. Special wiring designs must sometimes be used to introduce phase reversals in order to produce partial cancellation of signal peaks.

## Compandors

The advantage in individual channel signal-to-noise performance gained by using syllabic compandors results from the fact that the range of signal amplitudes is substantially reduced for transmission over the medium. The reduction, called the compression ratio, is usually on the order of 2 to 1; a signal amplitude range of 50 dB is reduced to 25 dB, and its standard deviation is also reduced by a factor of 2 to 1 from 6 to 3 dB. The average signal amplitude is a system design parameter. Thus, in a given cable carrier system using compandors, the value corresponding to $V_0$ for a non-compandored channel must be selected by the designer to optimize performance. The optimization must take into account the fact that the use of compandors generally results in a higher average power per channel.

Sometimes, when the multiplexed signal of a compandored carrier system is applied as a portion of the signal to another system of higher capacity which does not normally use compandors, precautions must be taken to avoid overload in the higher capacity system. Two effects must be taken into account: one is the higher average power in a voice channel due to the compandor action; the other is the high power represented by the transmitted carriers in most Bell System compandored systems. The requirements of the high-capacity system may be met by reducing its channel capacity, by lowering the amplitude of the total applied compandored signal load, or by lowering the relative amplitudes of the carrier components. In some instances, some of the compandor advantage may be lost. This loss is usually not important because the high-capacity systems are less noisy (by design) than the systems for which compandors are provided.

## TASI

Time assignment speech interpolation, discussed briefly in Chapter 2, has as its principal effect on broadband signal characterization an increase in the activity factor for each speech circuit. The amount of increase is a function of the TASI system design, which must take into account the number of trunks between terminals, the number of lines being served, the ratio of the two, the allowable degradation due to freeze-out, and the syllabic content of the language being used. (Freeze-out is an effect that leads to clipping of initial speech bursts due to the fact that all trunks are active and therefore busy.) In TASI systems, the trunk activity factor may increase from 25 or 30 percent to 90 or 95 percent.

## Microwave Radio Systems

The Federal Communications Commission specifies that the frequency deviation of a frequency-modulated microwave carrier be confined to the allocated band [3]. Modern microwave systems, while designed to carry a long-term average per-channel signal power load of −16 dBm0, are tested by noise loading techniques with a load equivalent to −15 dBm0 per 4-kHz channel. This approach provides some margin against peak excursions of composite signals which produce the extremes of the microwave frequency deviations. At the same time, signal-to-noise objectives are met for the system when operated at a load value equivalent to −16 dBm0 per channel.

### REFERENCES

1. McAdoo, Kathryn L. "Speech Volumes on Bell System Message Circuits — 1960 Survey," *Bell System Tech. J.*, Vol. 42 (Sept. 1963), pp. 1999-2012.

2. Bennett, W. R. "Distribution of the Sum of Randomly Phased Components," *Quarterly of Applied Mathematics*, Vol. 5 (Jan. 1948), pp. 385-393.

3. Federal Communications Commission. *Rules and Regulations*, Vol. 2 (Aug. 1969).

Section 4

Impairments and Their Measurement

The next six chapters contain descriptions and definitions of impairments suffered by telecommunication signals as they are transmitted through various channels and media. These descriptions, qualitative for the most part, are related to the sources of impairment, the manner in which the impairments are measured, and the units in which the measurements are expressed.

Signal transmission is subject to impairment by a number of imperfections in channels; thus, signal impairment may be regarded as resulting from channel impairment. The channel imperfections include interferences induced from external sources, interferences that are signal-dependent and caused by nonlinear channel input/output characteristics, distortions of the channel transmission characteristics, and indirect effects such as timing and synchronization errors.

Transmission irregularities affect various types of signals differently. Consider, for example, the transmission of a variety of signals over imperfect voiceband channels. In one case, carrier signal generating equipment may produce jitter which is scarcely noticeable in speech signal reception but which causes disastrous impairment of voiceband data signals. In another case, the channels under consideration may have impedance discontinuities that produce intolerable echoes from the point of view of speech signal transmission but which may have negligible effects on data signal transmission.

Sometimes impairments may be dealt with in the design process. For instance, if a particular type of signal is intolerant of frequency shift, the effect can be eliminated by using a method of transmission that permits recovery of the transmitted carrier frequency and phase. Signals sensitive to frequency shift impairment should usually not be transmitted by suppressed-carrier methods.

410

Occasionally, somewhat degraded performance may be tolerated in time of trouble. Such a situation would be typified by an increase in noise, for example, while major troubles are being corrected. The ability to furnish some service, even though below normal, may be preferable to furnishing no service at all.

Some impairments are evaluated subjectively and others objectively. Sometimes the same impairment may be evaluated both ways depending on the type of signal to be transmitted. For example, random noise must be evaluated in terms of its annoying effect in speech or video signal transmission and in terms of the number of errors it causes in digital data signal transmission. In either case, the ultimate expression must be in terms that can be stated quantitatively so that meaningful values may later be established for objectives and requirements.

Chapter 17 deals with various types and sources of noise and crosstalk. Some forms of noise are induced in transmission channels from external sources by a number of different coupling mechanisms. Some have their sources within the channel of interest. Some forms of noise are independent of the signal transmitted, while others are functions of the transmitted signal.

Signal transmission may be seriously impaired by departures from desired amplitude/frequency channel characteristics. These characteristics and some departures from ideal are discussed in Chapter 18 in relation to typical signals and channels.

A discussion of timing and frequency synchronizing relationships is found in Chapter 19. Impairments that relate to such functions are seldom controlling in baseband systems. They appear in carrier systems or when signals are otherwise processed by time or frequency functions.

Chapter 20 discusses the effects of echoes in a telephone message channel. Echoes cause serious impairment to speech and telephotograph signals. The impairing effects are a complex combination of echo amplitude and the amount of delay difference between the signal and its echo.

Delay distortion has a number of adverse effects on signal transmission, particularly on video and data signals. Chapter 21 treats the impairments caused by delay distortion in channels of any bandwidth that may carry signals sensitive to delay distortion.

imate deterioration of signals occurs when transmission
by a failure. Chapter 22 considers the design, construction,
ıd operation of the transmission plant from the point of
its reliability. Protection switching arrangements, emer-
toration, and diversity of routing are subjects of discussion.
tion of transmission need not be total, however. Poor main-
ractices and/or inadequate support equipment and support
ıay bring about a gradual circuit deterioration that causes
ıts to increase with time. These subjects are also discussed
· 22.

Chapter 17

# Noise and Crosstalk

The transmission of telecommunications signals is degraded by the limitations of practical channels and by the existence of various types of interference in the channel of interest. Interference may be induced from a source *outside* the channel of interest (e.g., power line noise picked up by a voice-frequency circuit), or it may be generated from *within* (e.g., intermodulation noise caused by non-linear input/output characteristics of repeaters in an analog transmission system).

The effects of some interferences depend on the type of signal affected. For example, bursts of impulse noise are usually of little consequence in the transmission and reception of speech signals because of the use of amplitude limiters and the relative insensitivity of the human ear to this type of impairment. However, impulse noise can seriously impair digital signal transmission. Some other types of interference, such as thermal noise, degrade the transmission of all types of signals. Unwanted signals and interferences, their sources, means of controlling the sources and coupling paths, the nature of the impairments incurred, and the methods of measurement are all basic to an understanding of impairments suffered in the transmission of signals in the telecommunications network.

## 17-1 COUPLING

Almost all transmission circuits are exposed to external influences and forces by virtue of proximity to other circuits. For example, a loop or trunk is usually physically close to other circuits in cables or on pole lines; multiplexed message channels share a wide bandwidth; any circuit passing through a central office is exposed to sizable switching transients; and many circuits have power transmission lines paralleling part of their routes. The exposure to electromagnetic fields created by the currents in these nearby circuits results in many possible interference coupling paths from a *disturbing* circuit to the circuit of interest, the *disturbed* circuit.

413

## Currents and Circuit Relationships

The extent of interference caused by coupling is dependent on the symmetry, or balance, of the disturbed circuit and on the type of current (longitudinal or metallic) that results from the coupling.

Currents that flow in the same direction in the two conductors of a pair of wires are called *longitudinal currents*, while currents that flow in opposite directions in the two conductors are called *metallic currents*. Both types are illustrated in Figure 17-1. The voltage sources, $E$ with internal impedance $Z_G$, are coupled to the transmission line by some form of coupling mechanism designated as $Z_c$. The resulting currents are transmitted through the central office, typically through common battery circuits which include impedances $Z_s$ as illustrated, to the load impedance, $Z_L$. The $Z_s$ impedances may represent a number of components such as supervisory relays, transformers, common battery supply leads, etc.

In Figure 17-1(a), the currents through $Z_L$ are exactly equal and opposite (net zero) if all the networks and the transmission line are exactly balanced, i.e., electrically alike and symmetrical with respect to ground. In this event, an interference voltage coupled as in Figure 17-1(a) causes no interference current in $Z_L$. However, if the $Z_c$ and $Z_s$ networks are not balanced, unequal currents flow in the two sides of the circuit. The difference between them is a metallic current that appears in the load, $Z_L$, as interference. This metallic component of the current can be represented as originating in an equivalent circuit like that of Figure 17-1(b), where $E$ may represent the source of unbalance current, some system-generated interference, or a wanted signal source.

The common battery does not cause appreciable unbalance because the internal impedance of a central office battery is extremely low, typically a small fraction of an ohm. However, other parts of the connection, those represented by impedances $Z_s$, are often unequal and and create an unbalanced circuit unless carefully controlled.

The circuits and currents of Figure 17-1 may be defined in terms of balanced and unbalanced conditions since the grounds located between impedances $Z_s$ and between impedances $Z_c$ provide references with respect to which circuit symmetry can be evaluated and direction of current flow can be determined. If the disturbed circuit has no ground or common return, a plane of symmetry cannot be established, and longitudinal currents are not generated.

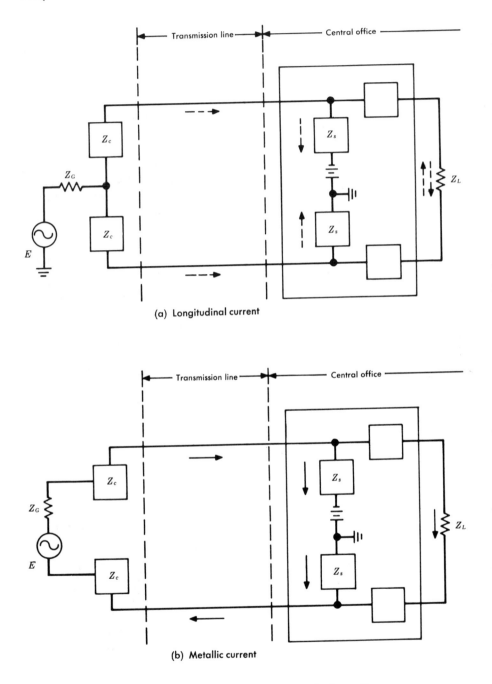

**Figure 17-1. Currents in a two-wire telephone line.**

## Coupling Paths and Their Control

The coupling path from a disturbing to a disturbed circuit may result from electromagnetic, electric (direct), or intermodulation phenomena. These phenomena and the resulting coupling paths are typical of interference problems found in telecommunications systems. Coupling path losses must be controlled so that transmission impairments may be held to tolerable values.

**Electromagnetic Coupling Paths.** A coupling path is involved when an electromagnetic field resulting from an alternating current carried in one conductor causes a voltage to be induced in another conductor [1, 2]. The induced voltage may result from either the magnetic or the electric field, whichever is dominant.

Figure 17-2 illustrates a simple and idealized case of *magnetic coupling* where A is one conductor of a disturbing circuit equidistant from the two conductors of disturbed circuit B. The magnetic field produced by current $I_0$ in conductor A induces voltages $E_1$ and $E_2$ in B. The resulting longitudinal currents, $I_1$ and $I_2$, in the two conductors of the disturbed circuit are exactly equal and of opposite polarity. Thus, they cancel one another, and there is zero net metallic current; i.e., $I_3 = 0$. Departures from the idealized conditions assumed in Figure 17-2 may cause the induced currents in the two

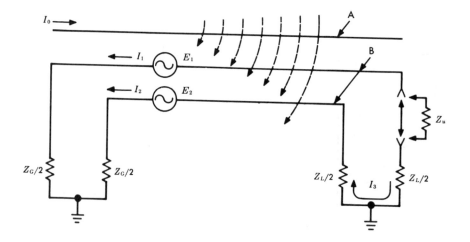

Figure 17-2. Circuit relationships — magnetic coupling.

conductors of the disturbed circuit to be different in magnitude, phase, or instantaneous direction of flow; the result would be a net metallic current in the disturbed circuit. This may occur if the couplings from conductor A to the conductors of circuit B are unequal or if the impedances to ground in the disturbed circuit are unequal, as would be the case if $Z_u$ were in the circuit. In either case, $I_3 \neq 0$.

Among the measures employed to control the magnetic coupling between circuits is shielding. Others include separation, orientation, and balance. Magnetic shielding is employed in braided or other forms of iron shields to cover the copper conductors to be used for the transmission of signals that are particularly susceptible to magnetically induced interferences. An example is the use of specially shielded wires for intracity television baseband signal transmission. More commonly, shielding is used on circuit packs having areas of high component density, especially where magnetic components (inductors and transformers) are used. Where transmission lines are shielded against magnetic coupling, the shields (such as cable sheaths) must be grounded at both ends.

*Electric field coupling* is the result of capacitance between adjacent parallel conductors. This form of coupling, resulting from the electric field produced by voltages in the disturbing circuit, is among the most important and most prevalent in communication systems. Capacitive coupling loss is a maximum at low frequency and tends to decrease at a rate of 6 dB per octave of frequency.

If the capacitance from a disturbing conductor to each of the two wires of a disturbed circuit is different, the current flowing in each of the disturbed conductors is different, and a resultant metallic current flows. Similarly, if the two currents are equal but the impedances of the two disturbed conductors are unequal (for example, due to a shunt impedance to ground on one conductor only), a resultant metallic current also flows.

Many of the guidelines that govern the relationships between disturbing and disturbed paths in magnetic coupling apply to electric field coupling also. Electrostatic shielding of conductors (by iron, copper, or aluminum shields), separation between disturbing and disturbed circuits, orientation of one circuit with respect to another, and balance of the impedances affect the control of electric field coupling paths.

**Electric Coupling Paths.** Electric coupling paths are those produced by impedances that are common to two or more otherwise independent circuits. These paths, which include common batteries and their supply leads and terminal multiplex equipment filters whose passbands are adjacent or overlap, are normally controlled in design; they are little affected by system application and maintenance.

The manner in which the common battery impedance can become a source of interference is illustrated by Figure 17-3. In Figure 17-3(a), two telephone station sets are shown with independent connections to trunks or other station sets. The two local station sets, STA 1 and STA 2, receive current from the common battery supply having internal impedance $Z_b$. The connections from the station sets are over loops and through central office battery supply and supervisory circuits designated $Z_s$. The transmission circuits are coupled by transformers. In Figure 17-3(b), the station set, loop, transformer, and supervisory circuit impedances of the two circuits are lumped together as $Z_1$ and $Z_2$. The battery impedance is shown as $Z_b$, and the transmitter at STA 1 is shown as a voltage generator, $E$. This simplified schematic shows clearly how STA 1 and STA 2 are coupled by $Z_b$. The following numerical example illustrates how interference from STA 1 to STA 2 is limited, or controlled, by maintaining $Z_b$ at a low value.

## *Example 17-1: Electric Coupling Path Control*

To simplify the example, assume that

    (a)  $Z_1$ and $Z_2$ in Figure 17-3(b) are both 1000 ohms.

    (b)  the interfering current, $I_2$, resulting from the voltage, $E$, must be at least 80 dB below the desired current, $I_1$; i.e., $20 \log I_1/I_2 \geqq 80$ dB.

What value of $Z_b$ will produce this result?

The voltage relationships in the right-hand mesh of Figure 17-3(b) may be written, by using Kirchoff's second law as shown in Equation (4-3), as

$$1000\, I_2 - (I_1 - I_2)\, Z_b = 0 \quad ,$$

or

$$\frac{I_1}{I_2} = \frac{1000 + Z_b}{Z_b} \quad .$$

If $20 \log I_1/I_2 \geqq 80$ dB, $I_1/I_2 \geqq 10{,}000$ .

Thus, $Z_b \lesssim 0.1$ ohm.

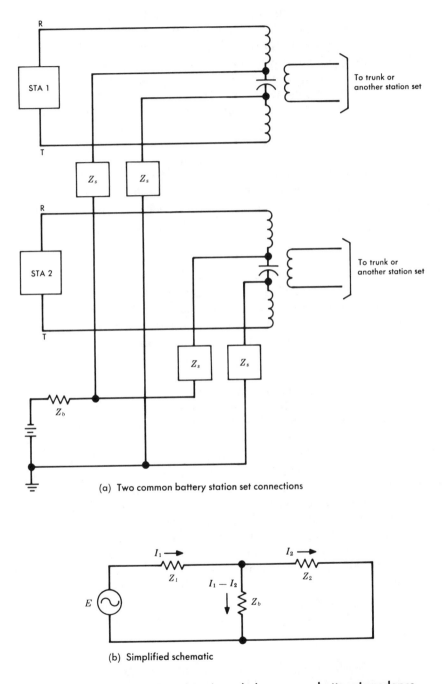

(a) Two common battery station set connections

(b) Simplified schematic

**Figure 17-3. Electric coupling through the common battery impedance.**

While somewhat oversimplified, this example illustrates how two circuits may be coupled by a common impedance such as $Z_b$ and how the coupling may be controlled in design by holding the common impedance to a much lower value than the coupled transmission circuit impedances. In a real case, the internal battery impedance tends to be very low, typically less than 0.01 ohm, and coupling becomes a problem only where long power leads are common to a number of otherwise independent circuits. Their impedance adds to $Z_b$ to give an effective value, $Z'_b$, that can introduce excessive coupling. In this case, the complex impedances of all the involved circuits must be taken into consideration. Sometimes, decentralized battery filters must be used to bypass the interference currents to ground. Although the common dc impedance remains high, these filters reduce the complex impedance at signal frequencies to low values.

In a large office, many interference currents such as $I_1 - I_2$ in Figure 17-3(b) are carried in the common battery. Even though each is small, there may be several thousand such signals simultaneously present. Together, they form a significant source of noise that must be carefully controlled by battery lead layout and appropriate filtering.

Figure 17-4 illustrates how carrier system terminal equipment filters may provide an electric coupling path. Two 4-kHz baseband input connections are shown at the left. The input signals modulate carriers at frequencies $f_{c1}$ and $f_{c2}$ (where $f_{c2} = f_{c1} + 4$ kHz) and produce double-sideband signals over bands ±4 kHz about the carriers. These double-sideband signals then pass through bandpass filters which pass their lower sidebands and suppress their upper sidebands. At the filter outputs, the two signals are combined. Note that the suppressed upper-sideband signal from Input 1 falls directly into the band occupied by the lower-sideband signal of Input 2. This form of coupling can sometimes be avoided by selecting carrier frequencies such that the overlap does not occur; however, bandwidth is wasted. Control is usually attained by designing the filters so that adequate suppression is obtained and interference currents are at an acceptably low amplitude, 50 to 80 dB below the wanted signal currents.

**Intermodulation Coupling.** The coupling that results from intermodulation among signals in an FDM carrier system cannot be described

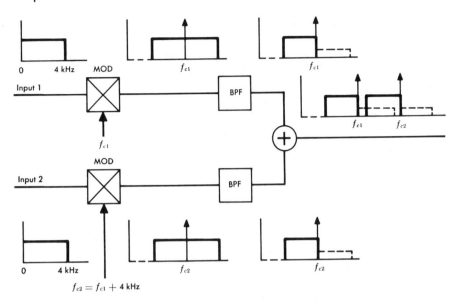

Figure 17-4. Electric coupling in multiplex terminal equipment.

in the same physical sense as the electromagnetic and electric coupling paths just discussed. Intermodulation is inextricably involved in the mathematics of the nonlinear input/output characteristics of the devices; therefore, a detailed discussion of this type of coupling is deferred to a later part of this chapter.

## 17-2  INDUCED NOISE AND CROSSTALK

The sources of many types of noise and crosstalk are outside the disturbed channel. Such interferences are coupled into the disturbed channel by coupling mechanisms and through the types of coupling paths discussed above. Here, some specific induced interferences and methods of control are described.

### Power System Noise

Problems involving inductive coupling arise where facilities for the power industry and the communications industry share the same underground or pole line environment. Communications channels carrying signals having components that extend down to or close to zero frequency are particularly susceptible to interference from the high-strength magnetic and electric fields generated by power systems.

**Nature of Impairments.** The characteristics of the 60-Hz wave present in most power distribution systems are high energy, high harmonic content (especially odd harmonics), and large differences between the currents carried by different conductors of the power line, i.e., high unbalance currents. The 60-Hz fundamental component is at such a low frequency that speech transmission is seldom impaired since most voice-frequency circuits and all carrier circuits have high attenuation at 60 Hz. However, high-amplitude odd harmonics of 60 Hz often cause an unpleasant hum in voice-frequency speech transmission systems.

The 60-Hz and harmonic components can also cause bar pattern interferences in television, PICTUREPHONE, or other video channels. Video signals require a flat attenuation/frequency response to essentially zero frequency; thus extraneous signals at low frequencies can cause picture impairment.

Data signals transmitted at baseband and requiring good response at low frequencies may also be seriously impaired by 60-Hz and harmonic interference. The impairment takes the form of increased error rate.

**Inductive Coordination.** This term is applied to the cooperative efforts of the power industry (represented by the Edison Electric Institute), other utilities, and the Bell System to solve problems that arise where facilities for different types of service share the same environment. In considering problems of interferences in communication circuits due to coupling from power circuits, three conditions are considered. These are influence, coupling, and susceptibility. *Influence* refers to those characteristics of power circuits and associated apparatus that determine the character and intensity of the fields they produce. *Coupling* covers the electric and magnetic interrelations between power and communication circuits. *Susceptibility* refers to the characteristics of communication circuits and associated apparatus, which determine the extent of any adverse effects from nearby power circuits. These three conditions form the basis of inductive coordination.

A large unbalance of the currents carried on the individual conductors of a multiphase power distribution circuit is a source of *influence* on communication circuits. A reduction of power line influence may be accomplished by transposing the power line conductors and by balancing load currents. The influence reduction results from a cancellation of fields caused by the unbalanced currents. In addition to reducing influence, load balancing makes power distribution more efficient and, therefore, more economical.

The distance between power and communication circuits and their mutual orientation are among the factors that must be considered in order to control *coupling*. On shared facilities the separation between the potentially interfering source and each communication circuit conductor must be as large as practicable, and the distance between the communication paired conductors must be as small as possible. When power and communication circuits must cross one another, a 90-degree crossing minimizes coupling between the lines.

The *susceptibility* of communication circuits can be reduced in a number of ways. The proximity of conductors subject to power line influence is effectively accomplished by twisting the conductors of each pair together where possible. Twisting of pairs is specified in the design of multiconductor cables and some open-wire pairs. This tends to equalize the distance from the conductor pair to each power conductor so that induced voltages are made nearly equal and metallic interference currents are minimized. Where twisted pairs cannot be used, the disturbed conductors are transposed, or frogged, at regular intervals to reduce susceptibility. Impedance balance to ground of the disturbed circuit is then an effective deterrent to the conversion of longitudinal to metallic current. Finally, good shielding, i.e., maintaining cable sheath continuity, offers some protection to communication circuits when they are in cables.

Some coordination problems are structural and, as such, may result in danger to personnel or communication equipment from high energy coupled from the power source into the communications circuits. While most systems are 60-Hz ac, some high-voltage dc power systems are also in use. Stray direct currents from the latter may cause noise or corrosion problems if structural problems and appropriate grounding arrangements are overlooked.

## Impulse Noise

Impulse noise consists of spikes of energy of short duration which have approximately flat spectra in the band of interest. The flat spectra are shaped by channel response characteristics so that, typically, the average spectrum of a large number of observed impulses approximates the frequency response of the channel on which the measurements are made [3]. Some of the more important sources of impulse noise are: (1) corona discharges in transmission lines simultaneously powering remote repeaters, (2) lightning, (3) electrical and electroacoustic transients associated with the termination of a call at a station set, (4) relay operations associated with switching and alarm functions, (5) microwave radio fading phenomena and the associated protection switching operations, and (6) many other transmission system operating and maintenance procedures.

In speech transmission, impulse noise causes little impairment. Above a certain threshold value, acoustic shock caused by impulse noise might be painful; however, circuits are designed to limit amplitudes well below the threshold, and acoustic shock is seldom experienced. At amplitudes below the limiting values, the human ear is quite tolerant of impulse noise.

In video transmission, the impairment takes the form of short-duration interferences such as small light or dark flickers in the picture (sometimes called pigeons) or, in extreme cases, a short-duration loss of synchronization that causes the picture to tear horizontally or to roll vertically. While all of these impairments are objectionable, such events are tolerated if they occur only occasionally.

The most serious effect of impulse noise is found in digital signal transmission. The interfering impulses are short compared with the time between them and, as a result, the receiving circuits resolve them as independent events. Depending on the impulse amplitude, polarity, duration, and time of occurrence, individual signal components or blocks of data symbols may be obliterated, resulting in errors in the received signal.

## Single-Frequency Interference

Single-frequency signals or wideband signals having discrete single-frequency components can be excessively annoying when coupled into a telephone channel. If they are of sufficient amplitude and fall between 200 and 3500 Hz, they may produce audible tones in the telephone receiver. Single-frequency interferences can also disturb SF signalling systems, produce bar patterns in video receivers, and cause high error rates in data transmission systems.

## Crosstalk

*Crosstalk* was initially used to designate the presence in a telephone receiver of unwanted speech sounds from another telephone conversation. The term has been extended in its application to designate interference in one communication channel or circuit caused by signals present in other communication channels. Consideration is limited here to the interference to one signal by another signal of the same general type—e.g., speech interfering with speech, video with video, digital with digital, etc. It should be pointed out, however, that crosstalk coupling of one signal type to another is often significant in establishing transmission level points for communication systems.

**Crosstalk Coupling and the TLP.** As discussed in Chapter 3, many of the complexities of system design and operation are made tractable by the concept of TLPs. The importance of any coupling path is strongly dependent on the relative magnitudes of the wanted signal and the unwanted interference. It is really a question of the signal-to-interference ratio, which is difficult to define in telephone practice; the TLP approach is found to be a useful way of dealing with signal-to-interference problems.

With this approach, the interference is defined in terms of its value at a specific TLP, and the coupling is expressed as equal level coupling loss (ELCL). The ELCL is defined as the ratio of signal power at some known TLP in the disturbing circuit to the induced power measured at an equal TLP in the disturbed circuit. The ELCL concept is illustrated in Figure 17-5 where two repeatered voice-frequency circuits are depicted as transmitting from left to right. A coupling path having 80-dB loss is shown from the output of the repeater in the disturbing circuit to the input of the repeater in the disturbed circuit. Thus, the coupling path loss must be adjusted by the gain of the repeater in the disturbed circuit (30 dB) to give a value of ELCL of 50 dB. Note that the ELCL is independent of signal amplitude and of the particular TLP used in its determination.

**Near-End, Far-End, and Interaction Crosstalk Coupling.** These forms of coupling, abbreviated NEXT, FEXT, and IXT, respectively, are subclasses of coupling modes that exist between communication channels or circuits. Although any of these forms of crosstalk coupling may cause impairment, NEXT and FEXT tend to be predominant. With NEXT coupling, the interference energy in the

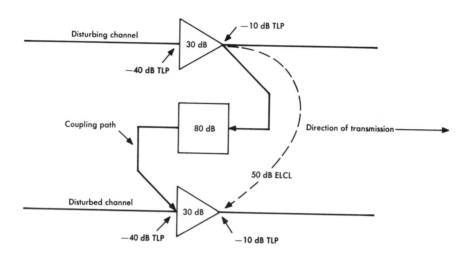

Figure 17-5. Coupling path loss and equal level coupling loss.

disturbed circuit is transmitted in the direction opposite to that of the signal energy in the disturbing circuit. With FEXT coupling, the signal and interference travel in the same direction. Interaction crosstalk occurs when energy is coupled to a tertiary path, propagates along that path, and then is coupled to the disturbed circuit. The two stages of coupling may be of the near-end or far-end type or a combination of both.

**Speech Crosstalk.** When an unwanted speech signal is coupled into another speech channel, the interfering signal may be intelligible or it may be unintelligible but have syllabic characteristics so that a listener thinks it may be intelligible. Such interferences are particularly objectionable because of the real or fancied loss of privacy. Even when they are clearly not intelligible, they tend to be highly annoying because of the syllabic content. Stringent objectives to minimize these interferences are applied in transmission systems.

When many unwanted speech signals appear in a disturbed channel simultaneously, each at such a low amplitude that neither intelligence nor syllabic variations are conveyed, the net effect may resemble random noise. In rare circumstances, the metallic coupling of large numbers of speech signals through common battery circuits in a central office might produce such an impairment. However, coupling losses through battery feed circuits are kept at high values by design.

**Video Crosstalk.** When a picture signal is coupled to another video channel, the interfering picture signal may be superimposed on the disturbed picture receiver. This form of interference is rare, however. The two signals are usually not synchronized, and the coupling path usually has a loss/frequency characteristic that distorts the interfering signal. A more common effect of such a coupling is known as the *windshield wiper* effect. The synchronizing pulses of the disturbing signal create a bar pattern across the disturbed picture. Since the two signals are normally unsynchronized, the bar pattern moves across the picture with a windshield wiper effect.

**Digital Signal Crosstalk.** As in most cases of interference to digital signals, crosstalk produces errors. Below some threshold essentially no errors are made, although a disturbing signal amplitude just slightly higher than the threshold value causes a very sharp increase in error rate.

In the design of digital transmission systems, the crosstalk due to the presence of the line signals of many systems in one cable is often the limiting factor in the spacing of regenerative repeaters.

## 17-3  SYSTEM-GENERATED NOISE AND CROSSTALK

Many sources of noise and crosstalk exist within a channel or within a transmission system. Such interferences are controlled by circuit design and, within the constraints of a particular design, by signal amplitude manipulation since the ultimate interfering effect is a matter of the signal-to-noise ratio. If interferences are independent of the signal amplitude, the signal-to-noise ratio is improved by raising the transmitted signal amplitude. If interferences are signal-dependent, the signal-to-noise ratio is generally improved by reducing the transmitted signal amplitude because this type of interference generally changes more rapidly than the signal amplitude. Thus, performance optimization in analog transmission systems involves the selection of optimum signal amplitudes because both types of noise are usually present.

### Random Noise

Random noise is an impairment that appears in all circuits as a result of physical phenomena that occur within the affected circuit or channel. The complete characterization of random noise types that are commonly found in transmission channels is beyond the scope of this chapter; but the important noise sources are briefly described, the resulting impairments to various telecommunication signals are discussed, and methods employed to measure and evaluate these types of interferences are described [4].

The terms *white noise* and *Gaussian noise,* often used to describe random noise characteristics, must be defined. The term white noise has become well established to mean a uniform distribution of noise power versus frequency, i.e., a constant power spectral density in the band of interest. The Gaussian noise distribution, discussed in Chapter 9, is the limiting form for the distribution function of the sum of a large number of independent quantities which individually may have a variety of different distributions. A number of random noise phenomena produce noise having this Gaussian amplitude distribution function.

By definition, Gaussian noise has a finite probability of exceeding any given magnitude, no matter how large. In practice, however, consideration can sometimes be limited to the magnitude attained 0.01 percent of the time. Thus, it is convenient to define the peak factor for random noise having a Gaussian distribution at 3.89 $\sigma_N$ where $\sigma_N$ is the rms value of the noise.* The peak factor is then 11.8 dB above the rms value (usually rounded to 12 dB for convenience). In cases where the value attained 0.001 percent of the time must be used, the peak factor is 13 dB.

Several types of signal independent random noise are encountered sufficiently often to warrant discussion. These include *thermal noise, shot noise, 1/f noise,* and *Rayleigh noise.* Since each of these types of noise has random characteristics, the total power in multiple sources, such as those encountered in multirepeatered analog transmission systems, may be computed simply by summing the powers. There is no amplitude or phase correlation between components from independent sources.

**Thermal Noise.** According to the kinetic theory of heat, electrons in a conductor are in a continual random motion which leads to an electrical voltage whose average value is zero but which has ac components of random amplitude and duration. The phenomenon produces an interference signal called thermal noise.

---

*The noise has an average value of zero, having random positive and negative excursions. The rms value can be shown as equal to the standard deviation, $\sigma_N$.

It is shown [5, 6] that for thermal noise the available noise power is

$$p_N(f) = kT \text{ watts/Hz} \qquad (17\text{-}1)$$

where k = Boltzmann's constant = $1.3805(10^{-23})$ joule/K, and $T$ is the absolute temperature of the thermal noise source in Kelvins. At room temperature, 17°C or 290 K, the available noise power is $p_N(f) = 4.0(10^{-21})$ watts/Hz or $-174.0$ dBm/Hz.

In theory, the thermal noise spectrum eventually drops to zero; actually, however, it is flat over all frequencies of practical interest from zero to the highest microwave frequencies used and can be termed white noise. Also, the available noise power is directly proportional to bandwidth and absolute temperature. Thus,

$$p_a = kTB \text{ watts} \qquad (17\text{-}2)$$

where $B$ is the bandwidth of the system or detector in hertz. This may be expressed in dBm as

$$P_a = -174 + 10 \log B \qquad \text{dbm.} \qquad (17\text{-}3)$$

While thermal noise has a flat power spectrum and a Gaussian amplitude distribution, it should not be concluded that white and Gaussian are synonymous; they are not.

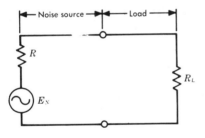

Figure 17-6. Equivalent circuit of a noisy resistor.

Sometimes it is desirable to determine the voltage generated by a thermal noise source. This may be accomplished by considering an equivalent circuit like that of Figure 17-6. The equivalent circuit assumes a noise voltage generator, $E_N$, in series with a hypothetically noiseless resistor of $R$ ohms. If this noise source is connected to a load resistor, $R_L$, and if $R_L$ is equal to $R$, the maximum power will be delivered to $R_L$. This maximum deliverable power is $p_a = E_N^2/4R$. As previously shown, the available noise power from a thermal noise source is $p_a = kTB$. Equating these two powers and solving for the rms voltage of the equivalent Thevenin generator yields

$$E_N = \sqrt{4kTBR} \qquad \text{volts.} \qquad (17\text{-}4)$$

**Shot Noise.** This type of random noise is found in most active devices. It is similar to thermal noise in that it has a Gaussian distribution and a flat power spectrum. However, it differs from thermal noise in the following two respects:

(1) The magnitude of thermal noise is proportional to absolute temperature, whereas shot noise is not directly affected by temperature.

(2) The magnitude of shot noise is proportional to the square root of the direct current through the device. Thus, the shot noise magnitude may be a function of signal amplitude if the signal has a dc component.

For fixed conditions in a particular design, it is often convenient to combine shot noise with thermal noise into a single equivalent noise source. The way in which the two combine depends on the particular circuit arrangement.

Shot noise may be computed as an rms current by

$$i = \sqrt{2qI} \qquad \text{ampere} \qquad (17\text{-}5)$$

where $q$ = charge of the electron = $1.6(10^{-19})$ coulomb, and $I$ is the direct current through the device.

**Low-Frequency ($1/f$) Noise.** This noise is associated with contact and surface irregularities in semiconductors and in the cathodes of electron tube devices. The noise has a Gaussian distribution, and in a given band (between $f_1$ and $f_2$) may be computed by

$$p = \int_{f_1}^{f_2} \frac{K}{f}\, df = K(\ln f_2 - \ln f_1) \qquad . \qquad (17\text{-}6)$$

Evaluation of the constant $K$ depends on specific devices and circuit conditions. The evaluation given by Equation (17-6) would result in infinite noise if the band were to extend down to zero frequency or up to infinite frequency. This is not to be expected, and the equation holds only for finite bandwidths which do not extend to either extreme.

**Rayleigh Noise.** When the bandwidth of the circuit or channel under consideration is small compared with its midband frequency, the noise in the band is considered as narrowband noise. If the noise has a Gaussian distribution, it appears to have the characteristic of a midband sinusoidal carrier modulated by a low-frequency signal whose highest frequency component is dependent on the bandwidth. The result is a noise which, when detected, has an envelope with a Rayleigh amplitude distribution.

Subjectively, there is little distinction between noises having a Gaussian or Rayleigh distribution. However, the peak factor of the Rayleigh distribution, the value exceeded 0.01 percent of the time, is 9.64 dB; this is more than 2 dB below that for a noise having a Gaussian distribution. This must sometimes be taken into account in circuit design or in system performance evaluation.

## Intermodulation Noise and Crosstalk

Intermodulation, caused by nonlinear input/output characteristics of analog system repeaters, may result in many different types of interference, all of which are in some way signal-dependent. The process is very complex and has many variables that need not be fully evaluated here [4]. However, a brief review of the principles is given.

The nonlinear characteristics may be expressed as a power series having an infinite number of terms. Usually, terms higher than third order are small enough to be ignored and the equation is written

$$e_o = a_0 e_i^0 + a_1 e_i^1 + a_2 e_i^2 + a_3 e_i^3 \quad . \qquad (17\text{-}7)$$

Consider an input signal, $e_i = A \cos \alpha t + B \cos \beta t + C \cos \gamma t$. If this signal is substituted in Equation (17-7) and expanded by trigonometric substitution, many interference frequencies are found in the output in addition to the wanted signals. All these components are given in Figure 17-7 except the dc term, $a_0 e_i^0$, which is usually of little interest; it is filtered out, in most cases, and causes no interference.

**Random Intermodulation Noise.** If all the signals involved in the intermodulation phenomenon are speech signals, the result is an interference very similar to random noise. If a signal is carried

**FREQUENCIES AND RELATIVE MAGNITUDES TO BE FOUND IN OUTPUT, $e_o = a_1 e_i^1 + a_2 e_i^2 + a_3 e_i^3$, FROM APPLIED SIGNAL, $e_i = A \cos \alpha t + B \cos \beta t + C \cos \gamma t$**

| | TERM 1 | TERM 2 | TERM 3 |
|---|---|---|---|
| dc | | $\frac{1}{2} a_2 (A^2 + B^2 + C^2)$ | |
| First order | $a_1 A \cos \alpha t + a_1 C \cos \gamma t + a_1 B \cos \beta t$ | | $\frac{3}{4} a_3 A (A^2 + 2B^2 + 2C^2) \cos \alpha t$ $+ \frac{3}{4} a_3 B (B^2 + 2C^2 + 2A^2) \cos \beta t$ $+ \frac{3}{4} a_3 C (C^2 + 2A^2 + 2B^2) \cos \gamma t$ |
| Second order | | $\frac{1}{2} a_2 (A^2 \cos 2\alpha t + B^2 \cos 2\beta t + C^2 \cos 2\gamma t)$ $+ a_2 AB [\cos(\alpha+\beta) t + \cos(\alpha-\beta) t]$ $+ a_2 BC [\cos(\beta+\gamma) t + \cos(\beta-\gamma) t]$ $+ a_2 AC [\cos(\alpha+\gamma) t + \cos(\alpha-\gamma) t]$ | |
| Third order | | | $\frac{1}{4} a_3 (A^3 \cos 3\alpha t + B^3 \cos 3\beta t + C^3 \cos 3\gamma t)$ $+ \frac{3}{4} a_3 \begin{cases} A^2 B [\cos(2\alpha+\beta) t + \cos(2\alpha-\beta) t] \\ A^2 C [\cos(2\alpha+\gamma) t + \cos(2\alpha-\gamma) t] \\ B^2 A [\cos(2\beta+\alpha) t + \cos(2\beta-\alpha) t] \\ B^2 C [\cos(2\beta+\gamma) t + \cos(2\beta-\gamma) t] \\ C^2 A [\cos(2\gamma+\alpha) t + \cos(2\gamma-\alpha) t] \\ C^2 B [\cos(2\gamma+\beta) t + \cos(2\gamma-\beta) t] \end{cases}$ $+ \frac{3}{2} a_3 ABC [\cos(\alpha+\beta+\gamma) t + \cos(\alpha+\beta-\gamma) t + \cos(\alpha-\beta+\gamma) t + \cos(\alpha-\beta-\gamma) t]$ |

*Note:* Observe that if in the applied signal $A = B$, then the amplitude of the $\alpha + \beta$ product, which is at the frequency $\alpha + \beta$, is 6 dB greater than the $2\alpha$ product. Similarly, $\alpha - \beta$ is 6 dB greater than the $2\alpha$ product, and $2\alpha - \beta$ (and similar terms) are 9.6 dB greater than $3\alpha$. If $A = B = C$, then the $\alpha + \beta - \gamma$ term and similar terms (but do not confuse with $2\alpha - \beta$ type) are 15.6 dB greater than $3\alpha$. The compression, or first-order component, arising from the $e_i^3$ term is at least 9.6 dB greater than $3\alpha$ and may be much greater, depending on the number of signals applied; for the three-frequency input given above, it is 23.5 dB greater. If the $a_n$'s are functions of frequency, the frequency effects must be added to the aforementioned effects to determine the amplitude differences between products.

Figure 17-7. Expansion of power series for three-sinusoid input.

in a speech channel, each fundamental signal (cos $\alpha t$, cos $\beta t$, etc.) may be considered as a band of energy 4 kHz wide. As a result, it can be seen that the frequency band of the intermodulation product (the interference) is 8 kHz wide for the second-order products shown in Figure 17-7 and 12 kHz for the third-order products. Thus, more than one channel can be disturbed by the interferences.

If a broadband signal has a large number of fundamentals, the number of disturbing products that can be produced is very large. For example, in a system of 10,000 channels, a disturbed channel may have well over one million third-order products. The probabilistic combination of the large number of contributors, together with the basic characteristics of each fundamental speech signal, generates an interference that is Gaussian in its amplitude distribution and has a flat power spectrum over the band of a disturbed channel.

Many other detailed characteristics of speech signals must be evaluated in determining the effects of random intermodulation noise. In addition to the speech signal characteristics, system characteristics must also be considered. For example, in analog cable systems, modulation products of different types accumulate from repeater to repeater according to different laws which are determined by the phase correlation between repeater sections. In microwave radio systems, the intermodulation phenomenon is as important as in cable systems but results from different basic causes. Intermodulation noise in AM systems is a function of signal amplitude, but in FM systems it is a function of the frequency deviation. In AM systems, the noise results directly from the nonlinear input/ output characteristic of amplifiers as illustrated by Equation (17-7). In FM systems, it results from gain and phase deviations in the transmission medium. The end result, provided the number of channels in the system is large, is essentially the same — a nearly flat spectrum of noise having a Gaussian distribution (see Chapters 12 and 16).

**Intermodulation Crosstalk.** The transmission of FDM signals over analog transmission systems produces interchannel coupling which may also yield intelligible crosstalk in a disturbed channel.

The nature of the coupling mechanism may be demonstrated by considering a simple illustration of two signals, $A$ cos $\alpha t$ and $B$ cos $\beta t$, transmitted simultaneously through a repeater whose input/output characteristic may be represented by the truncated power series of

Equation (17-7). The first term on the right side of Equation (17-7) is a dc term. The second term yields the wanted output signal, a reproduction of the input signal, $e_i^1$, multiplied by the gain, $a_1$. The third term (together with higher order terms) represents the inherent nonlinearity of analog repeaters. In this illustration, it is the third and fourth terms which can produce intelligible crosstalk through intermodulation.

For example, if the input signal is

$$e_i = A \cos \omega_1 t + B \cos \omega_2 t \quad , \tag{17-8}$$

the third term on the right-hand side of Equation (17-7) becomes

$$e_3 = a_2 e_i^2 = a_2 (A \cos \omega_1 t + B \cos \omega_2 t)^2$$

$$= a_2 (A^2 \cos^2 \omega_1 t + 2AB \cos \omega_1 t \cos \omega_2 t + B^2 \cos^2 \omega_2 t). \tag{17-9}$$

Trigonometric expansion of the cosine terms in Equation (17-9) results in the following expression:

$$e_3 = \frac{a_2 A^2}{2} + \frac{a_2 A^2}{2} \cos 2 \omega_1 t + a_2 AB \cos (\omega_1 + \omega_2) t$$

$$+ a_2 AB \cos (\omega_1 - \omega_2) t + \frac{a_2 B^2}{2} + \frac{a_2 B^2}{2} \cos 2 \omega_2 t \quad . \tag{17-10}$$

It can be seen then that unwanted signals have been produced at four frequencies [$2 \omega_1$, $(\omega_1 + \omega_2)$, $(\omega_1 - \omega_2)$, and $2 \omega_2$], all different from the input signals at $\omega_1$ and $\omega_2$. These unwanted signals are interferences to signals transmitted at the corresponding frequencies.

If it is now assumed that the signal at radian frequency $\omega_2$ is a single-frequency sinusoid and if the signal represented by frequency $\omega_1$ is a speech signal, the interference may take the form of intelligible crosstalk, or it may be inverted in frequency and seem to be intelligible, or it may be more nearly like thermal noise. Its characteristics depend on the signal components that form the intermodulation product, the frequency orientation of the product, and the number of other such interferences falling simultaneously into the disturbed channel.

Computation of interference signal amplitudes and evaluation of higher order terms in the power series expression, Equation (17-7), are covered in Volume 2. Control of this mode of coupling is primarily a matter of system design to suppress nonlinearities or their effects or to avoid them by a suitable choice of frequency allocations, followed by proper operation and maintenance.

## Digital Signal Noise Impairments

The various impairments and coupling modes discussed apply to digital systems and signals as well as to analog systems and signals. The nature of the impairments may differ somewhat, but the basic phenomena of interference generation and coupling are similar.

Many interferences to the digital signal of a digital transmission system are essentially nullified by the process of regeneration discussed briefly in Chapter 14. In this process, each pulse of the line signal arrives at a regenerative repeater with various impairments produced in one repeater section only. The function of the repeater is to restore the pulse to its original form and amplitude and thus eliminate the impairments incurred. When this is accomplished with few errors, system performance is good. As errors increase, system performance deteriorates very rapidly; therefore, adequate margin must be provided to keep error rates low.

One type of noise impairment is unique to the transmission of analog signals over digital systems. The noise, called quantizing noise, is introduced during the process of digitally encoding an analog signal such as a speech signal. It results from the assignment of a finite number of quantum steps chosen to limit the number of codes needed to represent the range of signal sample amplitudes that must be transmitted. A sample is transmitted precisely only when its value corresponds exactly to a quantum step value. Otherwise, the transmitted value may be in error within $\pm V_s/2$, where $V_s$ is the quantum step amplitude range. The noise can be reduced to an arbitrarily small value by reducing $V_s$ (and thus increasing the number of code steps). However, this increases the required bit rate (and bandwidth) or decreases the capacity of a fixed bit-rate system. Thus, it is economical to allow quantizing noise to be as large as tolerable.

When multiple terminals can be connected in tandem to establish a connection, each coding-decoding process encountered produces quantizing noise that increases with the number of tandem terminals. The total noise allowed must be allocated among the tandem-connected terminals, in effect limiting the allowable noise per terminal or, in another sense, limiting the number of terminals that may appear in a built-up connection.

Another aspect of quantizing noise to consider in the design of terminal equipment is the size of quantum steps relative to the range of amplitudes to be encoded. If uniform steps are used, the percent quantizing error is greater for small signals than for large signals, thus degrading the relative signal-to-noise ratio for small signals. It is desirable to use an increasing number of quantum steps of decreasing size as the analog signal amplitude decreases so that the percent error remains relatively constant over the expected range of amplitudes. This may be accomplished either by using a complementary nonlinear encoder-decoder arrangement or by using a linear encoder-decoder preceded by a compressor and followed by a complementary expandor. Practical systems now in use employ the former method, which is, in effect, the application of an instantaneous compandor.

Since quantizing noise is only present when a signal is present, it must be measured in the presence of a signal. The technique used is similar to that previously described for compandored systems (C-notched noise measurement) in which a holding tone is transmitted and attenuated at the input to the noise measuring instrument [7, 8].

Several other forms of distortion arise in the terminal equipment as a result of coding processes. These include harmonic distortion, which may be caused by overload or by poor compandor tracking, and foldover distortion, which may occur if the high-frequency channel cutoff is set at too high a value [9].

## 17-4 NOISE AND CROSSTALK MEASUREMENTS

In the measurement of interferences and coupling path losses, many factors must be considered; these include the purpose of the measurement, the parameter to be measured, the units in which the results are to be expressed, the instrumentation, and the procedure to be followed. Ultimately, measurements of impairments must be related meaningfully to the objectives or requirements that have been established.

There are two general purposes for measuring interferences coupled from sources outside the channel of interest. The first purpose is to determine the magnitude of the interference in the disturbed circuit or circuits, irrespective of the source or of the coupling mechanism. This type of measurement might be made to evaluate power hum, common battery supply noise, or impulse noise. Until the magnitude of the problem is evaluated, the mode of coupling and means for reducing the interference are of secondary importance. The second purpose is to determine the coupling loss between a disturbing and disturbed circuit. This measurement establishes the fact that a suspected source of interference involves a particular combination of disturbing and disturbed circuits and determines the increase in coupling loss needed to cure the problem.

### Parameters and Units — Noise Measurements

In evaluating interferences, electrical power is most commonly measured although voltage or current is occasionally measured. As discussed in Chapter 3, power measurements are usually expressed in decibels relative to one milliwatt (dBm) or in decibels relative to reference noise, weighted or unweighted (dBrn or dBrnc). Further, such expressions are often referred to 0 TLP and are expressed as dBm0, dBrn0, or dBrnc0. Often, the measurement of a single-frequency interference, such as a power-frequency harmonic, is made in dBm and later translated into dBm0 or dBrnc0. Some wave analyzers designed for such measurements are calibrated directly in dBrn. Some interferences which cover a broad spectrum are measured in the voiceband in dBrnc and translated into dBrnc0. The measurement may be made in dBm if the interference is being evaluated for wideband signal impairment. If the interference has impulse noise characteristics, the measurement must account in some way for interference amplitude and frequency of occurrence. The measurement is often expressed in counts per minute, an evaluation of the average number of impulses measured in excess of a threshold value. The threshold depends on the type of signal for which the interference is being evaluated and, of course, on the TLP at which the measurement is made.

## Speech Crosstalk Measurements

Because of subjective effects special consideration is given to crosstalk between speech circuits. The many parameters to be evaluated include the number of exposures to crosstalk, the volumes of interfering speech signals, the coupling loss for each coupling path, the gains and losses of each of the involved circuits (which led to the concept of *equal level coupling loss*), and the hearing acuity of the listener in the presence of noise.* Of these, only the coupling loss is subject to design or operating control.

**Coupling Loss Measurements.** The measurement of crosstalk coupling loss, as the name implies, is a loss measurement, one usually expressed simply in dB.† As implied, a test signal must be applied in the disturbing circuit at a known frequency and amplitude and then measured in the disturbed circuit. It is also implicit that disturbing and disturbed circuits can both be uniquely identified. Unless the coupling is intermodulation, the frequency is the same in both; if the coupling is intermodulation, the frequency may be shifted. Thus, measurements of the received test signal amplitude at the shifted frequency may be necessary, and suitable signal generation and detection equipment is required. Coupling loss measurements are often made across the spectrum of interest because the coupling loss is often a function of frequency.

The concepts of NEXT, FEXT, and IXT couplings are applied most often to crosstalk problems arising from parallel transmission lines (e.g., pairs in the same cable). In voice-frequency circuits, where the predominant coupling is usually FEXT and capacitive, the coupling loss tends to decrease at a rate of 6 dB per octave of frequency. It has been found that where smooth coupling of this type exists, a single-frequency measurement of coupling loss may be made at 1 kHz; from this measurement a good approximation to the effective coupling loss over the voice band may be determined by subtracting

---

*The efficiencies of the telephone transmitter and receiver are implied in the measurements of crosstalk volume and listener acuity.

†A unit occasionally used in crosstalk computations is the dBx; it is equal to 90 minus the measured coupling loss. The use of this unit is sometimes considered convenient because, as coupling becomes tighter (lower loss), the number ⋅⋅ increases rather than decreases. Thus, as crosstalk increases (less coupling values in dBx increase.

2 dB from the measured value [10]. Coupling loss of the FEXT type is also a function of the transmission line length. The loss decreases directly with the length of exposure.

The NEXT coupling loss tends to be independent of path length and decreases with frequency at a rate of 4.5 dB per octave. The effects of multiple couplings of the same combination of circuits, of multiple couplings from different disturbing sources, of frogging, and of several other coupling types of lesser importance are covered elsewhere [11].

The crosstalk coupling between speech circuits in FDM carrier system terminal equipment and that resulting from intermodulation in analog system repeaters tends to be relatively constant across a speech channel and, as a result, a single-frequency measurement is often sufficient to determine the coupling loss.

**Crosstalk Index.** Speech crosstalk coupling parameters and objectives have been condensed into *generalized crosstalk index charts* shown in Figures 17-8 through 17-14. These charts permit graphical solutions to crosstalk problems. The charts are plotted in terms of the probability of intelligible crosstalk (the crosstalk index), several arbitrarily defined parameters (symbolized $M$, $R$, and $B$), the number of disturbers, and an assumed activity factor of $\tau = 0.25$ for the disturbing circuits. A separate chart is used for each value of the number of disturbers. The parameters $M$, $R$, and $B$, which have no physical significance, are related to the mean and variance of the various factors entering into the crosstalk phenomenon and are defined by the following equations:

$$M = \frac{M_v - M_I}{\sigma_I} \quad , \tag{17-11}$$

$$R = \frac{\sigma_v}{\sigma_I} \quad , \tag{17-12}$$

$$B = \frac{5}{\sigma_v} \quad , \tag{17-13}$$

where

$$M_v = M_{TV} - M_{l1} - M_{Cl} - M_{l2} - M_{l0} \quad , \tag{17-14}$$

$$M_I = M_{INT} + M_N - 6.0 \quad , \tag{17-15}$$

$$\sigma_I = \sqrt{\sigma_{INT}^2 + \sigma_N^2} \quad , \tag{17-16}$$

and

$$\sigma_v = \sqrt{\sigma_{TV}^2 + \sigma_{l1}^2 + \sigma_{Cl}^2 + \sigma_{l2}^2 + \sigma_{l0}^2} \quad . \tag{17-17}$$

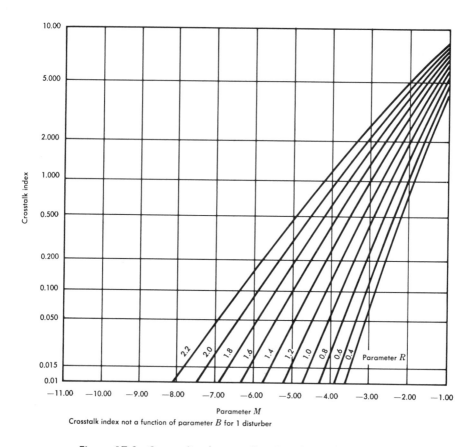

Figure 17-8. Generalized crosstalk index chart, one disturber.

In these expressions, $M$ represents mean values, and $\sigma$ represents the standard deviations of a number of measured parameters, some of which have been measured in the field and some in the laboratory. The subscripts refer to particular parameters as follows:

$v$, the crosstalking speech volume in vu at some defined reference point in the disturbed circuit.

$TV$, talker volume in vu at a convenient, well-defined reference point in the disturbing circuit.

$l1$, the loss in the disturbing circuit between the point at which $TV$ is measured and the point at which the crosstalk coupling occurs or is assumed to occur.

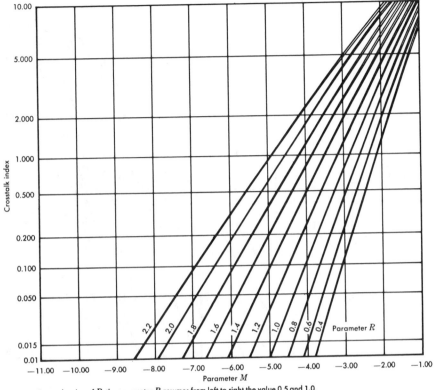

For each value of $R$, the parameter $B$ assumes from left to right the value 0.5 and 1.0.

Figure 17-9. Generalized crosstalk index chart, two disturbers.

$l2$, the loss in the disturbed circuit between the point at which crosstalk coupling occurs, or is assumed to occur, and some convenient intermediate point at which the interference is conveniently measured, not necessarily the final reference point.

$l0$, the loss, in the disturbed circuit, between the point at which $l2$ is measured and the final reference point of the computation.

$Cl$, the coupling loss between the disturbing and disturbed circuits.

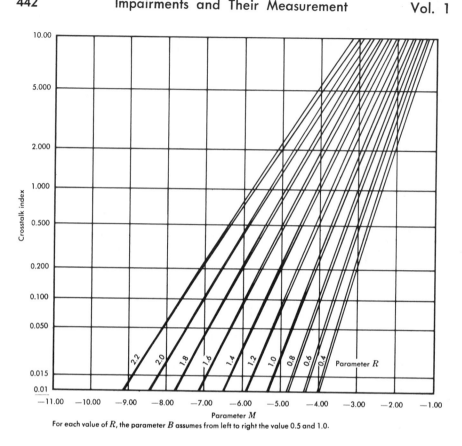

For each value of $R$, the parameter $B$ assumes from left to right the value 0.5 and 1.0.

Figure 17-10. Generalized crosstalk index chart, five disturbers.

$INT$, the listener acuity, without noise, referred to the reference point of the computation.

$N$, the noise measured at the reference point of the computation.

## Example 17-2: Use of the Generalized Crosstalk Index Charts

As an example of the use of these charts, consider a group of 101 trunks utilizing the same cable facility in which it is judged that far-end crosstalk may be controlling. The problem is to determine the mean value of FEXT coupling loss that would yield a crosstalk index of 1 (1 percent probability of intelligible

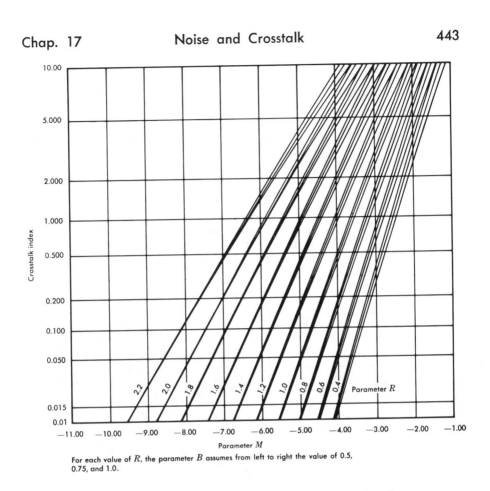

For each value of $R$, the parameter $B$ assumes from left to right the value of 0.5, 0.75, and 1.0.

Figure 17-11. Generalized crosstalk index chart, ten disturbers.

crosstalk) for 100 disturbers. The following values, chosen to be illustrative and not necessarily representative of a real problem, are given:

| Parameter | Mean | Sigma |
|---|---|---|
| Talker volume — outgoing switch — class 5 office | $M_{TV} = -16.8$ vu | $\sigma_{TV} = 6.4$ dB |
| Loss of trunks between class 5 offices | $M_l = 8.0$ dB | $\sigma_l = 3.0$ dB |
| Coupling loss | $M_{Cl} = $ Unknown | $\sigma_{Cl} = 4.5$ dB |
| Equivalent loop loss | $M_{l0} = 3.8$ dB | $\sigma_{l0} = 2.0$ dB |

| Parameter | Mean | Sigma |
|---|---|---|
| Noise at listener's station set | 20 dBrnc | 3.0 dB |
| Equivalent loop noise | $M_N = 22.0$ dBrnc | $\sigma_N = 3.0$ dB |
| Listener acuity without noise | $M_{INT} = -89.0$ vu | $\sigma_{INT} = 2.5$ dB |

The equivalent loop noise, $M_N$, is a combination of the measured circuit noise and the room noise which must be added to it. The total may be determined by the following:

$$M_N = [(N_c - 6.0) \text{ "+" } (N_R - 39.5) \text{ "+" } 6.3] + 6.0 \quad \text{dBrnc}$$

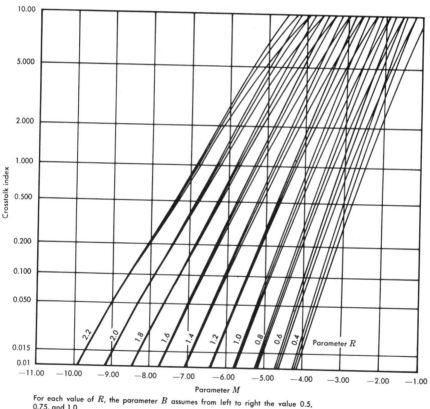

For each value of $R$, the parameter $B$ assumes from left to right the value 0.5, 0.75, and 1.0.

Figure 17-12. Generalized crosstalk index chart, 20 disturbers.

where $N_c = 20$ dBrnc is the circuit noise and $N_R$ (taken here as 50.0) is the room noise in dBt, an acoustic unit. The constants are empirically derived.

The disturbing circuit activity factor is taken as 0.25. The parameters $M$, $R$, and $B$ may now be computed by using Equations (17-11) through (17-17). For this example and the values given above, it should be noted that $M_l = M_{l1} + M_{l2}$ and

$$\sigma_l = \sqrt{\sigma_{l1}^2 + \sigma_{l2}^2}.$$

Using Equations (17-12), (17-16), and (17-17),

$$R = \frac{\sqrt{(6.4)^2 + (3.0)^2 + (4.5)^2 + (2.0)^2}}{\sqrt{(2.5)^2 + (3.0)^2}} = 2.20 \quad .$$

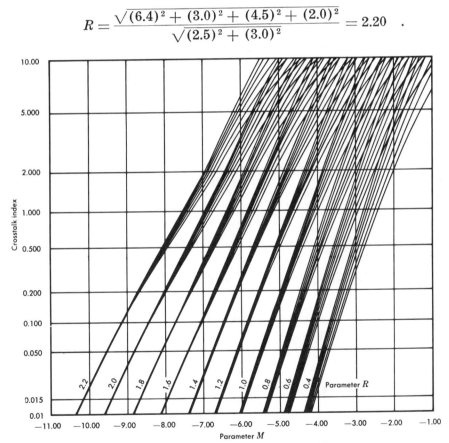

For each value of $R$, the parameter $B$ assumes from left to right the value 0.5 to 1.0 in steps of 0.1.

Figure 17-13. Generalized crosstalk index chart, 50 disturbers.

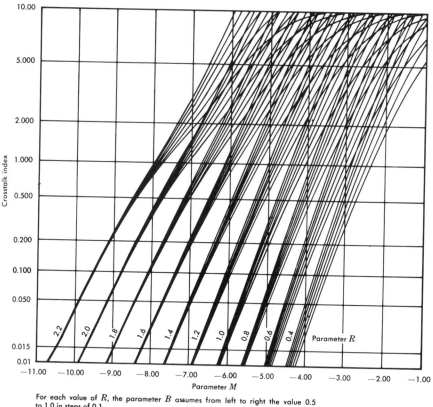

For each value of $R$, the parameter $B$ assumes from left to right the value 0.5 to 1.0 in steps of 0.1.

Figure 17-14. Generalized crosstalk index chart, 100 disturbers.

Using Equations (17-13) and (17-17),

$$B = \frac{5}{\sqrt{(6.4)^2 + (3.0)^2 + (4.5)^2 + (2.0)^2}} = 0.580.$$

For a crosstalk index of 1, 100 disturbers, and the above values of $R$ and $B$, the value of $M$ may be determined from Figure 17-14 as $-7.95$.

The required mean coupling loss may now be determined from Equations (17-11), (17-14), (17-15), and (17-16):

$$M_{Cl} = -M_l - M_{l0} + M_{TV} - M\sigma_I - M_{INT} - M_N + 6.0$$

$$= -8.0 - 3.8 + (-16.8) - (-7.95)(3.9) -$$
$$(-89.0) - 22.0 + 6.0$$

$$\approx 75.4 \text{ dB}.$$

Thus, if the mean value of the FEXT coupling loss equals or exceeds 75.4 dB, a crosstalk index of 1 or lower is realized.

### Digital Measurements

In the transmission of digital signals, whether digital data signals or digital carrier system line signals, the methods of measuring impairments other than coupling loss tend to be somewhat different from those used in analog transmission. There are four commonly-used methods of evaluating digital impairments: (1) impulse noise measurements, (2) error rate measurements, (3) studies of an eye diagram, and (4) P/AR meter measurements.

Impulse noise measurements are made as previously described. In evaluating the results in terms of digital transmission, calculations are often made in terms of the effects on error rate.

A direct measurement of error rate involves the transmission of a digital signal of known information content. The receiving equipment used in such an evaluation has stored within it the expected signal. It then compares the received signal, symbol by symbol, with the stored signal to provide the operator with a knowledge of the errors incurred during transmission. When data are collected, they are often plotted as illustrated in Figure 17-15. This figure, not representative of any real measurements, shows how the error rate increases as the signal-to-noise ratio is reduced. The first curve is *ideal* in that it represents the performance (measured or computed) in the presence of no impairment other than Gaussian random noise expressed as a signal-to-noise ratio. The other curves illustrate how performance may be degraded by two different impairments. The curves are usually similar in shape but displaced towards a better signal-to-noise ratio for the impaired conditions; i.e., the signal-to-noise ratio must be improved in the presence of the impairment in order to achieve the same error rate. The amount of the displacement is called the *noise impairment*.

The *eye diagram* and *P/AR meter* methods of test and evaluation, are particularly helpful in evaluating the total effect of all signal impairments. It is difficult to evaluate a single impairment by these methods because of the difficulty of separating effects.

A graphic illustration of an eye diagram of a ternary signal made up of raised cosine pulses is given in Figure 17-16. The figure shows

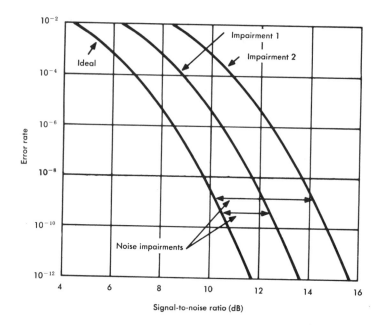

Figure 17-15. Error rate curves for ideal and impaired transmission.

two signalling intervals and a superposition of all possible undistorted pulse shapes. The decision area, or eye, for each of the two decision amplitudes is evident.* The horizontal lines (+1, 0, and −1) correspond to the ideal received amplitudes; the vertical lines (−T, 0, and +T) separated by the signalling intervals, correspond to the ideal decision times.

The decision-making process that must be implemented in the equipment receiving a series of such pulses can be related to the crosshairs shown in each eye. The vertical hair represents the de-

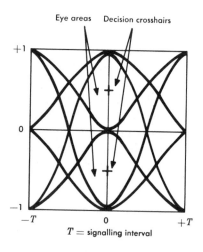

Figure 17-16. Eye diagram for a ternary signal.

*In an m-level system, there are m-1 separate eyes.

cision time, while the horizontal hair represents the decision amplitude. To permit accurate detection of the signal, the eye must be open (i.e., a decision area must exist) and the crosshairs must be within the open area. The practical effect of impairments of the pulses is to reduce the size of the eye. A measure of the margin against error is the minimum distance between the crosshair and the edges of the eye [12].

## Random Noise Measurements

The most common type of measurement of random noise is that of measuring the noise power in a given band. Such a measurement must be made at a known TLP (for telephone circuits), must cover the band of interest, and must include appropriate weighting factors or characteristics if applicable. The results are expressed in units appropriate to the measurement — dBrnc for telephone circuits and dBm for other types of circuits. The effect on digital transmission is usually expressed in terms of error performance.

Two transmission system-related measuring techniques are of interest here—the measurement of noise in compandored telephone circuits and the measurement of analog system performance by noise loading.

**Noise in Compandored Circuits**. Random noise that appears in telephone circuits equipped with syllabic compandors must sometimes be evaluated from two points of view, one when the circuit is used for speech signal transmission and the other when the circuit is used for digital signal transmission. In the case of speech signal transmission, the impairment, with or without a compandor, is greatest during silent periods when the noise can be heard best. When speech energy is present, the noise is subjectively far less interfering. Thus, in a compandored system, requirements for noise measured at the expandor input are somewhat less stringent than for an equivalent noncompandored system. However, at the expandor output, where much less noise is measured in the absence of signal, 5 dB must be added to the measured noise to account for the subjective effect of noise during quiet intervals.

For digital data signal transmission, the noise must be evaluated with signal present. Compandor action in the presence of signal usually results in an increase in noise at the expandor output. To accomplish such a measurement, a single-frequency signal, called a holding tone, is transmitted over the channel at about 2800 Hz* at an amplitude of −13 dBm0, to simulate a data signal. At the channel output, a band-elimination filter is used to suppress the holding tone. The noise, measured in dBrn or in dBrnc and translated to the 0 TLP as dBrn0 or dBrnc0, is called *notched noise.*

**Noise Loading.** The performance evaluation of broadband analog cable and radio systems is difficult, from an analytical point of view, because of the large number of parameters to be dealt with and, from a measurement point of view, because of the lack of control over a true telephone system load. Activity, type of signal transmitted, the percent of system equipped for service, and other important parameters are hard to determine or control. In such cases, a technique called noise loading is frequently used to evaluate system performance.

A band of flat Gaussian noise, limited to the spectrum normally occupied by transmitted signals, is applied to the system at a point where the normal multiplexed signal would be applied in practice. The magnitude of the applied noise is adjusted to simulate the loading effect of a normal signal.

In order to measure intermodulation noise, quiet channels (carrying no signal) are ordinarily used. To simulate quiet channels in a noise loading measurement, one or more band-elimination filters must be used to suppress the noise signal over small portions of the band at the output of the noise generator. At the system output, bandpass filters suppress all of the impressed noise signal. The passbands allow the noise that falls in the measurement bands, the quiet channels, to be passed on to the noise measuring equipment. This noise is due to intermodulation and other phenomena in the transmission system.

The noise measurement arrangements are illustrated in Figure 17-17. The output of the noise generator is depicted as covering the band of interest, $f_0$ to $f_t$. The input attenuator is used to adjust the signal amplitude to the desired value. The band-elimination filter suppresses the noise signal to create a quiet channel between $f_1$ and $f_2$ as

---

*The exact frequency depends on the design of the filter used at the receiver. Consideration is being given to the future use of 1000 Hz for all measurements requiring holding tones and filters.

illustrated. The noise signal is transmitted over the system under test (the system must be out of service, carrying no traffic) and then passed through the bandpass filter at the output. The only band that is passed is that of the quiet channel between $f_1$ and $f_2$. It contains all of the noise components (thermal, shot, intermodulation, etc.) that have been accumulated in the system. The output attenuator is used to adjust the gain of the measuring set-up so that the detector always measures noise at the same TLP, taken here as 0 TLP.

As previously mentioned, the noise loading test arrangement results in a measurement of all accumulated noise. The intermodulation noise can be separated from the other random noise sources by varying the noise signal amplitude over a range of values below the system overload point. The resulting curves, called V-curves, are plotted as in Figure 17-18 where the signal is varied over the range from about $-8$ dB to $+9$ dB. The reference, 0 dB, is arbitrarily defined as that value of signal amplitude which produces minimum noise.

These signal amplitude adjustments are made by adjustment of the input attenuator in Figure 17-17. For each such adjustment, a compensating adjustment must be made in the output attenuator in order that the overall gain from noise generator to detector remain constant and the detector always measure noise at the same TLP. The V-curves may now be interpreted in terms of the segments labeled A, B, C, and D.

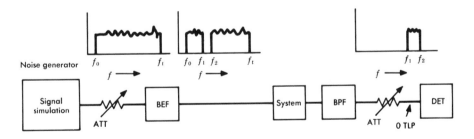

Figure 17-17. Noise loading test arrangements.

As the signal amplitude is reduced from its reference, 0 dB, the intermodulation noise is reduced and becomes insignificantly small. Other random noise components, such as thermal and shot noise, are signal independent. They appear to increase because, for each dB the signal is reduced, the output attenuation must be reduced, and the signal independent noise appears to increase at 0 TLP. Thus, the A segment of the V-curves has a straight-line constant slope of 1 dB per dB change in signal amplitude.

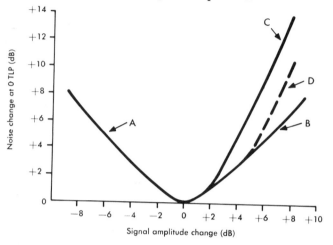

Figure 17-18. Noise loading V-curves.

As the signal increases above the 0-dB reference value, its relation to the total noise becomes dominated by intermodulation noise. Consider segment B of the V-curves. This illustrates a situation in which intermodulation products in the system are predominately second-order [derived from the $a_2e_i^2$ term of Equation (17-17)]. If the noise signal amplitude is increased 1 dB, the second-order noise increases by 2 dB. The output attenuator, however, has been adjusted to give 1 dB more loss to compensate for the 1-dB signal increase. Thus, the 2-dB noise increase is reduced to a 1-dB increase at the measuring point, 0 TLP. Segment B, then, has a slope of 1-dB increase in noise for each dB increase in signal amplitude.

Segment C of the V-curves is similar to segment B except that for C the intermodulation noise is dominated by third-order ($a_3e_i^3$) products. In this case, the noise increases 3 dB for each dB of signal amplitude increase, but the effect is reduced 1 dB by the adjustment of the output attenuator. The result is a 2-dB noise increase per 1-dB signal increase.

Finally, segment D illustrates a situation in which intermodulation noise is at first dominated by second-order products. The noise change follows curve B up to about +5 dB on the signal amplitude scale. Then third-order products predominate, and segment D follows the 2:1 slope of segment C.

The noise loading technique has three major uses in system test and evaluation. The first is simply to check performance against predicted or specified values. The second is to optimize signal-to-noise ratio by determining the drive level at which the signal-to-noise ratio is a maximum. The third is to provide information as an adjunct to trouble identification and isolation. In the latter case, measured results are compared with a predicted V-curve to determine if there is an excess of thermal or intermodulation noise.

The source of excessive noise can often be determined from such a comparison. In microwave radio systems, for example, excessive intermodulation noise shown by a V-curve at high frequency may be caused by waveguide or RF cable echoes, a defective RF amplifier stage, or a defective IF filter. Excessive modulation noise at low frequency may be caused by nonlinearity in an FM transmitter or receiver or in a baseband amplifier. Excessive low-frequency thermal noise may have as its source a defective local oscillator.

### REFERENCES

1. Rogers, W. E. *Introduction to Electric Fields* (New York: McGraw-Hill Book Company, Inc., 1954).

2. Binn, K. J. and P. J. Lawrenson. *Analysis and Computation of Electric and Magnetic Field Problems* (New York: The MacMillan Company, 1963).

3. Fennick, J. H. "Amplitude Distributions of Telephone Channel Noise and a Model for Impulse Noise," *Bell System Tech. J.*, Vol. 48 (Dec. 1969), pp. 3243-3263.

4. Technical Staff of Bell Telephone Laboratories. *Transmission Systems for Communications*, Fourth Edition (Winston-Salem, N.C.: Western Electric Company, Inc., 1970), Chapters 7, 8, and 10.

5. Johnson, J. B. "Thermal Agitation of Electricity in Conductors," *Physical Review*, Vol. 32 (1928), pp. 97-109.

6. Nyquist, H. "Thermal Agitation of Electric Charge in Conductors," *Physical Review*, Vol. 32 (1928), pp. 110-113.

7. Technical Staff of Bell Telephone Laboratories. *Transmission Systems for Communications*, Fourth Edition (Winston-Salem, N.C.: Western Electric Company, Inc., 1970), pp. 166-168 and Chapter 25.

8. Bell System Technical Reference PUB41008, *Analog Parameters Affecting Voiceband Data Transmission — Description of Parameters* (American Telephone and Telegraph Company, Oct. 1971), p. 24.

9. Bell System Technical Reference PUB41005, *Data Communications Using the Switched Telecommunications Network* (American Telephone and Telegraph Company, May 1971), pp. 10-11.

10. Sen, T. K. "Masking of Crosstalk by Speech and Noise," *Bell System Tech. J.*, Vol. 49 (Apr. 1970), pp. 561-584.

11. Technical Staff of Bell Telephone Laboratories. *Transmission Systems for Communications*, Fourth Edition (Winston-Salem, N.C.: Western Electric Company, Inc., 1970), Chapter 11.

12. Technical Staff of Bell Telephone Laboratories. *Transmission Systems for Communications*, Fourth Edition (Winston-Salem, N.C.: Western Electric Company, Inc., 1970), pp. 630-635.

13. Bell System Technical Reference PUB41004, *Transmission Specifications for Voice Grade Private Line Data Channels* (American Telephone and Telegraph Company, Mar. 1969).

14. Bell System Technical Reference PUB41009, *Transmission Parameters Affecting Voiceband Data Transmission — Measuring Techniques* (American Telephone and Telegraph Company, Jan. 1972).

Chapter 18

# Amplitude/Frequency Response

In the transmission of telecommunication signals, the fidelity of signal reception is strongly influenced by the frequency response characteristic of the channel involved. The amplitude/frequency response is the variation with frequency of the gain, loss, amplification, or attenuation of a channel or transducer. If a channel is linear and time invariant, its frequency response may be expressed as a ratio of output signal to input signal. This ratio, involving amplitude and phase relations, reflects the gain (or loss) of the channel and the phase shift through the channel. The desired relationships may be specified by a function that is flat with frequency or shaped in accordance with some specific rule. Whether it be a low-pass, bandpass, or high-pass function, the concern here is with departures from the function specified and from the definitions of linearity and time invariance.

The effects of bandwidth limitations, gain (or loss), gain variations with time, and attenuation/frequency distortion are all related to transmission impairments that may affect speech signals, voiceband or wideband data signals, and video signals. In some cases, unique methods are used to measure these impairments. The impairments are often related to transmission system design problems in significant ways.

## 18-1 TELEPHONE CHANNELS — SPEECH SIGNAL TRANSMISSION

In normal operation, a connection between two telephones may involve as little as two loops, one switching system, and the two station sets. On the other hand, the connection may be substantially more complex. It may contain several trunks between central offices and may be routed through a number of additional switching machines. It may be entirely at voice frequency, or it may involve a number of links utilizing analog or digital carrier systems. The two telephone speakers may be only a few feet apart or may be halfway around the world from one another. Such diversity in the makeup of connections makes it important to define and control the frequency response characteristic of each possible part of the connection and makes it

455

difficult to define and control the overall response characteristic of telephone connections.

## Channel Bandwidth

The development of frequency division multiplex (FDM) equipment during the 1930s made necessary the determination of the bandwidth to be assigned for the spacing of telephone channels in the spectrum. The problems of designing filters and channel combining and separating networks for the FDM equipment made it necessary also to determine what useful band was to be provided within the assigned channel band, i.e., the roll-off characteristics that could be tolerated.

The bandwidth of a telephone channel (established by subjective testing) is conveniently described in terms of the 4-kHz spacing of channels in the standard FDM equipment used in the United States (3 kHz in some submarine cable transmission systems). This description, however, is inadequate because it does not account for any of the effects which produce an effective bandwidth of less than 4 kHz, nor does it give the criterion used to define band edges. The band-narrowing effects include the loss characteristics of transmission media (loaded or nonloaded cable pairs on loops and trunks) and the frequency response of the filters used in terminal equipment, battery supply repeat coils, telephone station sets, etc. In addition, the tandem connection of multiple links may introduce a cumulative reduction of effective bandwidth.

The useful band of a telephone channel is defined as that between the 10-dB points on the loss/frequency characteristic of the channel, i.e., the points at which the loss is 10 dB greater than that at 1000 Hz, usually taken as the reference frequency. Without considering local loops, the bandwidth varies on switched telecommunications network connections from somewhat more than 3000 Hz to about 2300 Hz [1]. This reduction in bandwidth results primarily from technical and economic design compromises made in channel terminal equipment, particularly in the design of some short-haul carrier system terminals such as those used in N-type carrier systems. The reduced bandwidth, however, has generally given satisfactory speech signal transmission.

Intertoll trunks in the switched network of the Bell System may be regarded as having bandwidths extending from about 200 Hz to about 3600 Hz. Figure 18-1 shows a typical characteristic for a channel in the FDM hierarchy. Loops and toll connecting and direct trunks are frequently carried on loaded cable facilities; their useful bands extend from 0 Hz to nearly 3500 Hz. Where repeat coils are used, the low-frequency cutoff is at about 200 Hz.

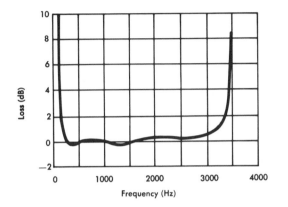

Figure 18-1. Typical channel loss/frequency characteristics of FDM equipment (toll quality).

Insofar as switched network circuits are concerned, the upper and lower cutoff frequencies (and therefore the bandwidth) are matters of transmission system design. For most systems, little can be done in system operations and maintenance that affects these parameters.

### Circuit Loss and Loss Variations

The losses in individual loops, trunks, and other transmission paths, such as those through switching machines, must be held to some maximum limit for two reasons. First, if the loss in a circuit or built-up connection is high, the received signal is low in volume and the listener either loses some of the transmitted information or is annoyed because he can not easily understand [2]. Second, if trunk losses are high, the contrast in received speech volume from call to call may be objectionable due to the many combinations of trunks that may be used. Consider, for example, successive calls to the same destination. On the first call the connection might be made over one intertoll trunk. On the second call, there might be several intertoll trunks in the connection because of alternate routing. If trunk losses were high, the resulting difference in volume between the two calls would be objectionable.

Ideally, all trunks should be operated at zero loss. This would permit the tandem connection of any number of trunks in a built-up connection, thus simplifying somewhat the problems of contrast, low volume, and noise. However, this mode of operation is impractical. Due to changes in terminating impedances, many circuits would become unstable (sing) or would be on the verge of singing and thus produce an unpleasant hollow effect in the received signal. Furthermore, echoes resulting from impedance mismatches would impair transmission. Thus, present circuit design is based on minimizing losses within the constraints of stability and echo control, a design concept called the *via net loss* design. Basically, the amount of loss in the toll portion of the network was determined by talker echo considerations. The allocation of loss to toll connecting trunks and intertoll trunks is based on the economics of supplying gain, the need for stability margins, and the transmission variations for alternate routing of calls (contrast).

The same reasons for controlling circuit loss apply to minimizing circuit loss variations. The control of cumulative losses to prevent low received volume, the prevention of excessive contrast between calls, and the need for controlling circuit stability and echo performance make it mandatory that loss variations with time be held to a minimum.

## Amplitude/Frequency Distortion

The transmission of speech signals is not seriously impaired by the type of inband amplitude/frequency distortion normally encountered in the switched network. The characteristics of loaded and nonloaded cable pairs tend to be smooth across the voiceband and, in general, not steeply sloped. There is, of course, a sharp roll-off at the high-frequency edge of the voiceband on loaded circuits. The characteristics of filters used in terminal equipment are also relatively smooth and introduce, except at the band edges, a gradually increasing loss as the frequency increases or decreases from the 1000-Hz reference frequency where transmission loss tends to be a minimum. Thus, the characteristics of inband distortion are usually expressed in dB of slope at 400 Hz and 2800 Hz, frequencies that are near the edges of the useful band. The slope is defined as the dB difference in loss at each of those frequencies relative to the 1000-Hz loss.

Except in the occasional instance of defective apparatus, the slope in a telephone channel is usually not a matter of field operating or maintenance control. The channel characteristics are established primarily by design, although some control may be applied by special equalization techniques.

## Measurements

Frequency response measurements are usually made by single-frequency measuring techniques involving a variable frequency oscillator and an adjustable detector. By this method several measurements at different frequencies must be made to establish the cutoff frequencies (10-dB points) and the slope at 400 and 2800 Hz.

Simplified evaluations of gain or loss and gain or loss variations with time are usually made at 1000 Hz. Individual loops and trunks are routinely measured at this frequency.

## 18-2  TELEPHONE CHANNELS — DATA TRANSMISSION

The specification and control of the frequency response characteristic of telephone channels is more critical for voiceband digital data signal transmission than for speech signal transmission because data signals are, in general, less tolerant of distortion in the frequency reponse than are speech signals. Such distortion may be described as a departure from the ideal amplitude/frequency response of a channel used for digital signals. The ideal response may be defined as that which, when combined with shaping or processing in the terminals, produces the minimum intersymbol interference in the received signal for the signal format employed as discussed in Chapter 14.

Where terminal equipment is designed to process data signals for transmission over the switched network, the processing (coding, rate of transmission, signal shaping, etc.) must result in signal characteristics that are compatible with the switched network channels. Where channels are dedicated to the transmission of data signals (private-line channels), the terminal equipment is usually designed so that the signal processing is coordinated and made compatible with the dedicated channel characteristics. *Conditioning*, the treatment of such channels to improve their amplitude/frequency response characteristics, involves the provision of fixed or adjustable equalizing networks.

The nature of amplitude/frequency distortion and the related effects of phase/frequency distortion often result in a need for more precise equalization than that provided by conditioning. Additional equalization may also be required because distortions vary with time, different types of facilities, and the different distances involved in successive connections. This additional equalization, provided by dynamic and adaptive equalizers, is usually designed in the form of tapped delay lines. Each tap is provided with an electronically controlled attenuator which automatically adjusts the delay line to approximate the inverse characteristic of the channel. The adaptive control is usually based on samples of transmitted pulses and an algorithm that uses statistical estimates of the sampled pulse response as control information [3]. This signal-dependent method of control combines amplitude/frequency and phase/frequency distortion correction and, in addition, can provide automatic gain control to compensate for changes in the overall gain of the channel.

Channel amplitude/frequency distortion, like most other digital signal impairments, causes an increase in the error rate. The effects are often expressed in terms of noise impairment, i.e., the dB improvement in the signal-to-noise ratio required to achieve the same error rate in a distorted channel as that achievable in the same channel when undistorted (see Figure 17-15).

## Available Bandwidth

As previously discussed, the bandwidth of intertoll trunks is about 3000 Hz (between 10-dB loss points); when toll connecting trunks and loops are included in an overall connection, the bandwidth is somewhat less. Within the band, the frequency response at high and low frequencies tends to roll off with increasing loss relative to the loss at 1000 Hz. Two questions now become apparent. How can the available band, with its roll-off characteristics, be most efficiently exploited? What is the sensitivity of the error rate to departures from the assumed channel characteristics? The answers depend on the chosen signal format and on the nature and magnitude of other forms of impairment.

Wherever possible, the available band is fully exploited by combining the known channel characteristic with the desired end-to-end transmission characteristic. For example, if the desired transmission characteristic is a raised cosine shape, the average expected channel characteristic is subtracted from the desired shape, and the difference is then supplied in the modem or data set so that the end-to-end characteristic matches the desired cosine shape as nearly as possible.

Where the channel is dedicated to a particular service, this match can often be made quite good. If transmission is over a switched network, the expected channel characteristics are quite variable, and compromise is necessary. Hence, the rate of transmission is sometimes significantly higher over a dedicated channel than on a switched channel.

If the data rate is too high relative to the bandwidth of the channel, serious signal distortion occurs because high-frequency components are attenuated. A simple example of this type of distortion may be seen in Figure 6-3.

## Loss and Loss Changes

The transmission loss between transmitting and receiving data stations may be controlled quite closely on dedicated channels, but may be quite variable on switched channels. This parameter is usually expressed in terms of the dB loss at 1000 Hz. On dedicated channels, the nominal loss is 16 dB $\pm 1$ dB. On switched channels, no such close control of the loss can be specified because the loss depends on the length of the connection, on the number of links in the connection, and on the loss of the loops at the two ends [4]. Thus, the design of receiving terminal equipment must take into account the variation of loss and the resulting variation of the received signal amplitude. Automatic gain controls are usually employed in the receiver to alleviate the problem of loss variation.

Loss (or gain) changes occur in telecommunication circuits for a number of reasons. *Slow* changes generally tend to be small and occur over a broad frequency range. They are generally caused by temperature changes or by the aging of active devices. Where carrier facilities are involved, these changes are usually compensated for by some form of automatic regulation. Such changes cause little impairment in the transmission of data signals. On the other hand, sudden gain changes occur sporadically as a result of faulty transmission components, substitution of broadband carrier facilities (protection switching), maintenance activities, or natural phenomena such as microwave radio fading. Even dropouts, i.e., momentary loss of signal, may occur. All such phenomena may cause signal impairment in the form of digital errors or severe analog signal anomalies.

## Inband Distortion

There are relatively few circuit elements that cause significant inband distortion of the frequency response characteristics in a telephone speech channel except for the roll-off. Slope distortion can be dealt with by providing adequate noise impairment margin (to permit satisfactory transmission over a switched network), by incorporating the channel characteristics in the overall amplitude/frequency response (in dedicated channels), or by conditioning (equalizing) some portion of the connection to satisfy requirements.

Two sources of distortion often occur in voice-frequency facilities as a result of trouble or oversight. Figure 18-2 illustrates the first of these, the deterioration of the insertion loss due to the presence of a *bridged tap* in a repeatered VF circuit. Cable layouts are often made with bridged connections at splice points to increase flexibility in circuit assignments. The bridged connection acts as a stub transmission line to produce an impedance irregularity. The second type of distortion, illustrated in Figure 18-3, occurs when a line is improperly loaded. This also produces serious amplitude/frequency distortion due to the impedance discontinuity.

### Measurements

A number of techniques are available for the evaluation of the performance of voiceband circuits for data signal transmission. These include the display of an eye diagram (Chapter 17), the measurement of errors in the transmission of a known message, and the use of the P/AR meter. The latter device measures the ratio of the pulse envelope peak to the envelope full-wave average for a closely con-

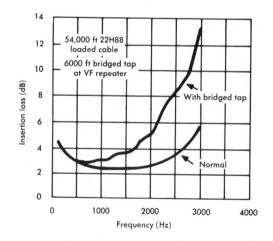

Figure 18-2. Effect of bridged tap on insertion loss of repeatered section.

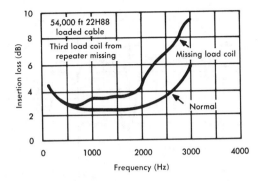

Figure 18-3. Effect of missing load coil on insertion loss of repeatered section.

trolled pulse stream transmitted over the channel under test. Distortions in the channel tend to disperse the energy in each pulse and to reduce the peak-to-average ratio [5]. All three techniques give an overall evaluation but do not provide a means for determining the specific cause of degradation. Thus, when such measurements indicate unsatisfactory performance, it is often necessary to resort to single-frequency measurements to determine the amplitude/frequency distortion in the band. These measurements may then be evaluated in terms of the equivalent noise impairment introduced by the frequency response characteristic.

## 18-3  WIDEBAND DIGITAL CHANNELS

The amplitude/frequency characteristics of wideband channels tend to have ripple components of higher amplitude than are typical of voiceband channels. Like voiceband channels, wideband channels display increasing loss toward band edges. However, wideband channels generally have less slope across the band. Figure 18-4 illustrates a typical wideband loss/frequency characteristic showing the cumulative ripple and nonuniform loss of about 1000 miles of an analog cable carrier system. The roll-offs at band edges do not appear in this figure because it does not include the effects of bandlimiting filters.

Nonuniform losses distort the digital signal spectrum and, hence, the desired waveform, resulting in a tendency toward increased errors. As with the voiceband channel, the nonuniform amplitude/frequency characteristic of the wideband channel is often corrected by fixed or adjustable networks designed to equalize the amplitude/frequency

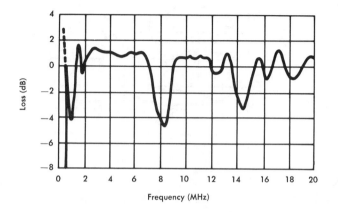

Figure 18-4. Typical amplitude/frequency characteristic of an equalized analog cable system (approximately 1000 miles long).

response. It is also sometimes necessary to provide adaptive equalizers for wideband digital transmission similar in concept to those used in voiceband transmission [6].

### Available Bandwidths

Terminal and transmission system equipment have been designed and wideband digital data services have been provided in a number of wideband channels. Standard tariff services correspond to building blocks in the FDM hierarchy. The building blocks used include the 48-kHz group band and the 240-kHz supergroup band. In addition, a half-group 24-kHz band is provided. In each case, the signal format is tailored in the terminal equipment (data stations and/or carrier system modems) to utilize the available bandwidth most efficiently, i.e., to provide the highest rate of transmission (bits per second per hertz of bandwidth) per unit of cost. Special facilities have also been provided from time to time to meet specific needs and to solve particular transmission problems.

Multilevel digital signals are transmitted on microwave radio systems at 1.5 megabits per second in a baseband extending from zero frequency to about 500 kHz [6]. Three-level signals are transmitted over wire pair cable facilities at 1.5 megabits per second and 6.3 megabits per second in the T1 and T2 carrier systems, respectively. In each case, the signal format has been designed to coordinate with the available bandwidth. The line signals may be composed of a variety of digital message signals combined by TDM techniques.

## Measurements

In the wideband systems and channels under discussion, point-by-point single-frequency measurements are often impractical because of their time-consuming nature. The evaluation of frequency response characteristics is therefore often made by the examination of an eye diagram or by an error rate measurement. The direct evaluation of the frequency response characteristic may be accomplished by sweeping the band with a sweep-frequency oscillator and then displaying the characteristic on an oscilloscope. A plot of the characteristic may be used to estimate the equivalent noise impairment caused by the distortion or it may be evaluated in terms of limits established for various portions of the band. These limits are defined so that, when not exceeded, the error rate objectives are met, provided other impairments are also held within limits. The terms commonly used to describe qualitatively the principal types of distortion are *slope, sag,* and *peak.* Slope describes the loss at the high end of the passband relative to that at the low end; sag describes the midfrequency bulge in the characteristic; peak describes the ripple components in the passband.

## 18-4   VIDEO CHANNELS

While wideband digital signal transmission has generally been adapted to available channel bandwidths, video signal transmission requirements have largely dictated the channel characteristics that must be provided* for television signal transmission and for PICTUREPHONE signal transmission. The significant impairments are those associated with bandwidth, cutoff characteristics, loss and loss changes, differential gain, and inband amplitude distortion.

## Bandwidth

As discussed in Chapter 15, the bandwidth required for video signal transmission is determined by the horizontal and vertical resolution to be provided in the received signal. These bandwidths

---

*An exception, of course, is telephotograph transmission, where the signal format has been tailored to the 4-kHz voice-channel spacing.

have been established at about 4.2 MHz for television signal transmission and about 1 MHz for PICTUREPHONE signal transmission.

For a picture generated with a specified number of scanning lines in a frame, the first noticeable impairment caused by a reduction in bandwidth is a loss of horizontal resolution, resulting in increasing difficulty in distinguishing between adjacent picture elements along a horizontal line. In addition to loss of horizontal resolution, color information is also lost as the bandwidth is reduced (recall that the color information is conveyed in television transmission by a carrier signal at about 3.58 MHz).

## Cutoff Characteristics

Another aspect of amplitude/frequency response is the nature of the video channel cutoff characteristics. At the low end of the band, it is necessary to provide good transmission essentially to zero frequency. Since frequencies near zero cannot generally be transmitted over analog facilities, the information in these components must be restored at the receiver. At the high end of the band, an essentially flat amplitude/frequency response must be provided to at least the color carrier frequency, 3.58 MHz. The channel loss above that frequency must be increased gradually because, if the band is cut off too sharply, a phenomenon called *ringing* may occur. A signal containing sharp transitions, when applied to such a channel, generates damped oscillations at approximately the cutoff frequency.

## Loss and Loss Changes

Video signals are generally not impaired by the overall loss or gain of a transmission system. The absolute value of gain or loss is set by the constraints of intermodulation, overload, crosstalk, and signal-to-noise ratio in the transmission system.

Television and PICTUREPHONE signals are impaired only slightly by gain or loss changes, provided these changes do not occur at a regular, low-frequency rate. When this does occur, the result is a flicker effect, a serious impairment of which viewers are quite intolerant. Telephotograph signals, on the other hand, are easily impaired by any loss or gain change. A gain change as small as 0.25 dB during picture transmission can be seen as a change of brightness in the received picture. Gain control circuits are often used to suppress such gain changes.

## Differential Gain

Video circuits generally require amplifiers. These amplifiers have nonlinear input/output characteristics that produce a signal-dependent form of amplitude distortion called differential gain, an impairment to which color television signals are particularly susceptible. Differential gain is the difference between unity and the ratio of the output amplitudes of a low-amplitude, high-frequency signal (simulating the color carrier) in the presence of a high-amplitude, low-frequency signal (simulating the luminance signal) at two different specified amplitudes of the low-frequency signal. The differential gain as defined may be expressed in percent by multiplying by 100, or in dB by taking 20 log the ratio of the two high-frequency signal amplitudes.

The effect of excessive differential gain on a color television transmission system is to cause undesirable variations in the saturation of the reproduced colors. The variations are a function of luminance signal amplitude.

## Inband Distortion

Departures from flat inband amplitude/frequency response cause a number of video signal impairments, two of which are called *streaking* and *smearing*. Both may be caused by transmission distortions in the frequency regions between about 60 and 1000 Hz and between 15 and 200 kHz. Both streaking and smearing cause objects in a picture to appear extended beyond their normal boundaries towards the right side of the received picture. With streaking, object extension appears undiminished; with smearing, the extension, which may be positive or negative in brightness relative to the object, diminishes substantially towards the right edge. The smearing impairment also tends to be more blurred than a streak.

If a channel has excess gain at high frequencies, sharp signal transitions may experience overshoot. The result is a black (or dark) outline to the right of a white object and a white (or light) outline to the right of a dark object.

Departures from flat response can, of course, be analyzed by Fourier techniques, and the loss characteristics may be expressed in terms of their Fourier components. If the departure from flatness is a simple sinusoid (gain or loss in dB versus frequency), the impairment can be shown to be a pair of echoes (low-amplitude duplicates of the signal displaced in time from the main signal) of the same polarity, one leading and one lagging the signal. Each of the Fourier components of the response characteristic produces such a pair of echoes.

The frequency band covered by one cycle of a Fourier component of a gain/frequency or loss/frequency characteristic may be defined as $\Delta f$. The relationship between $\Delta f$, the ripple frequency of the sinusoidal amplitude distortion, and $T$, the time displacement of the echo, is $T = 1/\Delta f$. Such distortions are often described in terms of coarse-structure and fine-structure deviations in the frequency domain. These terms have been arbitrarily defined relative to 555 kHz. If $\Delta f$ is less than 555 kHz, the distortion is called fine-grained; if $\Delta f$ is more than 555 kHz, the distortion is called coarse-grained. A coarse-grained ripple with $\Delta f = 2$ MHz produces a pair of echoes displaced 0.5 microsecond from the originating signal element, about 0.13 inch on a 17-inch wide television screen. A fine-grained ripple in which $\Delta f = 200$ kHz produces a pair of echoes displaced 5 microseconds, or about 1.3 inches from the signal.

The subjective effects of echoes due to distortion in the frequency response characteristic are dependent on the amount of time displacement of the echo and on the magnitude and shape of the distortion. The effects are also related to the presence of other echoes due to loss/frequency or phase/frequency distortion. These combined effects are given an echo rating, a method of assigning a single-number evaluation of a complex echo impairment [7].

## Measurements

Discussion of amplitude/frequency response measurements in relation to video signal transmission must be related individually to each of the three types of signals in use, telephotograph, PICTUREPHONE, and television.

Telephotograph Impairments. Telephotograph signals are transmitted in the voiceband. Because of their susceptibility to many of the impairments found on switched network voice channels, most telephotograph service is provided over dedicated facilities. Frequency response characteristic measurements are usually made on a point-by-point single-frequency basis. Overall circuit quality is judged by the transmission of special test pattern signals which can be viewed on an oscilloscope or printed as a picture for study.

**PICTUREPHONE Impairments.** This service has not yet developed to an extent requiring standard test procedures. The amplitude/frequency response of PICTUREPHONE channels is evaluated by sweep or point-by-point single-frequency techniques and by the transmission of special test patterns [8, 9].

**Television Impairments.** Point-by-point single-frequency measurements to determine the characteristics of a television channel are time-consuming, and therefore impractical, because of the wide channel bandwidth. While sweep techniques are sometimes used, special test signals are most commonly employed to evaluate a video channel. One such test signal is the test pattern shown in Figure 18-5.

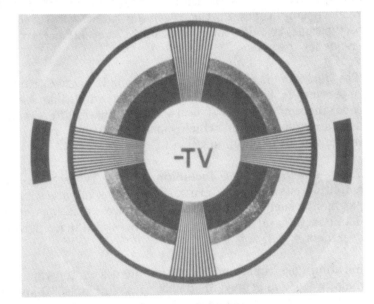

Figure 18-5. Typical television test pattern.

This test pattern may be used for many purposes including the lineup and adjustment of television receiving sets. It may also be used for gross evaluation of a channel and may be examined for a number of the amplitude/frequency response impairments previously discussed. For example, loss of horizontal resolution due to insufficient bandwidth would be evidenced by an inability to resolve the vertical lines in the striped wedges of the pattern. Ringing, smearing, streaking, and overshoot would all be easily seen by examination of such a pattern.

More objective measurements are made possible by the transmission of a variety of other test signals that are, in some cases, transmitted on an in-service basis with the video signal. In-service signals, sometimes called vertical interval test signals (VITS), are usually transmitted during the vertical blanking interval, and so they are not seen on the receiver. The waveform is displayed on an oscilloscope for examination and interpretation. Following are brief descriptions of some of the test signals used:

(1)  The *multiburst* is a frequency domain signal transmitted in-service. It is formed of brief impulses of 0.5, 2.0, 3.0, 3.6, and 4.2 MHz waves transmitted at equal amplitudes. Their relative amplitudes at the receiving point, measured on an oscilloscope, provide an evaluation of the channel amplitude/ frequency response.

(2)  The *stairstep* is also a frequency domain signal transmitted in-service. It is a signal of increasing amplitude formed of equal-increment steps. It is used to evaluate differential gain, excessive amounts of which cause departures from equality in the amplitude step sizes.

(3)  *Time domain signals,* including several types of pulse and amplitude step signals, are also used to evaluate various impairments such as unwanted luminance variations in large-detail sections of a picture, smearing, streaking, ringing, and overshoot.

The duration, rise time, and transition shapes of the time domain test signals are often defined in terms of a $\sin^2$ pulse shape. Pulse durations are defined in terms of the time between half-amplitude points, and the time of transition is defined as the Nyquist interval [10, 11, 12]. Recent work on video channel testing has favored the use of time domain signals because they appear to give a more direct measure of circuit quality. Frequency domain signals, such as the multiburst signal, are no longer in common use.

## 18-5  TRANSMISSION SYSTEMS

Since a channel may be comprised of a number of different combinations of baseband and carrier facilities, its amplitude/frequency response is to a degree dependent on that of the carrier system of which it may be a part. Some qualitative relationships may be used to show how channel characteristics may be affected by carrier system characteristics.

All of the channel frequency response characteristics previously discussed may be influenced in some way by the type of facility used. Included are bandwidth, cutoff characteristics, inband distortion, and loss variations with time.

## Bandwidth

Where a transmission medium or facility is dedicated to providing a single channel, the frequency responses of the facility and the channel coincide. However, where a facility or transmission medium is shared by many channels which have been multiplexed by FDM or TDM techniques, the relationship between the amplitude/frequency responses of the medium and the channel is not so clear-cut. In the latter case, the bandwidth limitations on the channels are most likely to be set by the filters in the multiplexing equipment.

A system characteristic may reduce the bandwidth of a channel where the channel is located near the edge of the band of the transmission system or near the edge of any band of the FDM hierarchy. At and near band edges, transmission response is most difficult to control, and an undesired roll-off that reduces the effective channel bandwidth is likely to be observed. The problem is primarily one of design; cutoff characteristics are minimally affected by operations and maintenance.

## Inband Distortion

The frequency response of a channel may or may not be impaired by the inband response of the transmission system. The departures from an ideal flat response in a broadband coaxial or microwave radio system tend to be broadband in nature. The inband distortion of a 4-kHz telephone channel is hardly affected by the system characteristic. For example, the maximum peak-to-peak deviation in Figure 18-4 is about 5 dB. This deviation occurs over a band of about 1 MHz, producing a slope of only 0.02 dB across a 4-kHz band. A broadband channel, on the other hand, may well be affected by inband system amplitude/frequency response distortion of the type shown in Figure 18-4.

## Loss-Time Variations

System and channel losses may vary due to temperature changes or due to operations or maintenance activities. Wherever possible, such loss changes are compensated by automatic gain control circuits called regulators. In some systems, these circuits operate in response to changes in the amplitude of a single-frequency signal, called a pilot, transmitted in the passband of the system or channel at a very precise frequency and amplitude. The regulator measures the received pilot amplitude, compares it to a reference, and then corrects the transmission according to the measured error by changing the loss of a network in the transmission path. The loss may be flat across the band of interest, or it may be shaped to compensate for a shaped loss variation. Pilot-controlled regulators are used in FDM equipment as well as in analog transmission systems.

Some systems, notably of the N-carrier type, regulate on the basis of total signal power rather than on a pilot. The system design is based on maintaining at the transmitting terminal a constant amount of signal power which is applied to the transmission line. The signal power is used for regulation in a manner similar to that described above for a pilot-controlled regulator.

### REFERENCES

1. Bell System Technical Reference PUB41005, *Data Communications Using the Switched Telecommunications Network* (American Telephone and Telegraph Company, May 1971).

2. Sullivan, J. L. "A Laboratory System for Measuring Loudness Loss of Telephone Connections," *Bell System Tech. J.*, Vol. 50 (Oct. 1971).

3. Lucky, R. W. "Techniques for Adaptive Equalization of Digital Communication Systems," *Bell System Tech. J.*, Vol. 45 (Feb. 1966), pp. 255-286.

4. Bell System Technical Reference PUB41007, *1969-70 Switched Telecommunications Network Connection Survey* (American Telephone and Telegraph Company, Apr. 1971).

5. Fennick, J. H. "Intersymbol Interference and the P/AR Meter," *Bell System Tech J.*, Vol. 49 (July-Aug. 1970), pp. 1245-1247.

6. Seastrand, K. L. and L. L. Sheets. "Digital Transmission over Analog Micowave Radio Systems," *IEEE* International Conference on Communications, June 19-21, 1972, Philadelphia.

7. Lessman, A. M. "The Subjective Effects of Echoes in 525-Line Monochrome and NTSC Color Television," *Journal of the SMPTE*, Vol. 81 (Dec. 1972), pp. 907-916.

8. Dougherty, H. J., E. B. Peterson, and M. G. Schachtman. "The PICTURE-PHONE System: Maintenance Plan," *Bell System Tech. J.*, Vol. 50 (Feb. 1971), pp. 621-644.

9. Favin, D. L. and J. F. Gilmore. "The PICTUREPHONE System: Line and Trunk Maintenance Arrangements," *Bell System Tech. J.*, Vol. 50 (Feb. 1971), pp. 645-665.

10. Schmid, H. "Measurement of Television Picture Impairments Caused by Linear Distortions," *Journal of the SMPTE*, Vol. 77 (Mar. 1968), pp. 215-220.

11. Davidoff, F. "Status Report on Video Standards; IEEE Video Signal Transmission Subcommittee 2.1.4," *IEEE Transactions on Broadcasting*, Vol. BC-15 (June 1969), pp. 27-32.

12. Schmid, H. "The $\text{Sin}^2$ Pulse and the $\text{Sin}^2$ Step in the NTSC TV System," *IEEE Transactions on Broadcasting*, Vol. BC-18 (Dec. 1972), pp. 81-84.

13. Fletcher, H. *Speech and Hearing in Communications* (New York: D. Van Nostrand Company, Inc., 1953).

Chapter 19

# Timing and Synchronization Errors

Baseband signals, defined here as signals not coded or processed by time or frequency functions, are usually transmitted without serious time or frequency impairment since the characteristics of most media and baseband apparatus tend to be stable and change very little with respect to frequency and timing relationships. However, where signal processing involves time-domain coding or frequency translation, impairments may result from lack of synchronization between the transmitter and the receiver or from deterioration of the timing signal itself. Some impairments to transmitted digital signals resulting from amplitude/frequency or phase/frequency distortions or impedance irregularities may lead to difficulties in timing signal recovery at regenerators or receivers.

Synchronization errors may be caused by incidental periodic, random, or discrete displacement of the carrier, resulting in unwanted amplitude, phase, or frequency modulation of the information-carrying signal. The discrete form of incidental modulation produces frequency offset (or shifting) of signal components in the received signal. Other forms of incidental modulation cause carrier signal impairments such as gain and phase hits, jitter, and dropouts.

In order to limit the impairments caused by synchronization problems, a national network distributes timing signals which synchronize analog systems and terminal equipment. This network, currently being modified, will also be used to synchronize digital systems.

## 19-1 FREQUENCY OFFSET

In most analog transmission systems employing suppressed-carrier FDM equipment, the output signal components may be offset in frequency from their proper values as a result of frequency differences between carriers in the transmitting and receiving terminals. The demodulation process must be controlled by carrier supplies at the receiving terminal, which must be synchronized with those at the transmitting terminal by some external means. Perfect sychronization can not be achieved and, to the extent it is not, frequency offset results.

In many short-haul FDM systems (such as N-type carrier), the carrier is transmitted with the signal, and its frequency and phase can be recovered with great accuracy to control the demodulation process. This practice is followed in designing for the transmission of many wideband digital line signals and video signals.

Consider first the offset phenomenon as it is produced in an AM analog transmission system. Assume $e_{in} = \cos \alpha t + \cos \beta t$ is the input signal which, at the transmitter, modulates a carrier, $\cos \gamma t$. Assume double-sideband suppressed-carrier transmission, where the signal components at the output of an ideal product modulator are

$$e_{mod} = (\cos \alpha t + \cos \beta t) \cos \gamma t$$

$$= \frac{1}{2} \cos (\gamma + \alpha) t + \frac{1}{2} \cos (\gamma + \beta) t + \frac{1}{2} \cos (\gamma - \alpha) t$$

$$+ \frac{1}{2} \cos (\gamma - \beta) t \quad .$$

Assume the lower sideband components are suppressed by filtering. Then, the single-sideband signal to be transmitted may be written

$$e_{SSB} = \frac{1}{2} \cos (\gamma + \alpha) t + \frac{1}{2} \cos (\gamma + \beta) t \quad . \tag{19-1}$$

If the demodulating carrier at the receiver is of the proper radian frequency, $\gamma$, the output of an ideal product demodulation process is

$$e_{dem} = \frac{1}{2} [\cos (\gamma + \alpha) t + \cos (\gamma + \beta) t] \cos \gamma t$$

$$= \frac{1}{4} \cos (\gamma + \alpha + \gamma) t + \frac{1}{4} \cos (\gamma + \beta + \gamma) t$$

$$+ \frac{1}{4} \cos (\gamma + \alpha - \gamma) t + \frac{1}{4} \cos (\gamma + \beta - \gamma) t \quad .$$

Then, if the first two terms, which are equal to $\frac{1}{4}\cos\ (2\gamma+\alpha)t$ and $\frac{1}{4}\cos\ (2\gamma+\beta)t$, are removed by filtering, the output signal is seen to be the input signal changed only by attenuation to one-fourth the amplitude of the original; that is,

$$e_{out} = \frac{1}{4}\ (\cos\alpha t + \cos\beta t) = \frac{1}{4}\ e_{in}\ . \tag{19-2}$$

The frequencies of the output signal are the same as those of the input signal.

If, at the receiving terminal, the carrier frequency is offset from that at the transmitter by $\Delta$ radians per second, the transmitted signal [Equation (19-1)] is demodulated to an output signal as follows:

$$e_{out} = \frac{1}{2}\ [\cos\ (\gamma+\alpha)t + \cos\ (\gamma+\beta)t]\ \cos\ (\gamma+\Delta)t\ .$$

Now, after the components containing $\cos\ (2\gamma+\alpha+\Delta)t$ and $\cos\ (2\gamma+\beta+\Delta)t$ have been filtered out, the output signal is

$$e_{out} = \frac{1}{4}\ \cos\ (\gamma+\alpha-\gamma-\Delta)t + \frac{1}{4}\ \cos\ (\gamma+\beta-\gamma-\Delta)t$$

$$= \frac{1}{4}\ \cos\ (\alpha-\Delta)t + \frac{1}{4}\ \cos\ (\beta-\Delta)t \neq \frac{1}{4}\ e_{in}\ . \tag{19-3}$$

Comparison of Equations (19-2) and (19-3) shows that in Equation (19-3) each signal component is shifted downwards by $\Delta$ radians per second; that is, the frequency shift is translated directly from the frequency error of the demodulating carrier to the baseband components of the output signal.

In carrier systems which transmit single-sideband suppressed-carrier signals at very high frequencies, the control of frequency offset imposes stringent requirements on the accuracy and phase stability (see Chapter 8) of the receiving terminal demodulating carrier. If the top frequency of a system is, for example, 100 MHz and if the frequency offset must be held to 1 Hz, the carrier at the receiver must be synchronized to within 1 Hz in 100 MHz or one part in $10^8$. Such accuracy requirements and stringent concomitant requirements on stability and reliability have made necessary the development of a national synchronization network and also have led to the development and use of very stable oscillators, highly sophisticated test equipment, and specialized methods of measurement and control.

## Speech and Program Signal Impairment

Frequency offset affects speech and program signals by reducing the naturalness of received signals. When music is transmitted, the effect is most objectionable to listeners with high aural acuity because many musical instruments produce sounds having high harmonic content. Consider a musical tone of radian frequency $\alpha$ having a strong second harmonic at radian frequency $2\alpha = \beta$. To preserve the natural harmonic relationship, the shift of $(\alpha-\Delta)$ radians per second should be accompanied by a shift of $\beta$ to $(\beta-2\Delta)$ radians per second. However, the effect of the frequency offset is to shift $\beta$ to $(\beta-\Delta)$ radians per second as in Equation (19-3). It is this type of discrepancy that causes the unpleasant subjective effect. It has been determined by subjective tests that frequency shift should be held to $\pm 2$ Hz to satisfy discerning listeners.

## Digital Data Signal Impairment

The manner in which digital data signals are impaired and the extent of the impairment are related to the signal format used. In many forms of digital data signal transmission, the timing signal used for signal decoding at the receiver is derived from the signal itself. In such cases, frequency offset is not a serious impairment, especially for the small offsets encountered when channels meet the requirements established for speech and program signal transmission.

Other digital data signals (particularly FSK) are prone to error in the face of frequency offset. For example, consider two received frequencies representing the space and mark signals common to telegraph signal transmission. A frequency offset in the received signal uses up margin with respect to threshold circuit recognition of the two conditions, thus making the receiver more prone to errors.

## Analog System Impairments

The most serious analog system impairment caused by frequency offset is the breakdown of system functions resulting from a large frequency offset that shifts signals outside the passbands of filters.

This is a rare occurrence when the 2-Hz requirement is met. In times of synchronizing system failure, however, substantial offsets may be occasionally experienced. Signalling may become impossible since the filters used for the single-frequency (SF) signalling system are relatively narrow. Large numbers of supervisory signals being shifted out of their passbands simulate a simultaneous call for service from a large number of callers and may cause a massive seizure of switching system equipment and breakdown of service. Frequency offset may also shift pilot frequencies and cause automatic gain circuits (regulators) to operate improperly. Such impairments are minimized by redundant designs of synchronizing arrangements used in the telephone plant to ensure reliability.

## Digital System Impairment

In most digital transmission systems, the signal formats are designed so that a timing signal can be derived from the line signal at regenerator and terminal stations. The timing signal recovery circuits are designed to operate within the normal range of frequency shift expected.

The most serious problem involving frequency offset and the synchronization of digital systems occurs where a number of such systems are to be interconnected and must in some way be synchronized with each other. At present this can only be accomplished by specially engineered arrangements where nearby terminals can be driven from a common clock or timing signal. Some designs of digital channel banks are provided with the capability of external transmit clock synchronization in order to be compatible with planned terminal synchronization arrangements and to facilitate the special engineering required.

Small differences between the frequency, or rate, of a received signal and that of a locally maintained clock signal can be detected and compensated for by buffer stores and bit stuffing. However, if the frequency difference persists, a buffer may overflow or underflow. The system is designed to reset the buffer when this occurs, thus causing deletion or repetition of bits from the output signal. The overflow or underflow and the resulting reset cycle of the buffer continues until the frequency offset is detected and corrected. The resulting impairments, called slips, cause serious deterioration of digital signal transmission.

## 19-2  OTHER INCIDENTAL MODULATION

In addition to frequency offset, synchronization and timing signals are subject to other forms of incidental modulation which may cause transmission impairments in the telecommunications channels they control. These include gain and phase hits, periodic or random jitter, and dropouts in the synchronization system.

Normally, these forms of incidental modulation have little effect on speech transmission, but they may introduce errors in digital signal transmission by reducing the noise impairment margin. Excessive impairment is also sometimes observed in the transmission of certain types of analog data, such as electrocardiograph signals.

### Gain and Phase Hits

Rapid changes in channel gain or phase result in signal impairments called gain or phase hits. These hits can be caused by timing signal aberrations or by transmission channel malfunction. The separation of these causes is difficult to determine by analysis or measurement.

One source of gain and phase hits is the switching of transmission facilities or multiplex equipment from working to standby facilities for trouble or maintenance work. If the facility that is switched carries a synchronizing signal or if the switch occurs within the synchronizing equipment, differences in phase or gain between working and spare equipment may cause a hit on the synchronizing signal, which may be extended through the working channels to the message signals. Or, the switching of transmission facilities may cause a hit directly on the transmitted signal as a result of the difference in attenuation or phase between the working and standby facility.

### Jitter

The generation of an absolutely pure single-frequency signal for use as a carrier is impossible; minute variations in amplitude, phase, and frequency always occur. These variations can usually be held to very small values, but from time to time they exceed acceptable limits and cause signal impairments. Continuously and rapidly changing gain and/or phase, which may be random or periodic, is defined as jitter. The principal sources of jitter have been in the power supplies and harmonic generators associated with analog system multiplex equipment.

Phase jitter, which tends to be a more serious impairment than gain jitter, is manifested as an unwanted change in phase or frequency of a transmitted signal. The problem results from some form of modulation of the wanted signal by another signal. A single-frequency signal that is so modulated has sidebands which may be discrete or random and noise-like, depending on the nature of the modulating signal. The amplitude of these sidebands relative to the wanted signal is one measure of the phase jitter suffered by the wanted signal.

Another useful measure of phase jitter is the time variation in zero crossings of a sine wave or of a pulse signal. Zero crossings are often used as decoding criteria by receiving logic circuits in digital signal transmission; thus, the variation in zero-crossing timing must be well-controlled. Variations in zero crossings of a sine wave are illustrated in Figure 19-1.

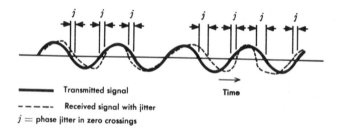

Figure 19-1. Effect of phase jitter on zero crossings of a sine wave.

Periodic forms of phase jitter sometimes are a result of modulation by power frequency or telephone ringing signal components and harmonics. Random forms of jitter may result from impulse noise or interfering signals having high-amplitude random components.

## Dropouts

Dropouts are short duration impairments in which the transmitted signal experiences a sudden drop in power, often to an extent that the signal is undetectable. They have been defined as any reduction in signal power more than 18 dB below normal for a period exceeding 300 milliseconds. Dropouts may be caused by facility or equipment switching or by maintenance activities. They usually occur rather infrequently but typically may be observed once an hour or somewhat more often.

## 19-3　THE SYNCHRONIZING NETWORK

The carrier frequencies used in analog systems in the United States are derived from and controlled by a single clock of extremely high accuracy and stability. The output of this clock is transmitted in a variety of ways to all parts of the country. At each location where synchronization is needed, a control signal derived from this clock is used as a master.

The distribution and transmission of the clock signal involve many intermediate links and pieces of apparatus. Many of the impairments described in this chapter occur as a result of impairments suffered by the clock signal in the process of transmission and distribution. Each dependent office has a clock or synchronizing signal source of its own. These local signal sources are controlled by the master clock as long as the master clock signal is available. Failure of intermediate transmission links or apparatus, however, can make the master clock signal unavailable. In such a case, the local clock is disconnected from the master and becomes free-running. Its frequency may deviate enough from that of the master to be a source of synchronization impairments.

The design and implementation of an improved synchronizing network is evolving to accommodate the interconnection and synchronization of new digital transmission systems, time division multiplex terminals, and time division switching systems. The introduction and construction of a new digital data network, the Digital Data System, impose new and more stringent accuracy and stability requirements on the synchronizing network; these requirements are also under study for application to the improved network.

### REFERENCES

1. Bell System Technical Reference PUB41008, *Analog Parameters Affecting Voiceband Data Transmission — Description of Parameters* (American Telephone and Telegraph Company, October 1971).

2. Bell System Technical Reference PUB41007, *1969-70 Switched Telecommunications Network Connection Survey* (Reprints of Bell System Technical Journal Articles), (American Telephone and Telegraph Company, April 1971).

Chapter 20

# Echo in Telephone Channels

Echo results when transmitted signal energy encounters an impedance discontinuity and a significant portion of the signal energy is reflected toward the energy source over an echo path. Echoes constitute one of the most serious forms of impairment in telephone channels, whether the channels are used for speech, data, or telephotograph signal transmission. The phenomenon is more difficult to control in switched networks, where terminating impedances may change with every new connection, than in dedicated private circuits, where the impedances are fixed and more nearly under the control of the circuit designer.

Transmission is impaired by echoes for both talker and listener on an established telephone connection. A frequently encountered source of echo, one that aptly illustrates the two forms of echo impairment (i.e. *talker echo* and *listener echo*), occurs at the junction of four-wire and two-wire circuits. This type connection and the resulting talker echo path and listener echo path are shown in Figure 20-1. The transitions between two-wire and four-wire modes of transmission are provided at each end of a four-wire connection by a circuit called a four-wire terminating set, designated HYB in the figure.

## 20-1 ECHO SOURCES

Consider the circumstances that make the interface between four-wire and two-wire circuits a frequent and difficult to control source of echo. Figure 20-2 may be used to review the relationship between the hybrid circuit and the impedances that must be matched for satisfactory operation. Two of these impedances, $Z_a$ and $Z_b$ in the four-wire circuit, are usually under design control, and there is little problem in achieving a good match between them. The match between $Z_c$, the impedance of the balancing network, and $Z_d$, the impedance

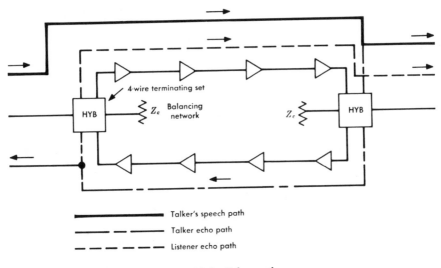

Talker's speech path

Talker echo path

Listener echo path

Figure 20-1. Echo paths.

of the two-wire connection, on the other hand, is not always under design control.

If the circuit of Figure 20-2 is to be used in a dedicated circuit, impedance $Z_d$ is under design control or at least has a known value. When this is so, the balancing network may be adjusted to match impedance $Z_d$ to any desired degree. Thus, in dedicated private line channels, echo is seldom a problem of great concern.

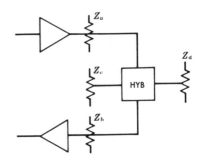

Figure 20-2. Simplified four-wire terminating set.

Next consider Figure 20-3. Here, office A may be a class 4 office (toll center) and office B a class 5 (end) office. Impedance $Z_d$ facing the four-wire terminating set is a function of the two-wire trunk impedance, $Z_T$ (a highly variable impedance depending on gauge, length, loading, etc.), and the terminating impedance, $Z_t$. For a particular trunk or group of trunks, impedances $Z_T$ and $Z_t$ can be controlled by using impedance compensators so that good balance

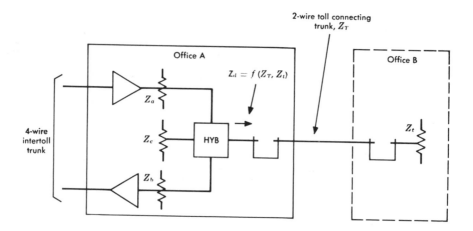

Figure 20-3. Terminating set impedances — four-wire trunk to two-wire trunk.

can be achieved. However, variations occur when loop impedances are substituted for $Z_t$.

Figure 20-4 illustrates the more practical situations that exist in switched network operation. This figure shows why it is so difficult to achieve good impedance matching at four-wire terminating sets and, therefore, why network echo performance is dominated by performance at these points. In Figure 20-4, the four-wire intertoll trunk terminating in office A may be connected to a distant office via two-wire toll connecting trunks having widely different impedance characteristics. At the distant class 5 office B, a connection may be made to a variety of loops designated $1, 2, 3 \ldots n$, each having a different impedance. Alternately, the four-wire trunk may be connected in office C to a variety of loops designated $4, 5, 6 \ldots m$. Thus, impedance $Z_d$, which should be equal to the balancing network impedance $Z_c$, is highly variable because it is a function of both the toll connecting trunk impedance and the terminating loop impedance (which is different for every established connection). It is impractical to control impedance $Z_d$ to a fixed value; hence, only a compromise value for $Z_c$ can be selected. Since $Z_c$ and $Z_d$ are generally not well matched, transmission loss across the hybrid (transhybrid loss) from $Z_a$ to $Z_b$ is reduced, and echo is returned to the talker and listener as illustrated in Figure 20-1.

Where the toll connecting trunk is two-wire and the four-wire terminating set is at the toll office, loop impedance variations are

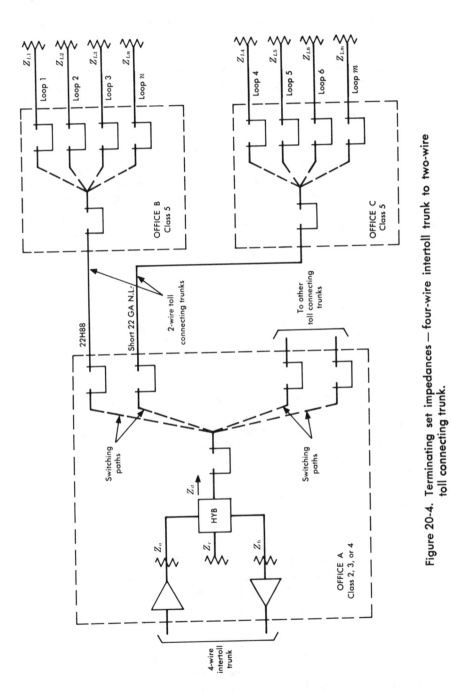

Figure 20-4. Terminating set impedances — four-wire intertoll trunk to two-wire toll connecting trunk.

somewhat masked by the controlled impedance of the toll connecting trunk. Where the interface between four-wire and two-wire circuits is at a class 5 office, more serious echoes occur because the great variability in loop impedances is not buffered by the toll connecting trunk. If an intertoll trunk is two-wire, the interface may occur anywhere in the network, but impedances are still controlled better than at the class 5 office.

While any impedance irregularity produces echoes, such irregularities are usually of little interest because the impedances are controlled by design; however, under trouble conditions, serious mismatch may occur. Examples include the impedance mismatch between a transmission line and electronic circuits caused by device failure, etc., and deteriorated structural return loss in a transmission line due to the mismatch caused by damage to the line or by a damaged or misplaced load coil.

Another source of echo is the crosstalk coupling between the two directions of transmission in a four-wire circuit. The measurement and control of this echo source are based on techniques discussed in Chapter 17. The nature of the impairment is the same as that caused by impedance discontinuity reflections.

## 20-2 NATURE OF ECHO IMPAIRMENTS

Transmission impairments caused by echo must be considered in relation to the type of signal involved. In some cases these impairments must be evaluated subjectively, as in speech signal transmission; in other cases the impairments must be evaluated objectively, as in data signal transmission.

### Speech Signals

It has been found that talker echo usually produces a more serious impairment to speech signal transmission than does listener echo; the latter is a result of a double reflection and is usually of low amplitude. Therefore, talker echo is stressed in the following.

**Talker Echo.** If the elapsed time is very short between the production of a speech signal at a station set and the reflection of that signal to the speaker's ear, the echo sounds like sidetone. Unless it is very loud, the speaker may not even be aware of the presence of the echo. On the other hand, if the elapsed time is long and the echo path loss is inadequate, the echo sounds very much like an echo resulting from acoustic reflection from an obstacle. In extreme cases, the telephone speaker may get the impression that the distant party is trying to interrupt him, and this can interfere with the speaker's normal process of speech. The overall effect of talker echo depends on (1) how loud it is (which is dependent on how loud the speaker talks and how much loss is in the echo path), (2) how long the echo is delayed in transmission, and (3) the speaker's tolerance to the echo phenomenon. All these are interrelated and all are best expressed in statistical terms.

It is important to note at this point that the impairment due to echo is related directly to the magnitude of the received echo signal, which can be reduced by increasing the loss in the echo path. Loss in the echo path can be increased by increasing the transhybrid loss in the four-wire terminating sets or by increasing the loss in the transmission path. Because of the high cost of modifying the millions of existing loops and trunks, it would be uneconomical to increase the average transhybrid loss. Thus, any increase in loss can be inserted only in the transmission path between the talker and the point of impedance mismatch. Such action, however, is accompanied by an unavoidable reduction in received volume, which may also lead to serious impairment. The echo problem must then be solved by a compromise between echo and volume loss impairments. This compromise, based on measured performance and its relationship to subjective evaluations of echo and loss impairments, is achieved in the via net loss (VNL) transmission design of the switched network [1, 2].

Now consider the evaluation of the three important aspects of talker echo—talker tolerance, magnitude, and delay. Talker tolerance to echo has been established by carefully controlled experiments whose results are summarized in Figure 20-5, which relates echo path delay to the echo path loss needed to satisfy the *average observer*. It was found that for any value of delay, the echo path loss (the measure of tolerance) had a normal distribution with a standard deviation, $\sigma_p = 2.5$ dB. These data apply to talkers on short loops only.

Echo magnitude is primarily a function of the losses in the echo path. The overall echo path loss is made up of two important components, *return loss* and *transmission loss*, each of which is a variable

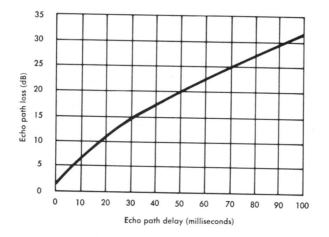

Figure 20-5. Talker echo tolerance, average observer.

having an average value and a standard deviation. Consider first the return loss. This parameter has, at class 5 offices, an average value of 11 dB and a standard deviation $\sigma_t = 3$ dB. These values have been determined by field measurements.

The transmission loss is, of course, highly variable. It is dependent on distance, number of trunks in the connection, types of facilities, plant maintenance capability, etc. The standard deviation for each trunk in a built-up connection is estimated to be $\sigma_v = 2$ dB for the round-trip loss, which takes into account differences in loss values for the two directions of transmission.

From the standard deviations just given for losses and observer tolerance, a value may be derived to represent the standard deviation of the minimum permissible echo path loss on connections made up of a number of trunks, $N$, and used by talkers of different echo tolerance,

$$\sigma_c = \sqrt{\sigma_p{}^2 + \sigma_t{}^2 + N\sigma_v{}^2} \quad \text{dB.} \quad (20\text{-}1)$$

The table of Figure 20-6 has been derived by substituting the previously given numerical values in Equation (20-1). The table shows the expected standard deviations of echo performance on several built-up connections.

| NO. TRUNKS | STANDARD DEVIATION, DB |
|:---:|:---:|
| 1 | 4.4 |
| 2 | 4.8 |
| 4 | 5.6 |
| 6 | 6.3 |

Figure 20-6. Standard deviations of minimum permissible echo path losses on built-up connections.

A relationship between loss and talker echo can now be demonstrated. It is convenient to express this relationship in terms of the minimum permissible *one-way* overall connection loss (OCL) that allows 99 percent of all calls to be completed without echo impairment. The value of the two-way OCL is the average echo tolerance reduced by the average return loss at the controlling point of echo generation and increased by 2.33* times the standard deviation determined in Equation (20-1). The result is divided by 2 to determine the permissible one-way OCL:

$$\text{OCL} = \frac{\text{Avg. Echo Tolerance} - \text{Avg. Ret. Loss} + 2.33\sigma_c}{2}. \quad (20\text{-}2)$$

The average echo tolerance in this equation must be adjusted to eliminate the effect of the transmitting loop loss. Equation (20-2) then yields the OCL between class 5 switching offices. Values of permissible OCL for various numbers of trunks are plotted in Figure 20-7. Linear approximations can be made for the curves of Figure 20-7 that also satisfy echo performance objectives on 99 percent of the connections experiencing the maximum allowable delay and even higher percentages on connections having less than the maximum delay. These approximates are used in the development of the via net loss plan for the message network.

**Listener Echo and Near-Singing.** In modern circuits, listener echo is usually negligible if talker echo is adequately controlled because, as shown in Figure 20-1, listener echo is suppressed by a second transhybrid loss (at the talker end of the four-wire circuit) and by the

---

*This value may be found by referring to Figure 9-14 or 9-15.

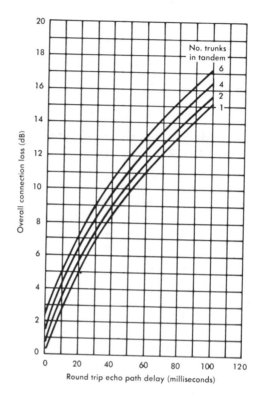

Figure 20-7. One-way loss for satisfactory echo in 99 percent of connections.

loss of one additional end-to-end transit of the four-wire circuit. An exception, relatively rare and yet important where encountered, is found in large multistation private line circuits. Here, listener echo is often controlling, and special care must be taken in designing this type of circuit.

There is a close relationship between listener echo and near-singing of a circuit. Both conditions are caused by currents circulating within a transmission path. Circuit instability, or singing, and singing margins were discussed in Chapter 4. Interestingly, transmission impairment occurs before singing actually takes place. If the singing margin is too low, the near-singing condition of the circuit causes voice signals to sound hollow, somewhat like talking into a barrel. To avoid this effect, singing margin is maintained at 10 dB or more in 95 percent of all connections and seldom, if ever, less than 4 dB.

## Digital Data Signals

When digital data signals are transmitted over telephone channels, there is greater concern with listener echo than with talker echo. Listener echo is usually very low in amplitude when talker echo is well enough controlled to satisfy speech signal transmission but some built-up connections in a switched network, public or private, may have poor return losses at both ends of a four-wire intermediate link. As a result, echo performance may be poor, and digital data echoes may be generated at high enough amplitudes to interfere with reception. Circuits designed for dedicated data signal transmission may be controlled so that listener echo is of low amplitude.

Data signalling rates between 1000 and nearly 10,000 bits per second are commonly used. Thus, a listener echo delayed by 0.1 to 1.0 millisecond or more appears at the receiver as an unwanted interfering signal having relatively little correlation with the wanted signal. The magnitude of such an interference may be high enough to be a serious source of data errors.

Talker echo is usually far less disturbing to data signal transmission than is listener echo. In certain instances, when a data terminal is switched from *send* to *receive*, a talker echo resulting from the end of a transmitted signal may appear as the leading edge of an unwanted received signal. This may be confused with the expected reply to the transmitted signal.

When data signal transmission equipment is designed for use on the channels of a switched network, the problems relating to echo must be solved by terminal design. Adequate signal-to-noise margin must be provided in the receiver to cope with listener echo effects. Timing circuits must be provided to avoid talker echo effects in terminal equipment which may serve as both transmitter and receiver. However, the amount of time delay allowed for in engineering design (called turnaround time) reduces the data transmission efficiency, or throughput rate. When this is critical, full duplex four-wire private line operation may be required or full duplex operation may be established over the switched network by using two separate connections.

## Telephotograph Signals

Telephotograph signals are very susceptible to echo impairment in the form of a "ghost" of the desired picture. The echo elements are detached from the signal elements by an amount proportional to the delay in the echo path. The echo may be positive or negative, depending on phase relationships between the signal and its echo. Since most telephotograph transmission is over dedicated channels where impedances can be better controlled, echo is a serious problem only when switched network channels are used.

## 20-3  ECHO MEASUREMENT AND CONTROL

The measurement of echoes in telephone channels is accomplished primarily by return loss measurements. Echo performance is controlled by impedance adjustments and by the use of echo suppressors on long circuits having delays in excess of 45 milliseconds. Private line, or dedicated, channels are often provided as four-wire facilities in order to avoid the echo problem. Even where some two-wire links are necessary, the performance of such channels can usually be made satisfactory by design control at the two-wire-to-four-wire interface.

### Echo Return Loss

Return loss, discussed in Chapter 4, is usually used as a measure of echo performance resulting from an impedance discontinuity such as that found at a four-wire terminating set in a telephone channel. Return loss is defined rigidly in terms of the ratio of the sum and difference of the complex impedances at the discontinuity. However, the complexity of phase relationships in the incident and reflected voltage or current waves makes it impractical to express return loss over a band of frequencies except by averaging the performance over the band of interest. A suitably weighted power average is used. This average, called echo return loss (ERL), is applied to the band from 500 to 2500 Hz.

This type of measurement is sometimes made by applying random noise to the 500 to 2500 Hz band, sometimes by using a sweeping frequency that covers this band and sometimes by measuring a number of single-frequency return losses across the band. The weighting is applied by the inclusion of appropriate networks in the test equipment or by the manner in which the results are processed. Several weightings are used (including flat weighting), none of which is universally accepted.

Echo return loss evaluation may be illustrated by considering return loss measurements at a number of single frequencies. Typically, measurements are made at frequencies $f_1 = 500$, $f_2 = 1000$, $f_3 = 1500$, $f_4 = 2000$, and $f_5 = 2500$ Hz. These measurements may be regarded as applying to the edges of four equal frequency bands, designated $B_1$, $B_2$, $B_3$, and $B_4$. The plane geometry rule for finding the area of a trapezoid may be applied by analogy to the determination of the average return loss over the four bands of interest. This rule, as applied here, yields an echo return loss value,

$$\text{ERL} = -10 \log \frac{B}{4B} \left( \frac{\text{RL}_{f1} + \text{RL}_{f5}}{2} + \text{RL}_{f2} + \text{RL}_{f3} + \text{RL}_{f4} \right) \text{ dB} \quad (20\text{-}3)$$

where $B$ is the bandwidth of each of the four bands of interest (500 Hz), and $\text{RL}_{fn}$ is the return loss at each of the five frequencies expressed as a *power ratio*. This equation has the effect of weighting the return losses at the band edges (500 and 2500 Hz) to one-half the effectiveness of the return losses at 1000, 1500, and 2000 Hz.

The following table, which gives return loss values that might be measured at the five frequencies of interest, provides a summary of how the data would be analyzed.

| FREQUENCY, Hz | RETURN LOSS, dB | RATIOS | | WEIGHTED POWER RATIO |
| --- | --- | --- | --- | --- |
| | | CURRENT | POWER | |
| 500 | 25.0 | 0.056 | 0.00316 | 0.00158 |
| 1000 | 30.0 | 0.032 | 0.00100 | 0.00100 |
| 1500 | 25.0 | 0.056 | 0.00316 | 0.00316 |
| 2000 | 22.0 | 0.079 | 0.00631 | 0.00631 |
| 2500 | 18.0 | 0.126 | 0.01583 | 0.00792 |

The last column, which represents the five return loss terms in Equation (20-3), totals 0.01997. When divided by 4, this yields a value of 0.00499. Then,

$$\text{ERL} = -10 \log 0.00499 = 23.0 \text{ dB}.$$

If a straightforward averaging of the five power ratios in the next-to-last column had been made, the echo return loss would be calculated as

$$\text{ERL} = -10 \log \frac{0.02946}{5} = 22.3 \text{ dB}.$$

While the difference between these approaches, 0.7 dB, appears to be small, it is significant and there is evidence that the weighted value more nearly represents the subjective effect of echoes as evaluated by return loss measurements.

In the switched network, ERL measurements are made at various switching offices; impedance adjustments are made to guarantee the echo performance at each office involved. The measurements and adjustments are usually made with some standard value of impedance as a reference. The measurements are called *through balance* or *terminal balance* measurements.

## Singing Return Loss

As previously discussed, margin must be provided against instability or singing. Singing return loss measurements, made to give assurance of the necessary stability, must be made at all frequencies at which a circuit might become unstable. Experience has shown that the important bands are those from 200 to 500 Hz and from 2500 to 3200 Hz. Below 200 Hz and above 3200 Hz, telephone circuits usually have sufficient loss to suppress any tendency towards instability. Frequencies in the 500 to 2500 Hz range are usually satisfactory from the standpoint of singing return loss if they meet echo return loss requirements.

The singing return losses or the singing margins are measured in the field by sweep frequency or random noise techniques or in the laboratory by point-by-point methods. Impedance adjustments are specified to guarantee satisfactory performance; the adjustments specified in echo return loss tests generally tend to improve singing return loss performance.

## Echo Suppressors

Figure 20-7 shows that the overall connection loss must be increased substantially as echo delay becomes greater. However, as transmission loss increases, talker volume at the listener's station set decreases; loss of volume may become a serious impairment.

Experience has shown that the loss associated with a 45-millisecond echo delay (one-way loss of about 9 to 11 dB) is about as much as can be tolerated and still produce satisfactory received volume.

Thus, circuits with echo delays in excess of 45 milliseconds are equipped with echo suppressors. These devices, used on four-wire trunks, insert high loss in the return direction when speech energy is being transmitted and permit a lower insertion loss on trunks that are so equipped.

## REFERENCES

1. Huntley, H. R. "Transmission Design of Intertoll Telephone Trunks," *Bell System Tech. J.*, Vol. 32 (Sept. 1953), pp. 1019-1036.

2. Clement, M. A. *Transmission* (Chicago: Telephony Publishing Corporation, 1969).

Chapter 21

# Phase Distortion

Some signals, such as digital and video signals, are particularly sensitive to departures from linear input/output phase characteristics in the channels over which they are transmitted. Speech signals are not adversely affected by these irregularities because the human hearing mechanism resolves signal components at different frequencies in a way that has little phase dependence; thus, little attention was originally given to phase irregularities in telephone channels. However, with the increased use of such channels for the transmission of other types of signals, the necessity for understanding and coping with phase-related impairments has continually increased. The characterization and control of wideband channels are also increasingly necessary because they are largely used by types of signals most sensitive to departures from linear input/output phase characteristics.

## 21-1 PHASE/FREQUENCY MATHEMATICAL CHARACTERIZATION

In Chapter 6 the breakdown of a square wave into its Fourier components was discussed, and the necessity of maintaining proper amplitude and phase relations among the signal components was mentioned. Impairments caused by departures from ideal (flat) amplitude/frequency response were considered in Chapter 18. Here, consideration is given to the impairments resulting from departures from the ideal (linear) phase/frequency characteristic in a channel.

### Departure from Linear Phase

Consider first, the simple square wave of Figure 21-1(a). As pointed out in Chapter 6, this wave can be synthesized from an infinite number of odd harmonics of its fundamental frequency. These harmonics and the fundamental must be controlled in both amplitude and phase. Figure 21-1(b) shows the dc, fundamental, and third-harmonic signal components of the idealized square wave of

**496**

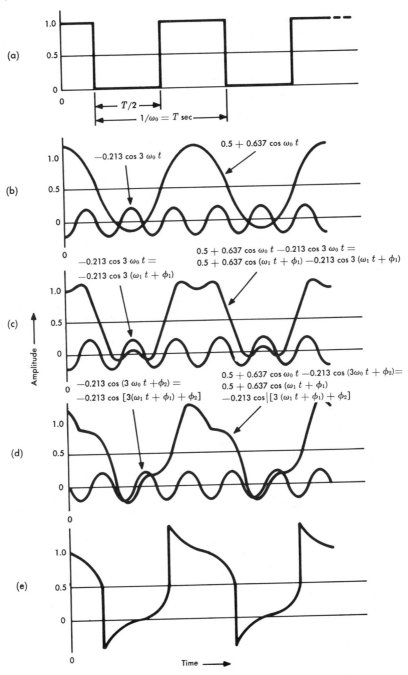

Figure 21-1. Waveforms with and without phase distortion.

Figure 21-1(a) in the phase and amplitude relationships they must bear to one another to form the original square wave. Figure 21-1(c) shows how these add to approximate the square wave.

If the channel over which these components are transmitted has a linear phase/frequency characteristic (for example, phase shift at $3\omega_0$ is three times that at $\omega_0$), the time relationship between components is unaffected by transmission. (For this discussion, the channel is assumed to have flat gain equal to 0 dB over the entire spectrum.) Since the delay is the same at all frequencies, the approximated square wave is identical at the output to that at the input (the absolute time delay is neglected).

Now consider the effect of phase distortion, i.e., a departure from linear of the phase/frequency characteristic. Assume that the phase shift at radian frequency $3\omega_0$ is not linearly related to that at $\omega_0$. The fundamental may be written $\cos(\omega_0 t + \phi_1)$, where $\phi_1$ is an arbitrarily assigned reference value of phase; the third harmonic is $\cos[3(\omega_0 t + \phi_1) + \phi_2]$, where $\phi_2$ represents the departure from a linear phase/frequency curve. In Figure 21-1(d), this relationship is illustrated for a value $\phi_2 = T/12$ seconds. The approximation to the square wave, initially shown in Figure 21-1(c), is seen in Figure 21-1(d) to be badly distorted.

Qualitatively, phase distortion changes the square wave so that it appears as in Figure 21-1(e). The exact resultant wave shape depends on the nature and magnitude of the departure from linearity of the phase/frequency characteristics.

## Phase Delay

*Phase delay, propagation time, group delay,* and *absolute envelope delay* are expressions used to define in various ways the time delay between a signal or its components at the input and at the output of a network or transmission line. These characteristics are usually functions of frequency and must be used with reference to a specific frequency. When the values of phase shift are plotted as a function of frequency, the plot is known as a *phase shift* characteristic.

As covered in Chapter 5, the ratio at any frequency of the input current, $I_1$, to the output current, $I_2$, of a four-terminal network or transmission line of unit length may be written

$$I_1/I_2 = e^{\alpha + j\beta} = e^{\alpha}\underline{/\beta} \qquad (21\text{-}1)$$

where $\alpha$ is the attenuation constant in nepers and $\beta$ is the phase constant in radians. This may also be written

$$I_2 = I_1 e^{-\alpha} \underline{/-\beta} \quad . \tag{21-2}$$

For convenience, a lossless network ($\alpha = 0$) is assumed in the following analysis. Thus, $I_2 = I_1 \underline{/-\beta}$. Equation (21-2) shows that an input signal, $I_1$, is shifted in phase by $\beta$ radians in transmission through a network. By virtue of the definitions, a positive shift, $\beta$, in the network causes a negative shift in phase of the output current, $I_2$, relative to the input current, and vice versa. A phase shift of $\pi/2$ radians between input and output is illustrated in Figure 21-2.

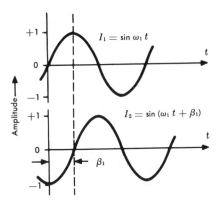

Figure 21-2. Phase shift and phase delay.

Phase delay is defined by

$$D_\phi = \frac{\beta_1 \text{ radians}}{\omega_1 \text{ radians/second}} \tag{21-3}$$

where $\beta_1$ is the phase shift at radian frequency $\omega_1$. If the phase shift characteristic is not linear, i.e., if the phase shift is not directly proportional to frequency, the phase delay characteristic is distorted. This is called *delay distortion*.

### Delay Distortion

Delay distortion is defined in terms of the delay at one frequency relative to that at another. In a telephone channel, the reference frequency is often taken as 1700 or 1800 Hz. In any channel, the reference frequency may be taken as the frequency of minimum delay.

If the phase shift characteristic is known, the delay distortion between two given frequencies may be calculated by

$$DD = \left( \frac{\beta_2}{\omega_2} - \frac{\beta_1}{\omega_1} \right) \text{ seconds} \tag{21-4}$$

where, as before, $\beta_2$ and $\beta_1$ are expressed in radians and $\omega_2$ and $\omega_1$, in radians per second. The delay distortion is usually expressed in microseconds. Note that if $\beta_2/\omega_2 = \beta_1/\omega_1$, delay distortion is zero. While the phase/frequency characteristic between $\omega_2$ and $\omega_1$ might thus appear to be linear, the ratio $\beta/\omega$ might vary considerably between the two frequencies. A more useful parameter, called *envelope delay*, is one that takes into account the rate of change of $\beta/\omega$.

## Envelope Delay

Although the phase characteristic can often be determined mathematically, it is sometimes more convenient to derive it graphically by measuring the area under the envelope delay curve, as illustrated in Figure 21-3.

Envelope delay, commonly used in describing and measuring phase characteristics of channels, is defined in terms of the slope of the phase characteristic; that is,

$$\text{ED} = \frac{d\beta}{d\omega} \text{ seconds.} \tag{21-5}$$

While phase distortion is difficult to measure or even to define explicitly, envelope delay can often be used directly in the evaluation of transmission quality.

In cases where phase delay is the quantity of interest, it can be derived in useful form from envelope delay. Where the envelope delay characteristic can be expressed mathematically, the phase shift at any radian frequency, $\omega_x$, is

$$\beta_x = \int_0^{\omega_x} \frac{d\beta}{d\omega} \, d\omega \text{ radians.} \tag{21-6}$$

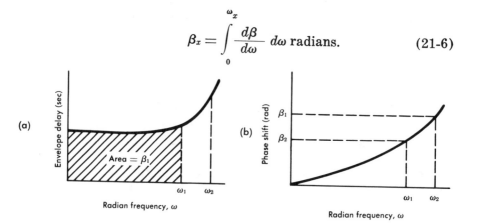

Figure 21-3. Phase shift derived from envelope delay.

## Envelope Delay Distortion

Envelope delay distortion, or relative envelope delay, is the differ-ence between the envelope delay at any frequency and that at a ref-erence frequency. The reference is usually taken as the frequency at which the envelope delay is a minimum. Thus, envelope delay distor-tion may be written

$$\text{EDD} = \frac{d\beta}{d\omega} - \left(\frac{d\beta}{d\omega}\right)_1 \text{ seconds} \qquad (21\text{-}7)$$

where $\left(\dfrac{d\beta}{d\omega}\right)_1$ is the envelope delay at the reference frequency.

## Illustrative Characteristics

Figure 21-4 illustrates the various expressions for the phase dis-tortion characteristics previously discussed. Three kinds of *channel* characteristic are illustrated, namely, an ideal linear phase charac-teristic (one that can only be approached in practice), a low-pass characteristic typically found in baseband cable transmission facili-ties, and a bandpass characteristic typically found in FDM carrier transmission facilities. The characteristics are displayed qualitatively to show their general shapes. In each sketch, the curve from $\omega = 0$ to $\omega = \omega_t$ illustrates the general shape of the inband channel char-acteristic. Above the top channel frequency, $\omega_t$, the trend of the out-of-band characteristic is illustrated for each case in a general sense. Exact characteristics vary widely according to design. The bandpass characteristic is illustrated in terms of its baseband equivalent.

## Intercept Distortion

Some signals transmitted over a channel having a certain type of phase characteristic may suffer from a form of distortion known as intercept distortion. This distortion may occur even though the phase characteristic is linear over the useful part of the band, i.e., the part carrying all significant components of the transmitted signal. This form of distortion may be illustrated mathematically with the help of Figure 21-5.

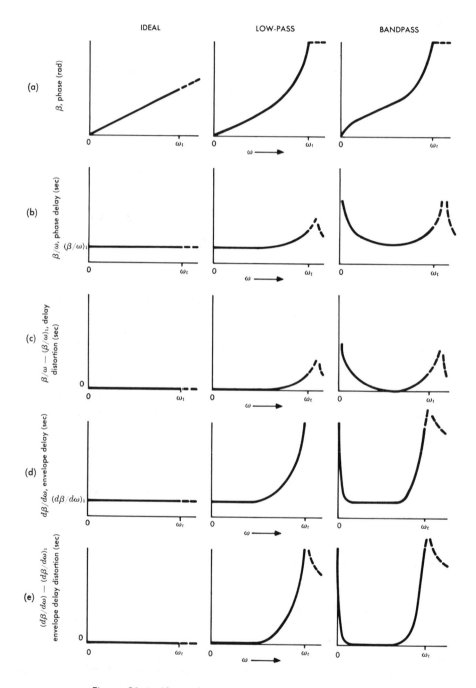

Figure 21-4. Phase distortion and related parameters.

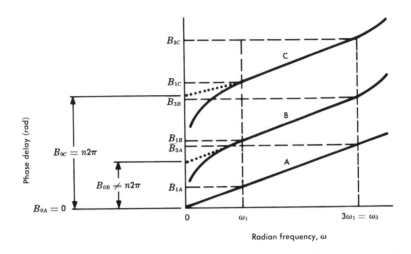

Figure 21-5. Phase delay characteristics to illustrate intercept distortion.

The three characteristic curves in Figure 21-5 are all linear in the region from $\omega_1$ to $3\omega_1 = \omega_3$. Curve A is an ideal linear phase curve; curves B and C are bandpass phase characteristics that typify transmission through a bandpass channel such as those found in FDM equipment. The characteristics have been translated (shifted in frequency) to their baseband equivalents. The dotted extrapolations of the low-frequency ends of curves B and C are straight-line extrapolations of the linear portions of those curves down to zero frequency.

The familiar analytic geometry expression for a straight line, $y = mx + b$, may be used as the equation for any of the characteristics, A, B, or C, at least up to frequency $\omega_3$. The expression, for present purposes, is

$$\beta = m\omega + \beta_0 \tag{21-8}$$

where $\beta_0$ is the zero-frequency phase intercept of the straight line with the $\beta$ axis, $\omega$ is the radian frequency, and $m$ is the slope of the straight line.

Now, consider two components of a simple input signal

$$A_1 \cos \omega_1 t + A_3 \cos 3\omega_1 t = A_1 \cos \omega_1 t + A_3 \cos \omega_3 t$$

where $\omega_3 = 3\omega_1$. At the output (if the circuit has a gain of unity), these signal components may be written

$$A_1 \cos (\omega_1 t + \beta_1) + A_3 \cos (\omega_3 t + \beta_3).$$

As previously pointed out, the signal is transmitted without distortion if

$$\cos (\omega_3 t + \beta_3) = \cos 3 (\omega_1 t + \beta_1). \qquad (21\text{-}9)$$

Now, write Equation (21-8) as

$$\beta_{3x} = m\omega_3 + \beta_{0x}$$

or

$$\beta_{1x} = m\omega_1 + \beta_{0x} \quad .$$

These relationships may be applied to the curves of Figure 21-5 by substituting A, B, or C, as appropriate, for $X$ in the subscripts. Substitute these values in Equation (21-9). Then the criterion for distortionless transmission is

$$\cos (\omega_3 t + m\omega_3 + \beta_{0x}) = \cos 3 (\omega_1 t + m\omega_1 + \beta_{0x})$$

or

$$\cos (\omega_3 t + m\omega_3 + \beta_{0x}) = \cos (3\omega_1 t + 3m\omega_1 + 3\beta_{0x}).$$

Since $\omega_3 = 3\omega_1$,

$$\cos (\omega_3 t + m\omega_3 + \beta_{0x}) = \cos (\omega_3 t + m\omega_3 + 3 \beta_{0x}). \qquad (21\text{-}10)$$

Now, examine the three characteristics of Figure 21-5. For the ideal characteristic A, $\beta_{0A} = 0$ and Equation (21-10) is satisfied. For C, $\beta_{0C} = n2\pi$, where $n = 1, 2, 3 \ldots$, and Equation (21-10) is also satisfied since the cosine of any angle $\alpha$ is equal to the cosine of any angle $(\alpha + n2\pi)$. For B, however, $\beta_{0B} \neq n2\pi$ and Equation (21-10) is *not* satisfied. The result is signal distortion similar to the distortion due to departure from linear phase.

Intercept distortion is a significant factor in producing signal impairment only when baseband digital signals are transmitted over a single-sideband AM carrier system and means are not provided for demodulating the received signal by a carrier properly related in frequency and phase to the signal received at the demodulator. The difficulty of providing such a carrier and the impairing effects resulting from failure to do so are among the reasons that such signals are so seldom transmitted over single-sideband AM facilities. Frequency shift, discussed in Chapter 19, causes impairment of the received signal by virtue of differences between input and output signal component frequencies. This impairment can be regarded as being due to a continual shift of the zero-frequency phase intercept, a shift that is further modified if the demodulating carrier drifts in frequency. Even if the demodulating carrier is synchronized exactly to the required frequency, it is difficult to maintain its phase so that the zero-frequency phase intercept is held to a value of $n2\pi$.

Generally, double-sideband AM, vestigial sideband AM, and PM signals are not impaired by intercept distortion; frequency and phase of the carrier is or can be transmitted with the signal for use at the receiver to control the frequency and phase of the demodulating carrier. Intercept and quadrature distortion both produce signal components in quadrature with the desired components. The quadrature components cause distortion of the carrier-frequency envelope waveform that is eliminated if the signal is demodulated by a carrier that is properly synchronized to the received signal. Thus, the impairment of the received signal may be large or small depending on the successful treatment of the signal in demodulation or detection at the receiving carrier terminal.

### Quadrature Distortion

When signals are transmitted by single-sideband or vestigial-sideband methods, quadrature components are generated at the transmitting terminal and appear in the carrier frequency signal to distort the waveform envelope. These components do not impair speech reception and so they are not generally eliminated by the design of receiving terminal equipment used in speech circuits. Data and video signals are impaired by quadrature components, however, and they must be eliminated or suppressed in the detection or demodulation process.

Figure 21-6 illustrates amplitude/frequency response characteristics for vestigial-sideband transmission in a manner that can be related to a mathematical demonstration of how quadrature distortion arises and how it may be evaluated. The amplitude/frequency response characteristic of the assumed channel is shown in Figure 21-6(a); signal components are transmitted from zero frequency to a top frequency of $\omega_u$. (The upper cutoff characteristic is not important in this discussion.) Figure 21-6(b) shows the amplitude/frequency response at the output of a double-sideband modulator, and Figure 21-6(c) shows the double-sideband signal as modified by a vestigial-sideband filter. The transmitted signal then has a principal lower sideband extending from the carrier frequency, $\omega_c$, to $(\omega_c - \omega_u)$ and a vestigial upper sideband extending from $\omega_c$ to $(\omega_c + \omega_v)$. Finally, as an aid to understanding quadrature distortion, the vestigial-sideband channel characteristic of Figure 21-6(c) may be regarded theoretically as being the sum of the characteristics shown in Figures 21-6(d) and 21-6(e). Note that the latter are both double sideband; Figure 21-6(d) displays even symmetry about $\omega_c$ and Figure 21-6(e) displays odd symmetry about $\omega_c$.

Now, consider a single-frequency component transmitted at radian frequency $\omega_i$. It appears in the waveform at various frequencies as shown in Figure 21-6. This signal component is translated to a double-sideband signal at carrier frequencies $(\omega_c + \omega_i)$ and $(\omega_c - \omega_i)$.

Let the signal component at frequency $\omega_i$ be represented by $a_i \cos \omega_i t$, and let the carrier be represented by $\cos \omega_c t$. The modulated signal may then be written

$$S = (A + a_i \cos \omega_i t) \cos \omega_c t \qquad (21\text{-}11)$$

where a dc component, $A$, has been arbitrarily added to the input signal component. Equation (21-11) may be expanded trigonometrically and rewritten as

$$S = A \cos \omega_c t + \frac{a_i}{2} \cos (\omega_c + \omega_i) t + \frac{a_i}{2} \cos (\omega_c - \omega_i) t. \quad (21\text{-}12)$$

To simplify the illustration of quadrature component generation, assume that the component at $\omega_i$ is in the flat portion of the baseband of Figure 21-6(a) so that the component at $(\omega_c + \omega_i)$ frequency is completely eliminated by the vestigial shaping filter of Figure 21-6(c). The remaining components then constitute a single-sideband signal with transmitted carrier,

$$S_{\text{SSB}} = \frac{A}{2} \cos \omega_c t + \frac{a_i}{2} \cos (\omega_c - \omega_i) t. \qquad (21\text{-}13)$$

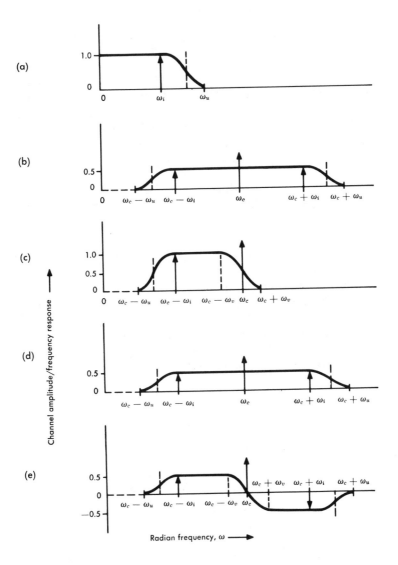

**Figure 21-6. Amplitude/frequency responses for vestigial-sideband transmission.**

The amplitude of the carrier component is halved by the vestigial filter. By further trigonometric expansion of the $(\omega_c - \omega_i)$ term and multiplication of the right side by 2 (assumed flat amplification), Equation (21-13) may now be written as

$$S_{\text{SSB}} = A \cos \omega_c t + a_i \cos \omega_c t \cos \omega_i t + a_i \sin \omega_c t \sin \omega_i t$$

$$= (A + a_i \cos \omega_i t) \cos \omega_c t + a_i \sin \omega_i t \sin \omega_c t$$

$$= [A + P(t)] \cos \omega_c t + Q(t) \sin \omega_c t \qquad (21\text{-}14)$$

where $P(t) = a_i \cos \omega_i t$ and $Q(t) = a_i \sin \omega_i t$. Note that $P(t)$, the wanted component, and $Q(t)$, the quadrature component, are both equal to the input signal but that $Q(t)$ is shifted in phase by $\pi/2$ radians relative to $P(t)$. Equation (21-14) also shows that $P(t)$ and $Q(t)$ may be regarded as modulating carriers of the same frequency, $\omega_c$, again separated by $\pi/2$ radians.

The signal components of Equation (21-14) are shown vectorially in Figure 21-7. Note that the amplitude of the resultant vector, equal to the envelope of the signal, is determined by

$$| S_{\text{SSB}} | = \sqrt{(A + | P |)^2 + | Q |^2} \qquad (21\text{-}15)$$

The distorting effect of the quadrature component is illustrated in Figure 21-8. The amplitude of the peak in the illustration can be estimated from Equations (21-14) and (21-15). If there is no dc component, $A$, the amplitudes of $P(t)$ and $Q(t)$ are equal and the peak excursion of the signal of Figure 21-8 is 1.7 times the amplitude of the step. If a dc component is added, the influence of the quadrature term

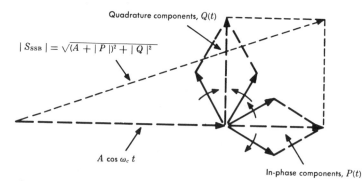

Quadrature components, $Q(t)$

$| S_{\text{SSB}} | = \sqrt{(A + | P |)^2 + | Q |^2}$

$A \cos \omega_c t$

In-phase components, $P(t)$

Figure 21-7. Analysis of SSB signal into in-phase and quadrature components.

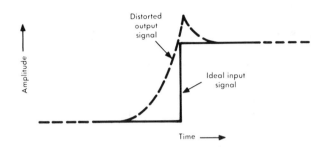

Figure 21-8. Effect of quadrature distortion on signal characteristic.

is reduced, as can be seen by examination of Equation (21-15). The shape of the modified step transition in Figure 21-8 is influenced by the ratio of the vestigial to the principal sidebands and the shaping of the characteristic through the vestigial region.

### Differential Phase

In a video transmission system, differential phase is defined as the difference in output phase of a low-amplitude, high-frequency sine-wave signal at two stated amplitudes of the low-frequency signal on which the high-frequency signal is superimposed. Since color television is the only signal significantly affected by the phenomenon, the definition of differential phase is in terms pertinent to color signal characteristics. Differential phase measurements are made using a 3.579545-MHz low-amplitude signal to simulate the color carrier in an NTSC color signal. The low-frequency signal used is usually a 15.75-kHz sine wave simulating the fundamental line scan frequency of the luminance component of a standard color signal. Other low frequencies are also sometimes used.

Mathematically, the generation of differential phase can be demonstrated by considering the intermodulation of the two test signals due to nonlinearity in the circuit being evaluated. Let $\beta$ represent the low frequency and $\alpha$ represent the high frequency. The input signal then may be represented by $e_{in} = A \cos \alpha t + B \cos \beta t$. Also, let the input/output characteristic of the circuit be represented by

$$e_{out} = a_0 + a_1 e_{in} + a_2 e_{in}^2 + a_3 e_{in}^3 + \ldots \qquad (21\text{-}16)$$

To simplify the illustration, consider only the wanted output term, $a_1 e_{in}$, and the distortion term, $a_3 e^3_{in}$. When the input signal is substituted in these terms, the output may be written

$$e_{1out} = a_1 \, (A \cos \alpha t + B \cos \beta t) + a_3 \, (A \cos \alpha t + B \cos \beta t)^3. \quad (21\text{-}17)$$

Now, with respect to the definition of differential phase, attention may be concentrated on just three of the terms that can be derived by expanding Equation (21-17).* These may be written

$$e_{2out} = a_1 A \cos \alpha t + \frac{3}{4} \, a_3 \, AB^2 \cos \, (\alpha + 2\beta) t$$

$$+ \frac{3}{4} \, a_3 \, AB^2 \cos \, (\alpha - 2\beta) t. \quad (21\text{-}18)$$

The signal of Equation (21-18) may be regarded as an amplitude-modulated version of the desired output component, $a_1 \, A \cos \alpha t$, with sidebands at frequencies $(\alpha + 2\beta)$ and $(\alpha - 2\beta)$. Amplitude and phase relations among the signal at frequency $\alpha$ and its sidebands determine the nature and magnitude of the impairment. If the sideband components combine in phase with $a_1 \, A \cos \alpha t$, the impairment is differential gain, discussed in Chapter 18; if the sideband energy is at $\pi/2$ radians relative to the wanted signal, the impairment is essentially all differential phase. If the sidebands combine and lie between 0 and $\pi/2$ radians relative to the wanted signal, the impairment is a combination of the two.

Differential phase is illustrated by the phasor diagram, Figure 21-9. If the phasors, $e_{(\alpha+2\beta)}$ and $e_{(\alpha-2\beta)}$, are small compared to $e_\alpha$, the magnitude of the resultant, $e_r$, is nearly constant and equal to $e_\alpha$. Also, the differential phase shift, measured by the angle $\phi_d$, is a

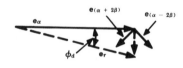

Figure 21-9. Phasor diagram illustrating differential phase.

maximum when $e_{(\alpha+2\beta)}$ is in phase with $e_{(\alpha-2\beta)}$. Thus, the differential phase is equal to

$$\phi_d = \tan^{-1} \frac{|\, e_{(\alpha+2\beta)} + e_{(\alpha-2\beta)} \,|}{|\, e_\alpha \,|} \quad . \quad (21\text{-}19)$$

* The complete expansion of a 3-component input signal through a nonlinear power series like that of Equation (21-16) is given in Chapter 17 (Figure 17-7).

## 21-2 PHASE DISTORTION IN TELEPHONE CHANNELS

All forms of phase distortion described so far may occur in standard 4-kHz channels used for telephone signal transmission. With the exception of differential phase, all may impair the signals typically transmitted in these channels. The sources of signal impairments are largely departures from linear phase/frequency characteristics, the resultant generation of echoes, and signal-dependent distortion such as quadrature distortion. The effects on various types of signals may be quite different depending on the nature of the transmitted information [1].

### Phase/Frequency Impairments

Phase/frequency impairments may result from many sources of distortion in a telephone channel. The two most common sources are those related to the transmission characteristics of loaded cable circuits and those related to the cutoff characteristics of filters in FDM equipment.

The transmission characteristics of loaded cable are similar to those of a low-pass filter. The phase/frequency characteristic of a low-pass filter is illustrated qualitatively in Figure 21-4. The exact shape and magnitude of the channel distortion and the extent of the impairment to signal transmission depend on many interrelated parameters such as the type of cable and loading, the lengths of line and end section, the impedance match of the source and load to the line, and the characteristics and susceptibility of the transmitted signal.

The effects of bandpass channel filters are also illustrated in Figure 21-4. The shape and magnitude of the channel distortions and the resulting signal impairments are significantly influenced both by the number of filters that may be connected in tandem in various situations and by the transmitted signal characteristics.

**Digital Signals.** As in the case of other forms of impairment, the effect of phase/frequency distortion on digital signals is conveniently expressed in terms of an increase in error rate or in terms of an equivalent signal-to-noise impairment. One simple form of signal impairment, illustrated in Figure 21-1, results from an assumed simple type of phase/frequency distortion. More complex forms of channel distortion cause more complex forms of signal distortion.

Some are amenable to theoretical analysis and evaluation, but in field situations they usually are most easily evaluated by error rate or eye diagram measurements. One useful type of analysis involves expressing the effect of channel impairments in terms of the generation of signal echoes. This work is particularly useful in the evaluation of analog signal distortion but can be adapted to digital signal impairment in some cases [2].

**Analog Signals.** While speech signals are not adversely affected by phase/frequency distortion, analog data signals and particularly telephotograph signals transmitted over telephone channels may be severely impaired by echoes that result from this form of distortion.

In the telephotograph picture, an echo appears as a low-amplitude reproduction of the picture or portions of the picture displaced from the picture signal itself. The amount of displacement and the faithfulness of reproduction in the echo depend on the magnitude and nature of the distortion. If the displacement is small, picture details may simply be blurred or distorted; if the displacement is large, the impairment may more nearly appear as a faint reproduction of the picture with positive or negative polarity.

Detailed analyses of echo phenomena have been made in order to provide quantitative evaluation of such impairments [3]. One type of analysis that has proved to be valuable for all forms of video signal transmission utilizes the concept of an *echo rating*. While applicable to any form of video signal transmission, the concept was first applied to television signal transmission and later used to analyze transmission over a PICTUREPHONE channel [4].

## Quadrature Distortion

As illustrated in Figure 21-8, quadrature distortion causes unwanted changes in a signal shape at a point where there is a sharp transition in the signal amplitude. Digital data signals and telephotograph signals commonly contain such sharp transitions and are thus particularly susceptible to quadrature distortion. In a carrier signal, the impairment must be evaluated in terms of possible overload effects. Peak factors such as those illustrated in Figure 21-8 may cause overload in carrier equipment and result in clipping of the signal peaks or in an increase in intermodulation or both. Intermodulation may cause deterioration of the signal itself or of other signals sharing the same medium.

While quadrature distortion is most evident in carrier signal wave-forms, its effects may also appear in the signal after detection. These effects may increase the error rate for digital signals or may cause blurring or smearing of portions of a telephotograph signal in areas where there are large and sudden changes in the luminance of the image.

## 21-3 PHASE DISTORTION IN BROADBAND CHANNELS

Broadband channels are provided primarily for the transmission of television, PICTUREPHONE, and wideband digital signals, all of which are particularly sensitive to phase distortion. All of the general comments and discussion relating to phase distortion in telephone channels apply equally well to transmission in broadband channels. In addition, quadrature distortion and differential phase have partic-ularly serious effects on the transmission of wideband signals.

### Phase/Frequency Distortion

Minimizing phase/frequency distortion in broadband channels through equalization techniques is made difficult by the multiplicity of phase/frequency characteristic shapes encountered across the band in a long transmission facility. This difficulty is a result of the many modulation stages and tandem application of connectors and/or blocking filters that may be found in long systems.

### Quadrature Distortion

Quadrature distortion effects on wideband digital signal trans-mission are similar to the effects on voiceband digital signal trans-mission. Signal peaking must be controlled to avoid clipping and transmission system overload. The loss of margin, or increase in error rate, caused by quadrature distortion of the detected signal usually can not be tolerated; therefore, this form of distortion must be carefully controlled in the received signal. Often some form of vestigial sideband modulation is chosen so that the carrier can be recovered in proper phase at the demodulator.

Quadrature distortion of PICTUREPHONE and television signals must also be carefully controlled to limit the blurring or smearing of the picture at sharp luminance transitions. There has been little PICTUREPHONE transmission by SSB or VSB techniques so that the quadrature distortion problem has not really been faced. Considerable effort was made to limit quadrature distortion in the transmission of television signals over L1 and L3 Coaxial Carrier Systems [5, 6].

## Differential Phase

Differential phase, as discussed previously, is of importance only in the transmission of NTSC color signals. When this distortion is encountered, the phase relationships among color signal components transmitted at high video frequencies vary with respect to one another. The result is a change in the hue of various colors. If the distortion is excessive, colors may change completely as the luminance of the picture varies.

## 21-4 MEASUREMENT, EVALUATION, AND CONTROL OF PHASE DISTORTION

Most of the impairments discussed in earlier chapters can be measured by relatively straightforward methods and by the use of conceptually simple test ·equipment such as signal generators and detectors. Phase distortion, however, is somewhat more difficult to measure and evaluate, partly because of the difficulty of defining and measuring a reference phase. As a result, most measurements are made in terms of relative phase. These observations apply equally to voiceband and broadband measurements; and while different test set designs are used for different bandwidths, the theory of operation is often independent of the bandwidth.

### Phase/Frequency Measurement and Evaluation

Phase/frequency measurements are seldom made in the field on voiceband circuits because of the inherent difficulty of establishing a reference phase. In addition, the complexities of measuring and evaluating even relative phase (delay distortion or envelope delay distortion) are great enough that time-domain measurements are usually favored. Time-domain measurements have the advantage of relative simplicity and speed, but they have the disadvantage that the effects of individual impairments are difficult to separate from one another. When poor performance is indicated, the time-domain measurements must usually be supplemented by frequency-domain measurements in order to identify specific sources of impairment. The common time-domain methods of measurement and evaluation include error rate measurements by means of pseudo-random data messages and eye-diagram presentations on a cathode-ray oscilloscope. Another method involves the use of the P/AR system of measurement.

**The P/AR System.** The peak-to-average ratio (P/AR) provides a single number measure of the overall quality of circuits used for the transmission of voiceband data signals [7]. A P/AR generator transmits a precise, repetitive pulse through the system. The pulses, dispersed by all of the impairments in the channel, are then delivered to the P/AR receiver which responds to the pulse envelope peak and the pulse envelope full-wave average. A meter in the receiver is used to indicate the ratio of these two parameters. The ratio, called the P/AR rating, appears on the meter in a form determined by

$$\text{P/AR rating} = 100 \left[\, 2\, \frac{\text{E (pk)}}{\text{E (FWA)}} - 1 \,\right] \qquad (21\text{-}20)$$

where E(pk) is the normalized peak value of the pulse envelope and E(FWA) is the normalized full-wave rectified average value of the envelope.

The P/AR approach to voiceband circuit evaluation is valuable because of its simplicity and speed. It also lends itself well to analytical evaluation based on measured or estimated amplitude/frequency and phase/frequency channel characteristics. Computer programs have been developed to accomplish this evaluation [8].

**Envelope Delay Distortion Measurements.** While phase measurements per se are seldom made, envelope delay measurments can be made without excessive complexity of test equipment, measuring technique, or interpretation. The results of these measurements can then be converted to a phase shift characteristic when needed. Usually, however, the envelope delay distortion curve suffices for evaluation of a circuit.

The usual method of measuring envelope delay can be described with the assistance of Figure 21-10, which illustrates the inband phase/frequency characteristic of a bandpass channel translated to baseband. Envelope delay can be measured to a close approximation at radian frequency $\omega_c$ by transmitting a carrier frequency signal

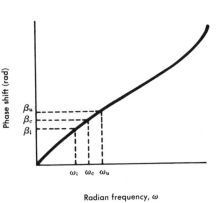

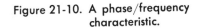

Figure 21-10. A phase/frequency characteristic.

at $\omega_c$ modulated by a signal which places sidebands at frequencies $\omega_l$ and $\omega_u$. If the modulating frequency is chosen so that the phase characteristic between $\omega_l$ and $\omega_u$ is *approximately* linear, the phase difference between the carrier and the two sideband frequencies may be written

$$\beta_u - \beta_c = \beta_c - \beta_l.$$

The envelope delay may then be determined as the slope of the phase curve at the carrier frequency, $\omega_c$, used for the measurement. Equation (21-5) may then be applied to this measuring technique:

$$ED = \frac{d\beta}{d\omega} \approx \frac{\beta_u - \beta_c}{\omega_u - \omega_c} \approx \frac{\beta_c - \beta_l}{\omega_c - \omega_l} \quad . \tag{21-21}$$

The modulating frequency must be quite low to make Equation (21-21) a reasonable approximation. In Bell System voice-band measuring sets, the modulating frequency is 83-1/3 Hz. Sets designed for wideband channel measurements use proportionately higher carrier and modulating frequencies.

Measurements of the type just described may be made on a point-by-point basis by adjusting $\omega_c$ manually and measuring the phase differences at each setting. The measurement may be automated by sweeping the band with a continuously varying $\omega_c$ and displaying the envelope delay on an oscilloscope or other recording device.

**Echo Observations.** Sometimes, gross evaluation of channel performance can be accomplished by transmission of test signals and examination of the received signal on an oscilloscope or, in the case of video circuits, on a television or PICTUREPHONE receiver. More precise analysis or measurement by other techniques is then necessary.

## Phase Distortion Control

The control of phase/frequency characteristics, quadrature distortion, and differential phase must all be considered in the design and/or operation of telecommunication facilities.

**Phase/Frequency Characteristic Control.** The necessity of controlling high-frequency and low-frequency cutoff characteristics has been previously discussed from the point of view of satisfying Nyquist's criteria. These cutoff characteristics must generally be designed to have linear phase/frequency characteristics, most nearly achievable if the roll-offs are gradual. Sharp channel cutoffs are to be avoided.

Where channels are to be used for digital signal transmission, their design may involve a conflict which must be resolved by carefully determined compromises. To make transmission efficiency as high as possible, it is tempting to design for the absolute maximum rate of transmission that a channel can theoretically accommodate. This design approach usually produces significant amounts of energy near the band edges where characteristics are hardest to control. Generally, it is advisable to restrict the bit rate to something below the maximum in order to provide operating margin.

Another means of controlling the phase/frequency characteristic is by equalizing the channel to produce a near linear phase/frequency characteristic. Many types of delay (phase) equalizers are used for this purpose. If a channel is dedicated to the transmission of a limited number of signal types, specific designs of equalizers may be used to equalize that channel for satisfactory performance. If the channel is partly dedicated and partly switched, fixed equalizers may be provided for the dedicated portion of the circuit and for some average characteristic representing the switched portion. Adjustable equalizers may also be provided in both dedicated and switched channels in order to compensate for seasonal or other time-varying phenomena. Finally, adaptive equalizers have been designed to adjust themselves automatically to gain and delay impairments on the basis of certain signal characteristics. These equalizers are usually of the tapped delay line type [9].

**Quadrature Distortion Control.** There are a number of design-related ways of controlling quadrature distortion. These include several that, of necessity, involve design compromises. For example, the wider the vestigial sideband, the less the quadrature distortion; however, the decrease in distortion can only be achieved at the expense of a greater total bandwidth or a reduction of the width of the wanted sideband.

The index of modulation is another design parameter that may be used to reduce quadrature distortion. The amplitude of the component $A \cos \omega_c t$ in Equation (21-14) is related to the index of modulation, as discussed in Chapter 8. The larger this component (low index of modulation), the more nearly the total transmitted signal is in phase with the wanted signal $P(t)$ cos $\omega_c t$; the quadrature component, $Q(t)$, then has less effect, as can be seen by examination of Figure 21-7. However, this means of reducing quadrature distortion can only be accomplished at the expense of overall signal-to-noise ratio. For a given maximum magnitude of transmitted modulated signal, the higher the index of modulation, the better the resulting signal-to-noise ratio.

The last important factor involves the use of product demodulation rather than envelope detection at the receiver. As previously described, quadrature distortion is first evident in the SSB or VSB carrier frequency signal. If this distorted signal is envelope-detected, the quadrature distortion is carried through the receiving equipment and appears in the output signal. If product demodulation is used at the receiver and the distorted carrier frequency signal has not been clipped or limited in any way, the quadrature distortion does not appear in the output signal, provided the phase of the demodulating carrier is very close to that of the received carrier [6].

**Differential Phase Control.** This form of distortion cannot be controlled in operation except by ensuring that signal amplitude does not exceed the design value. This distortion can only be controlled in the integrated signal-system design process, which must result in satisfactory performance—both signal-to-noise and differential phase performance—in the face of natural nonlinearities in the channel transmission equipment.

### REFERENCES

1. Bell System Technical Reference PUB41004, *Transmission Specifications for Voice Grade Private Line Data Channels* (American Telephone and Telegraph Company, Oct. 1973).

2. Mahoney, J. J., Jr. "Transmission Plan for General Purpose Wideband Services—Appendix A," *IEEE Transactions on Communications Technology*, Vol. COM-14 (Oct. 1966), pp. 647 and 648.

3. Wheeler, H. A. "The Interpretation of Amplitude and Phase Distortion in Terms of Paired Echoes," *Proceedings of the IRE* (June 1939), pp. 359-385.

4. Crater, T. V. "The PICTUREPHONE System: Service Standards," *Bell System Tech. J.*, Vol. 50 (Feb. 1971), pp. 255-258.

5. Morrison, L. W. "Television Terminals for Coaxial Systems," *Transactions of the AIEE*, Part II, Vol. 68 (1949), pp. 1193-1199.

6. Rieke, J. W. and R. S. Graham. "The L3 Coaxial System—Television Terminals," *Bell System Tech. J.*, Vol. 32 (July 1953), pp. 915-942.

7. Campbell, L. W. Jr. "The PAR Meter: Characteristics of a New Voiceband Rating System," *IEEE Transactions on Communication Technology*, Vol. COM-18 (Apr. 1970), pp. 147-153.

8. Fennick, J. H. "The PAR Meter: Applications in Telecommunications Systems," *IEEE Transactions on Communication Technology*, Vol. COM-18 (Feb. 1970), pp. 68-73.

9. Lucky, R. W. "Techniques for Adaptive Equalization of Digital Communication Systems," *Bell System Tech. J.*, Vol. 45 (Feb. 1966), pp. 255-286.

Chapter 22

# Maintenance and Reliability

This chapter relates the basic principles and general application of maintenance and reliability to transmission and service impairments. Serious impairment to transmission or service can occur when poor maintenance design practices are followed, when maintenance procedures are inadequate, or when unreliable equipment, apparatus, and facilities are used. Maintenance and reliability are interrelated; in the extreme, poor maintenance can lead to the ultimate impairment, system failure.

While maintainability and reliability must be carefully planned during the design, development, and installation of all equipment and systems, maintenance must also be a continuing concern throughout the service life of each system or item of equipment so that performance standards continue to be met. Awareness of and familiarity with all facets of maintenance systems, maintenance support systems, and test equipment are major elements in the control of network transmission performance.

The reliability aspects of transmission systems vary widely in accordance with such factors as accessibility, availability of protection switching and emergency broadband restoration facilities, and the impact of service outages on the kinds of circuits to be routed over the system. A balance must be sought among such factors as the degree of reliability improvement obtained, the cost of the improvement, the time allowance for temporary outage deemed acceptable to the customer, and the cost of service restoration when outages do occur.

Economics plays a large role in the design, development, and operation of maintenance and reliability aspects of equipment and systems. One example may be found in submarine cable system design and operation. The cables and repeaters in these systems are placed in a highly isolated and stable environment, the ocean floor.

520

However, when failure occurs, the recovery of cables and repeaters for repair is a very time-consuming and costly operation. The revenue lost during system outages can also represent a substantial financial penalty. For these reasons, it is economical in submarine cable systems to spend large sums to provide high system reliability and accurate fault-location equipment in spite of the favorable environment.

## 22-1 MAINTENANCE

Maintenance work is carried out either to correct an existing trouble or to minimize or avoid the occurrence of trouble. In the first case, there are various indications of trouble conditions, which alert maintenance personnel to the need for repairs. The indications may come directly from a customer, an operator, or other observer of a malfunction; or trouble may be indicated by local or remote alarms or measurements which reveal that some parameter fails to meet requirements.

The second case, preventive maintenance, is performed on a routine basis for the purpose of recognizing, limiting, or preventing the deterioration of transmission performance and minimizing the likelihood of service failure. Preventive maintenance activities may involve only measurements; if no trouble is indicated by the measurements, further action may be unnecessary.

Many transmission parameters, such as noise, loss or gain, balance, etc., are measured periodically under various environmental conditions. The results of the measurements are reported to a central point where the data are analyzed and combined. From these analyses, indices are derived and published both as a means of comparing performance with other organizational units for which similar indices are derived and as a means of determining trends in performance. By using these indices as guides, it is often possible to see where and when preventive maintenance routines are not being followed or are inadequate or incomplete and where their application must be strengthened. In many cases, routine maintenance procedures are prescribed in which a system is temporarily removed from service at specific intervals so that it can be realigned for optimum performance.

Requirements for preventive maintenance and periodic performance measurements have greatly increased since direct distance dialing has become widespread. In the past, most long distance calls and many local calls were established with the help of an operator. When transmission was unsatisfactory or when there was a service failure, it was usually possible for the operator to identify the defective circuit and to report it to maintenance personnel. Now, the customer often fails to report troubles, particularly those of a marginal nature; and even when a report is made, it is difficult to identify the source of trouble. In addition, the tremendous plant growth and the need to improve the productivity of plant maintenance personnel has added continuing emphasis to the need for preventive maintenance routines.

## Sources of Deterioration and Failure

Causes of performance deterioration are numerous. As devices age, they often perform less efficiently and cause changes in critical parameters. Aging effects are, of course, more pronounced in equipment employing electron tubes than in equipment employing solid-state devices; but all devices, active and passive, display some form of deterioration with age. This is most apparent where high mechanical, thermal, or electrical stresses exist.

Where moving parts are involved, electrical performance and reliability deteriorate due to mechanical wear. This type of deterioration is most often found in electromechanical switching systems; also, transmission paths through relay and plug-in unit contacts, etc., are often adversely affected by increases in noise or loss.

Each year, millions of telephone connections are changed because people move, equipment is rearranged, and facilities are changed. The resulting plant rearrangements often cause performance deterioration since undesired bridged taps may be left on cable pairs, impedance relationships may change at interface points to produce changes in return loss and echo, and defective workmanship can cause unwanted grounds or circuit crosses.

Weather changes may also cause deterioration of transmission performance. Seasonal changes make it necessary in some carrier systems to readjust equalizers to compensate for changes in transmission characteristics. Moisture due to rain or humidity can produce trouble conditions in the loop and trunk plant, particularly if there are numerous open-wire circuits or cable sheaths that have deteriorated so that they are no longer waterproof.

Finally, equipment defects are also a source of impairment or unreliability. These may result from poor quality control or manufacturing errors, poor workmanship (including installer errors) in the field, damage in transport or in service, unusual stress, and many other causes.

## Maintenance Systems and Equipment

Maintenance arrangements are provided for transmission systems as required to meet best the demands of satisfactory cost and service relationships. Sometimes the maintenance equipment is built into the transmission system as a subsystem. In other cases the maintenance or monitoring equipment may be centralized and applied to several transmission systems. Maintenance operations may be automatic or manual, locally or remotely controlled, and may involve the use of fixed or portable test equipment. All these options depend on specific applications to the purpose to be served.

**Integrated Designs.** Where continued satisfactory operation of a transmission system depends on the frequent adjustment of system components or where efficient fault-location procedures must be provided in order to minimize the cost of service failure and repair, maintenance equipment is often built in as a subsystem of a transmission system.

An example of integrated maintenance equipment is found in the L5 Coaxial Carrier System. In L5, a transmission surveillance center is provided at certain main repeater stations. This equipment provides the capability of measuring remotely the gain-frequency characteristics of a large number of coaxial transmission systems or selected parts of systems. In addition, fault-location equipment may be activated from the surveillance center to assist in the identification and isolation of troubles in remote repeaters. Protection switching functions can also be activated from the surveillance center.

**Adjunct Designs.** Many designs of maintenance equipment and complete maintenance systems have been provided as adjuncts to transmission systems or to the switched networks. These are maintenance facilities that interconnect manually or automatically with transmission systems or with large groups of trunks. Their functions range from simple manual or automatically sequenced measurements of loop-to-ground resistance to complex series of automatic loss and noise measurements of both directions of transmission on interoffice trunks. Special test bays are provided to measure automatically or manually the performance of circuits dedicated to the transmission of specialized types of signals such as digital data and video signals.

The provision of adjunct test facilities has been stimulated by the expanding plant. The large number of circuits that require testing has led to considerable automation; time, cost, and manpower limitations simply do not permit manual testing. The expansion of types of services has created a demand for well designed test facilities and orderly procedures for their use. For example, the increased use of telephone channels for data transmission and the increased interconnection of customer-provided data transmission equipment have led to the design of test facilities that can be located in a telephone central office and yet test the circuits in both directions of transmission ("loop-around testing"). This mode of testing has the added advantages of minimizing the number of visits that must be made to remote locations to test such circuits and minimizing the amount of portable test equipment that must be carried to customer locations, thus making the maintenance job more economical.

A wide variety of portable test equipment is also required for various phases of the maintenance task—prevention, identification, isolation, and repair. Portable adjunct equipment includes signal generators and detectors, noise measuring sets, and delay distortion measuring sets. They may be used sporadically in investigating trouble situations or periodically in preventive maintenance procedures. Most central office transmission paths, for example, are tested periodically for noise and, against standard terminations, for adequate return loss.

**Maintenance Support Systems.** Systems that may be regarded as in the maintenance support category are those that provide communication service for maintenance personnel (order wires), local and remote alarm and telemetry arrangements, and system features that are adapted to the maintenance function. These support arrangements may be integrated into the transmission system or may be provided as adjuncts.

Consider first the communication facilities needed by maintenance personnel. In some instances, the equipment may be simply a telephone connected to the public switched network to permit direct dialing by the maintenance man at a remote location to his home base or to another remote location. In other cases, typically in coaxial system operations, private channels (called order wires) are installed. These use interstitial wire pairs in the cable or may even be assigned separate facilities in order to increase reliability. Order-wire systems have become quite sophisticated and may include switching arrangements and alternate use for data transmission. Such facilities may also be integrated with the transmission system. An example is found in the TH radio system which includes an order-wire circuit in the basic system design.

Alarms are provided in every system to alert maintenance personnel to real or incipient trouble conditions. This type of maintenance support equipment also varies widely in design and application. An alarm may be as simple as a local alarm actuated when a fuse operates to light a lamp and/or to sound a bell or buzzer. Alarm information may be extended from remote, unattended locations to a manned central location. The extension of the alarm may again be very simple—for example, the connection of the alarm indicator to the remote location over a pair of wires—or the connection may be over a data transmission system that collects alarm information from many remote locations and forwards it to the central location. This type of system may even provide for the remote control from the central location of certain maintenance functions at the remote stations. For example, in certain versions of the TH-3 Microwave Radio System, equipment is built into the radio equipment bays to provide for some remote control maintenance functions and to report to a central location the existence of alarms and other forms of trouble indication at the remote radio repeater stations.

Automatic protection switching systems are often provided so that a hot standby facility is switched into service in the event of failure of a working system. These switching facilities often provide a maintenance support function. When it is necessary to perform maintenance, measurement, or repair of a working system, service may be temporarily transferred to the spare facility while the maintenance work proceeds. Even where protection switching facilities are not available, this function is sometimes accomplished by patching to spare facilities.

**Documentation.** The description of maintenance equipment and the specification of tests and testing intervals are important aspects of maintenance operations. For tests to be meaningful, the test equipment must be properly calibrated and personnel must be trained in its use.

In the Bell System, many internal operating and maintenance documents are devoted to descriptions of all types of test equipment and to their calibration and use. Sections devoted to descriptions of transmission systems and their operation have parts which contain directions and suggestions on system maintenance procedures and intervals. When these guides are not faithfully followed, system deterioration may accelerate.

## 22-2 RELIABILITY

Reliability may be considered with respect to a device, a circuit, a transmission system, or service to the customer.* The reliability of a device is defined as the probability that the device will continue to function satisfactorily during some specified interval, normally its useful life. Where repair and replacement of failed devices (i.e., maintenance) is feasible, reliability is defined for a system comprised of discrete devices as the percentage of time the system is expected to operate satisfactorily over a given time interval.

The opposite of reliability is the probability of failure during a specified time interval. For systems, this measure of unreliability is often expressed as the *outage* time over a given time period. Short-term and long-term outages and intervals between which outage times are measured may all be important, as in the case of microwave radio systems where short-term outages due to fading differ in their effect from longer outages due to gross equipment failures. Typically, the objectives for system outage are expressed as minutes per year. Since systems are comprised of many devices, overall system failure rates and reliabilities are functions of many complex combinations of individual device reliabilities. The laws of probability, discussed in Chapter 9, are used to evaluate these combinations. In general, combinations of devices in series are more unreliable than the least reliable device; parallel combinations are more reliable than the most reliable device. Increased reliability of parallel com-

---

* An even broader term, survivability, is used to describe the ability of the network to function in the event of enemy attack on the contiguous 48 states. This subject is not covered here since it is only indirectly related to transmission.

binations is the justification for providing diversity for critical systems or components of systems.

## Sources of Failure

The sources, causes, and mechanisms of service failure may be categorized in many ways. The principal categories can be defined as external and internal; within each of these, there are natural and man-made categories. These categories may be considered briefly and their effects on the deterioration of service and signal transmission may be qualitatively evaluated.

**External Sources.** Among the most common external sources of failure are the effects of weather and other natural phenomena. Lightning, in its direct impact, causes serious damage even to well-protected cables and equipment. Indirectly, it is a source of impulse noise, static in radio transmission, and induced currents in wire circuits that can cause damage and system outage. Ice, snow, wind, and water can also be destructive. Ice on microwave system antennas causes serious deterioration of transmission performance. Ice, snow, and wind often bring down aerial wire and cable. They make access difficult where remote equipment and facilities are necessarily exposed. Water does tremendous damage when flooding occurs, but even relatively light rain or humidity can cause deterioration of service where insulation is exposed and weakened by age and the elements. Rain attenuation of some microwave radio signals is a serious source of impairment. Atmospheric layers not broken up by convection or winds are a source of refractive fading in microwave radio systems.

Other natural phenomena, such as sunspot eruptions and the aurora borealis, can create earth currents that temporarily disable cable system operations or create inoperable conditions in high-frequency radio transmission. Finally, communications systems are in no way immune to the devastation caused by earthquakes, landslides, and fire.

Among man-made sources of failure are the environmental hazards created by nearby power transmission systems. These systems may induce interference currents into communication circuits or, in the event of certain power system faults, may produce damaging currents and expose personnel to high voltages. If communications circuits are exposed to dc power systems, still found in traction company operations, the damaging effects of electrolysis must be considered.

Construction, installation, and maintenance are also frequent causes of failure. Outside plant may be damaged by workmen pursuing highway or building construction activities or service may be disrupted inadvertently during normal outside plant operations. Sometimes, service outages are a result of automobile accidents.

While little damage has occurred to Bell System plant as a result of enemy action in time of war, this potential source of failure must be of great concern to all those responsible for the design, development, and operation of all parts of the plant. Little can be done to protect against direct hits of even conventional weapons. The possibility of direct damage and the effects of electromagnetic pulses due to near misses of atomic explosives are given much consideration in the design of portions of the present-day plant which carry critical services.

**Internal Sources.** Within systems, circuits, and devices there are a number of natural or man-made stresses that may be causes or sources of unreliability. Among these stresses are high voltage (which may produce noise or failure by breakdown), heat (which accelerates the aging process and may cause fire), and mechanical stress (which may cause fatigue failure or breakage due to mechanical shock or long-term vibration in transport or service). Natural aging is, of course, also a source of performance degradation and, ultimately, failure.

Defects due to manufacture, handling, design, or improper installation may cause failure or deterioration. Such defects are sometimes hard to control because they are so unpredictable.

Another form of internal stress is that of overload. At least two forms of overload can cause transmission impairments. One form, which results when signal amplitudes exceed design values, produces serious transmission performance impairments due primarily to intermodulation and, in the extreme, can cause transmission system failure. This form of overload sometimes occurs when test signals are misapplied or when high noise amplitudes are introduced by a feeding system that has failed. The second form of overload, excessively high traffic, has its greatest impact on switching system operation. This form of overload causes blocking of calls and a breakdown of service. It can impair transmission performance only indirectly by imposing higher than average loads on affected systems.

## Designs for Reliability

Reliability in a telecommunication network hinges on the design of all elements in the network. Examples are cited of apparatus, circuit, system, and manufacturing designs that have had a direct impact on reliability performance.

**Apparatus.** Certain types of apparatus are commonly used to protect circuits and systems from failure due to high-voltage that may result from lightning or contact with power systems, or excessive currents caused by power system faults. Protective apparatus includes carbon-block, gas-tube, or solid-state (diode) protectors which break down when subjected to excess voltage and carry fault currents safely to ground. These devices are themselves sometimes sources of transmission impairments; when lightning or other faults cause partial breakdown, low resistance to ground or high resistance or opens in one conductor path may result, impairing the circuit by excessive attenuation, noise, distortion, or crosstalk.

Heat coils are used on many circuits to protect personnel and equipment from power sources which have voltages too low to operate carbon-block or gas-tube protectors and which produce currents to ground (through office equipment) too low to operate normal fuses or circuit breakers. Heat coils are designed to operate when fault currents exceeding specified values continue to flow on a communications conductor for time periods sufficient to cause excessive heating and/or fire in the equipment. Operation of the device serves to ground the offending conductor permanently. The heat coil must be replaced when the fault has been cleared.

Fuses and circuit breakers are devices which also operate to protect personnel and equipment from excessive voltage or current. These devices operate to open the offending circuit.

**Circuits.** A few examples of the many circuit arrangements furnished to provide reliable operation are cited here. Among the most important is the central office battery arrangement that is used to furnish common battery to all or most switching, signalling, and transmission systems associated with each office. The battery supply circuits are designed so that in normal operation the load is supplied from the primary commercial source with a charging current supplied to the battery. The size and capacity of the battery are determined by the load that must be carried by the battery alone in the event of primary power failure and by the amount of time the battery alone must carry the load without service failure.

Power distribution circuits within communications systems must also be designed to guarantee maximum overall system reliability. The design problems involve size of conductors, division of load among the battery feed conductors, and the location and capacity of fuses. The circuits must be arranged so that a fault in one part of a system is contained and the whole system is not taken out of service.

Grounding of cable sheaths, apparatus cases, and other outside plant items must be carefully controlled to minimize corrosion effects due to electrolysis, especially where dc power systems are used.

**Systems.** Perhaps the most common feature of system design for reliability involves the "hot spare," i.e., the provision of spare equipment that is powered and ready to operate. Service from a failed line or piece of equipment is transferred to the spare. The transfer may take place automatically, by action of a switching arrangement designed to recognize failure and to substitute the spare facility, or manually, by patching in the spare equipment. Coaxial transmission systems, microwave radio systems, and frequency division multiplex terminal equipment are usually provided with automatic switching facilities. Reliability improvement hinges on the reliability of the protection switch, which may lie idle for long periods of time before it is called into action. New designs of high-speed digital transmission systems are also provided with sophisticated monitoring and switching arrangements. Some short-haul carrier systems and most of the voice-frequency trunk plant are provided with flexible patching arrangements. The degree of protection in each case is determined by the reliability of the component parts of the system and the resulting effect on the overall end-to-end reliability of the circuits routed over the system.

*Emergency broadband restoration* is a procedure designed to maintain service on a broader scale than the switching and patching arrangements just discussed. Restoration arrangements are set up primarily to provide temporary service over spare facilities in the event of any kind of failure that is not automatically restored, including failure resulting from the cutting of a coaxial cable or the loss of a microwave repeater tower. Restoration patching and switching bays have been provided in many locations so that transmission systems may be patched together in a flexible manner. Bandwidths and transmission level points are critical parameters in the design of these bays.

Arrangements are also provided for physically replacing damaged plant on an emergency basis. Portable microwave radio repeaters and towers, coaxial cable lengths and splicing arrangements, gasoline-driven power supplies, and trailer-mounted PBXs and central office switching machines are all kept in storage, available on quick demand to furnish emergency service.

"Hardened" systems have also been installed in recent years to increase reliability of the network, especially in the event of enemy attack. Cables, structures, and buildings have been built or installed to meet stringent blast-resistant requirements, and equipment is often shock-mounted. Shielding is used on cables, structures, equipment, and buildings to minimize the possibility of service failure in the event of exposure to electromagnetic pulses that accompany nuclear blasts.

**Manufacturing Designs.** The reliability of apparatus, circuits, and systems is related to the design of manufacturing processes used. Mechanical, thermal, or electrical stresses can often be avoided by proper design of the manufacturing process. Reliability can be improved by proper test and inspection methods. All these, however, must be brought into economic balance. Manufacturing costs are increased by more stringent reliability requirements, and they can only be justified by savings realized in field operation—reduced maintenance, less outage time, reduced cost of repairs, etc.

## Network Operating Methods and Procedures

Reliability of service is related finally to network operating methods and procedures in the field environment. An important element in the layout of facilities for reliable operation is the provision of diverse routes and facilities. In the long distance plant, for example, diversity of trunking between distant cities is often achieved by dividing the trunks between coaxial systems and microwave radio systems so that some service will remain in the event either type system fails. Such diversity may well be further increased as domestic satellite systems are brought into operation.

The alternate routing features of the local and toll portions of the message network provide a great measure of reliability. If a route is blocked as a result of trouble or excessive amounts of traffic, alternate routes can usually be found to satisfy most service needs.

Many features of route layouts can be selected to maximize reliability. Hardened long-haul transmission systems are laid out so that the backbone route that carries the bulk of the traffic bypasses large cities. These routes are thus less vulnerable to damage by enemy attack. Service into the cities is carried by sideleg systems which are usually smaller in capacity and less well-protected against damage.

In certain environments, the provision of appropriate maintenance vehicles is an important element in system reliability. Access to outside plant facilities may be hampered by snow or other vagaries of the weather, long water crossings, or mountainous terrain. Trucks, snowmobiles, barges (for river work), and helicopters all find their places in route maintenance and reliability work, not only for repair activities but also for patrolling so that new construction work or other sources of trouble along the route may be anticipated.

Another environmental factor influencing reliability is out-of-sight plant; cable may be buried directly or placed in conduit. In recent years, there has been increased emphasis on the part of the public to improve and beautify our environment, one result of which is increased desirability of out-of-sight plant. While in many cases higher capital costs have resulted, some added benefits in reliability and lower maintenance costs have been realized. Generally, out-of-sight plant is less susceptible to damage by people, ice, snow, wind, rain, and lightning. It has also resulted in somewhat fewer outages due to cable damage. Offsetting the latter advantage, however, is the fact that outages tend to be of somewhat longer duration.

## Section 5

## Objectives and Criteria

The design, installation, operation, and maintenance of transmission facilities must be based on logically and scientifically established objectives that can be applied throughout the useful life of the facilities. The objectives must also be realistic in that, when met, they lead to customer satisfaction at a reasonable cost. Objectives are dynamic. They must be changed to accommodate changing customer opinion and the introduction of new services. However, the degree of change and adjustment of objectives must be tempered by economic considerations. If objectives are too stringent, excessive costs may be incurred for new system designs and for maintenance of existing systems. If objectives are too lenient, performance may be so poor that customer satisfaction may be low, and an excessive number of service complaints may be received. To avoid either extreme, objectives are continually re-examined and re-established.

There are occasions when relaxation of objectives must be considered for economic and/or plant reasons. The importance of the service, a complete understanding of the basis of the objective, and a firm plan to correct the situation are required before relaxation can be implemented. It is sometimes tempting, for example, to apply objectives lower than optimum in the case of a new service on the basis that the new service is temporary or limited in application. If this approach is taken, there is danger that concentrations of the new service may occur to the extent that damaging effects cannot easily be overcome. There is also the danger that the demand for the service will increase and that problems first introduced as a result of poor judgment will proliferate and then require years of effort to correct.

Many objectives are determined by a process of subjective testing because impairment judgments are based on the sight or hearing mechanisms of the users. Subjective testing is not much used in the operating companies; it is carried out primarily under controlled laboratory conditions. A knowledge of subjective testing techniques and methods is desirable, however, in order that results can be properly interpreted and used. Information about this type of testing is covered in Chapter 23.

Customer satisfaction with the transmission performance of the switched network is conveniently expressed in terms of *grade of service,* a measure of the expected percentage of telephone users who rate the quality of telephone connections excellent, good, fair, poor, or unsatisfactory when the connections include the effects of a given class of transmission impairments. The grade-of-service concept is described in Chapter 24, and several applications of the concept are discussed.

Transmission objectives are subject to considerable manipulation to make them applicable to various operational situations. After the objectives have been determined, a number of ways of interpreting them must be considered to account for such factors as variability of an impairment and the probability of its occurrence. Also, objectives must be translated into firm requirements for system or circuit performance; then the requirements must be allocated to different parts of the network and to different impairments. These methods of treating objectives are considered in Chapter 25.

Chapter 26 discusses a number of specific transmission objectives. As related to the message network, many of these have been accepted for general application. Others, including many that apply to other services, have only provisional status; and in some cases, objectives have not yet been established. The chapter contains some general discussion to indicate the nature of the problems involved when objectives are not established.

The economic tradeoffs that must be considered as engineering compromises in the design, application, and operation of transmission facilities are interrelated and involve judicious application of transmission objectives. These relationships are discussed and illustrated by significant examples in Chapter 27.

The North American telecommunications network must interconnect with and operate with the networks of many other countries and thus may be regarded as part of a world-wide telecommunications network. The coordination of the many facets of international operation is guided primarily by the International Telecommunication Union (I.T.U.), a specialized agency of the United Nations for telecommunications. The International Telegraph and Telephone Consultative Committee (C.C.I.T.T.), and the International Radio Consultative Committee (C.C.I.R.) are two permanent organs of the I.T.U. and are concerned with international coordination of operations. Chapter 28 describes some of the significant ways in which the Bell System and the telecommunications industry of the U.S.A. interact with these international bodies, stressing the relationships that have evolved in respect to transmission objectives as they are expressed in the Bell System and by the international organizations.

Chapter 23

# Subjective Testing

Clearly defined transmission objectives must be established for use in setting performance standards and maintenance limits for existing systems and in setting design and development requirements for new systems. The performance standards and maintenance limits must be adjusted to achieve economically feasible performance that yields satisfaction for the majority of customers.

Various methods of subjective testing have been developed for the purpose of measuring customer opinion as to the disturbing effects of transmission impairments. The data thus acquired can be used by the application of statistical principles to establish relationships between transmission impairment measurements, the subjective effects of the impairments, and overall customer satisfaction. These relationships can then be applied to the establishment of transmission objectives.

Transmission objectives should be reviewed frequently and brought up to date so that they reflect changes resulting from the introduction of new transmission technology, the introduction of new signals or services, and the slowly evolving customer responses to these changes. If customer opinion and objectives are not reviewed regularly, there is the danger that they may be accepted as a matter of habit and thus become traditional.

Some objectives, whether old or new, can be established or revised in a discrete or quantitative manner because they relate only to machine and equipment performance; some, such as the signal-to-noise objectives for digital signal transmission, can be stated discretely because performance thresholds are sharply defined; other impairments, however, must be judged by subjective testing; and in many cases objectives and performance must be expressed statistically.

In these instances, one of a number of available test methods must be selected, the purposes of the tests must be well-defined, and the

536

test environment must be well-controlled. Also, it must be possible to quantify and express the results in useful terms for application to system design and operation. In addition, objectives are also related to transmission grade and grade of service for use in transmission management. Subjective testing does not provide the answers to many of the questions relating to objectives or grade of service, but it is often the starting point.

## 23-1 SUBJECTIVE TEST METHODS

Subjective tests of communication system phenomena fall generally into one of three categories: (1) threshold tests to determine threshold values of impairment; (2) pair-comparison tests to compare interfering effects of two different forms of impairment; and (3) category judgment tests to establish subjective reactions to a wide range of impairments (including intelligibility of telephone circuits), a range that spreads from threshold values to unusable values. Circumstances determine selection of the category to be used.

### Threshold Testing

In the determination of an impairment threshold, the test procedures are arranged so that each participant, or observer, is given the opportunity to establish a value of impairment magnitude at which the stimulus is "just perceptible" or "just not perceptible." Threshold measurements are often made at the beginning of more extensive test procedures in order to establish a base from which other work may proceed. Threshold measurements are also valuable in determining the sensitivity of observers to a particular type of impairment, i.e., to determine if the variation of reactions shows a large or small standard deviation. This information is useful in assessing the importance of other parameters that may affect the result by masking or enhancement.

Threshold measurements may also be used to determine the importance of some newly observed impairment phenomenon. If, in the normal course of development or operation, the impairment is well below the threshold value, it may sometimes be safely ignored; if it is at or above the threshold value, its importance is increased and more extensive testing is indicated.

Threshold testing is usually easier to carry out than other types of subjective testing programs. In many cases, the broad judgments involved and the fact that designs are seldom based on threshold values make it unnecessary to perform threshold tests with the attention to detail required for useful results in other types of testing. Threshold measurements also need only a few test subjects since opinions are not involved.

## Pair-Comparison Testing

Occasionally, a new form of impairment can be evaluated by a test procedure, called pair-comparison testing, designed to establish a magnitude that makes the impairment being tested as disturbing as another type of impairment for which objectives are well defined. Two typical arrangements for this type of testing, also called isopreference testing, are illustrated in Figure 23-1. In Figure 23-1(a) the

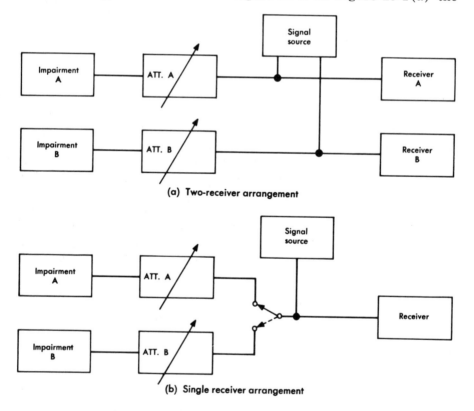

Figure 23-1. Experimental arrangements for pair-comparison testing.

signal source is connected to the two receivers which are adjusted to be as nearly alike as possible with no impairments present. Attenuator A is then adjusted to impress on receiver A an impairment such as random noise. The test subject is then asked to adjust ATT B so that in his judgment the disturbing effect of impairment B (for example, a single frequency interference) is as disturbing on receiver B as impairment A is on receiver A.

In Figure 23-1(b) the same procedure is followed except that only one receiver is used. The test subject adjusts ATT B in the same manner as previously described but makes the comparison between impairments by switching back and forth between the two disturbing sources. In the second procedure the uncertainty of the equivalence of the two receivers is removed. However, other uncertainties exist such as the effects of switching transients and, in telephone testing, the coupling-decoupling between the ear and telephone receivers. Hence, there is little choice between the procedures.

The pair-comparison method of testing may be used for either visual or hearing tests. The results are generally considered more valid, or at least more useful, than threshold test results; however, all the statistical aspects of observer reactions are not included. Impairment evaluation is most valid when the impairment is rated relative to some standardized scale that can be applied to all kinds of impairments.

## Comment Scale Testing — Category Judgments

The two types of service for which subjective testing has been conducted widely are telephone and television. For each of these, there has evolved a mode of subjective testing which involves a scale that permits quantitative evaluation of judgments of the disturbing effects of impairments. The two scales are different, though they have certain similarities, and they are used somewhat differently.

**Telephone Impairment Testing.** Bandwidth limitation was one of the first types of telephone transmission impairment for which objectives were established by subjective testing. The test method that evolved, known as articulation testing, measured intelligibility of received speech. The test procedure involved the preparation of stimuli in the form of standard lists of vowel and consonant sounds, syllable sounds, real and nonsense words, and sentences. These stimuli were transmitted by a number of speakers to various listeners who recorded what they heard. Errors in their records were used as the basis of evaluating the effects of variations in the high-frequency or low-frequency cutoffs or in the overall bandwidth of the circuit [1, 2, 3, 4, 5, 6]. Such lists are still used in some related types of testing [7].

Category judgment evaluations have been applied analytically to the results of articulation tests. At present, other kinds of impairments to the transmission of speech signals over message channels are often evaluated directly by subjective tests in which participants rate specific transmission conditions *excellent, good, fair, poor,* or *unsatisfactory*. The tests are conducted in such a manner that various impairments are rated under listening conditions selected to be as representative as possible of operating conditions.

Usually, these tests are carried out to evaluate a specific impairment — for example, a single frequency tone. Other impairments either are suppressed or are impressed on the listening circuits at values typical of those found in practice. Ultimately, objectives must be established for combinations of impairments which are evaluated to establish a *grade of service*.

**Television Impairment Testing.** Subjective testing of television impairments has followed the complete cycle of threshold testing, comparative testing, and comment scale testing. The comment scale shown in Figure 23-2 was adapted for use in the Bell System and was selected after many efforts to find suitable terms that would adequately describe a wide range of impairments.

| COMMENT NUMBER | COMMENT DESCRIPTION |
|:---:|:---|
| 1 | Not perceptible. |
| 2 | Just perceptible. |
| 3 | Definitely perceptible, but only slight impairment to picture. |
| 4 | Impairment to picture, but not objectionable. |
| 5 | Somewhat objectionable. |
| 6 | Definitely objectionable. |
| 7 | Extremely objectionable. |

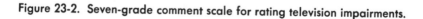

Figure 23-2. Seven-grade comment scale for rating television impairments.

In this type of test, a minimum of ten observers are usually used. The observers are persons (usually technically trained) having good eyesight, experience in judging television picture impairments, and proven consistency in their evaluations over a reasonable period of time.

## 23-2   TEST PLAN AND PROCEDURES

The design and administration of a subjective test program involves careful planning and preparation. To have maximum value, such tests must have well-defined goals and must be carefully controlled throughout. The nature of the goals often influences the choice of test method, determines the details and sophistication of test arrangements, establishes the importance of providing appropriate and well-designed environmental test conditions, and helps to establish the number and qualifications of observers.

### Setting Goals

Some carefully considered questions — by definition, the *right questions* — are involved in the determination of the goals of a subjective test program. The determination of the right questions is sometimes an iterative process, because they are not automatically known. However, if the attempt is not made to ask the right questions, the risk is high of getting the right answers to the wrong questions and then to set off in the wrong direction. To illustrate, consider the problem of determining how much frequency offset can be tolerated in FDM channels. If a subjective test program is conducted in which only speech signals are used, the results would indicate listener tolerance of tens of hertz to this impairment. Early considerations, however, should have included a preliminary investigation to determine what type of signal is most susceptible to frequency offset. The preliminary tests would quickly have shown that certain types of music are most susceptible, and a subjective test program would then have been designed around the criteria of satisfying critical listeners to musical programs; speech signal testing could have been limited to that which would assure that speech transmission would be acceptable when objectives for music were satisfied.

Who is to be pleased and to what degree are two other questions that must be answered. For television signals, transmission objectives that satisfy home viewers may not satisfy the broadcasters or the advertisers. Transmission objectives must be set to satisfy the most critical of these groups, and it is often necessary to set objectives at threshold or near-threshold values.

## Test Locale

Subjective tests may be conducted in the field or in the laboratory, with the choice normally determined by the test goals and by the relative advantages and disadvantages of each locale.

**Field Tests.** The principal advantages of conducting tests in the field are that the normal, uncontrolled environment of the operating plant is used and that the test subjects (observers) are the customers who normally use the service. Thus, realistic appraisals of various phenomena can be made under operating conditions.

The same parameters that are judged to be advantages of field testing are also the greatest disadvantages when considered from another point of view. The inability to control the test environment and the observers makes it difficult to achieve accuracy and precision in the test results. Service impairments such as *slow dial tone* and *all circuits busy* conditions can be evaluated best in the field environment. The subjective evaluation of transmission impairments, however, seldom falls within the framework of broad judgments and evaluations of service.

**Laboratory Tests.** Most subjective testing of transmission impairments is carried out in the laboratory where test conditions can be controlled and where observers can be selected and trained. Sometimes the entire test program is carried out in the spirit of a laboratory experiment, with all facets of the program (impairment simulation, environment, procedure, etc.) carefully designed and controlled.

In other cases, a laboratory test program may be designed to simulate the field environment but in a controlled manner. One such test program, for example, involved the introduction of computer-controlled time delays (with and without echo suppressors) into working telephone circuits for the purpose of simulating very long transmission circuits and evaluating the effect of echo. The observers, selected laboratory personnel, evaluated the effects of the impairment during actual conversations and reported to the computer by prescribed station set dial operation [8].

Normally, laboratory tests of transmission impairments are greatly preferred over field tests. Better simulation of the impairment under test can usually be made in the laboratory, test results are less likely to be influenced by external factors, and better control can be exercised over the program. Also, laboratory testing is usually less expensive than field testing.

## Test Conditions

After the initial test plans have been formulated and the procedure has been decided upon, the next step is to establish all test conditions and facilities. It must be decided whether the tests are to be made under laboratory or field conditions; a source of the impairment (real or simulated) must be provided; the circuits required for conducting the test must be selected or designed; and the necessary test equipment must be procured.

**Laboratory Environment.** The process of laboratory subjective testing must involve careful control of the test environment, including both the physical environment in which the test is conducted and the environment induced by circuit conditions and arrangements. In both cases, the laboratory environment must realistically simulate the environment in which the impairment actually occurs and must be consistent with the stated goals of the test program.

*Physical Environment.* It is not possible to specify the nature of control necessary over the physical environment in any given situation; however, the physical environment must be consistent with the goals of the test. If, for some reason, the threshold value of a stimulus must be determined under the most stringent conditions, external distractions must be minimized. For example, if a listening test is required, the environment must be something in the nature of a soundproof room; if a visual test is required, the environment must be a darkened room.

On the other hand, if a stimulus is to be evaluated by category-judgment-type testing with normal observing conditions, it may be desirable to create a noisy environment by playing recorded street sounds or room noise at appropriate sound levels. If the tests involve television viewing, appropriate ambient lighting may be used.

*Circuit Conditions.* Circuit conditions that are provided for subjective tests are perhaps even more variable than physical conditions, but they must also be consistent with pre-established goals. Some examples may be given.

If telephone listening tests are to be made, a number of questions must be answered. First — would the purposes of the tests be served in the presence of impairments other than the one under test (multiple impairment testing) ; and if so, how loud should they be? For example, if echoes are being evaluated, should there be noise on the test circuits or should they be as quiet as possible? Impairments are interdependent; while subjective test results may apply to one or a combination of several impairments, other interrelated impairments must always be kept in mind. Second—should there be a normal signal present, and if so, what kind of signal? If a particular noise impairment is under test, for example, should the observers listen to a simulated conversation while evaluating the noise? Or would it be better to have the observers listen to continuous speech? Or perhaps there should be no speech signal on the circuit at all.

If television viewing tests are involved, the same kinds of questions must be answered. For significant results, should a picture be present or should the screen be blank? Should the picture be one with large flat areas of constant brightness or should it be "busy" with a lot of high-frequency components present in the signal? Should other impairments be present in the picture during the tests?

These questions have no general answers and therefore must be considered both separately and collectively. The answers can often be determined from the results of preliminary testing carried out to establish procedures and to determine which parameters affect the results sought.

**Source of Impairment.** Introduction into the test circuit of the impairment to be evaluated is, of course, a prerequisite to subjective testing. Sometimes this is straightforward, particularly when the impairment is well-defined and easy to simulate. For example, a signal generator to produce a single frequency interference or a random noise generator to introduce random noise into the test circuit may well suffice.

On the other hand, it is often necessary to record the impairment and then to use the recording as the source during the test. This approach is used in cases where the impairment is intermittent or has some other unusual characteristic that is not easily reproduced except under carefully controlled conditions. An example is the evaluation of interference falling into a telephone channel due to cross-modulation between multifrequency signalling tones and television signal components where combined system (telephone-television) operation is contemplated. In this situation, the interference falls at different frequencies in the disturbed channels, has highly variable amplitude characteristics, is randomly intermittent, and has a complex and variable spectrum. Recordings of such an impairment are the only way that the impairment can be adequately simulated for test purposes, and great care must be taken that the recordings adequately represent the variables mentioned.

**Test Circuits.** The principal requirement of test circuits used in subjective testing is that they must be capable of delivering signals and impairments to the test area without introducing distortion that might mask the results of the test. It is essential, therefore, that all transmission, distribution, and control circuits be thoroughly tested under all conditions to which they will be subjected in the test program.

**Test Equipment.** The selection of test equipment is as important as the selection of test circuits. The test equipment must be available before the test program is begun, and effort must be expended to assure that its capabilities and accuracies are appropriate to the task. Consideration must be given to the human engineering of the tests so that the selected test equipment can be used conveniently. Consideration must also be given to the need for and availability of automated measurements and recordings of measurements.

### Test Procedures

The actual conduct of the test program finally must be worked out in detail. Only general guidelines can be given here because the procedure may be different for each test. In most situations, ten or more observers (sometimes expert and sometimes non-expert) are asked to participate, particularly when comment scale testing is to be used. For these purposes, an *expert observer* is defined as a person with good vision (for television) or hearing (for telephone) with experience in judging impairments, who has exhibited consistency in his evaluations over a reasonable period of time. Frequently, television testing involves the use of expert observers; for telephone testing, nonexpert observers are chosen to be a representative sample of all users.

Usually, a test program is begun by training the observers. The training involves first an exposure to the impairment under test so that each observer knows what to look for or listen for. If test arrangements are such that other impairments are present and cannot be eliminated, the observers may be told to try to ignore them and judge only the impairment under test. The observers then are given the scale of comments to be used in judging the impairment. Often, a few trial runs are used to give the observers a sense of understanding of the test process.

Some pitfalls are to be guarded against. One is the introduction of observer bias such as the bias that might result from presenting an ordered sequence, like best-to-worst or worst-to-best test conditions, to all observers in all test sequences. Randomizing the sequence of presentation is desirable. Another pitfall to be avoided is observer fatigue. This condition can sometimes be identified in preliminary testing; experienced observers may become inconsistent after a period of observing, and that period of time may then be used for the duration of final testing.

Before actual testing begins, every step of the procedure should be rehearsed, and every effort must be made to ensure that all observers are exposed to the same or very similar viewing or listening conditions. The circuits, test equipment, television pictures or telephone signals, and impairment sources should be checked and calibrated before each test in order to eliminate unwanted and unexplained variations. In short, the entire procedure must be conducted with great care and precision to ensure valid results.

## 23-3  DATA ANALYSIS

For a subjective test program to produce useful results, the accumulated data must be analyzed and presented so that they may be related to performance criteria. For these purposes the subjective test data and performance criteria are often expressed in terms of mean values and standard deviations.

Methods of analysis have improved as subjective testing procedures have become more scientific and as the importance of test results has become more widely appreciated. The progress made with respect to the evaluation, analysis, and presentation of the results of subjective testing of television impairments is a case in point. In the middle and late 1940s, when the television industry was growing rapidly, most of the development and research emphasis was placed on gaining an understanding of the fundamentals of television camera, transmission, and reproduction processes. Later, experiment and analysis were devoted to the evaluation of various impairments.

One study exemplifying this type of analysis involved the evaluation of random noise impairment of television pictures [9]. In this study, the similarity of random noise effects to the graininess of photographs was noted, and the number of spots or grains that could be expected from noise was compared with the graininess of photographic pictures. The analysis had to take into account the size of the grains (due to the duration of noise bursts), all of the television signal processes (camera and viewing tube spot sizes and shapes, scanning process, brightness-voltage relationships), viewing distance, etc.; furthermore, all of these parameters had to be expressed mathematically. The analysis was closely tied to earlier observations that the ratio of the minimum perceptible change of a stimulus to the value of the stimulus tends to be constant over a wide range of stimulus values (the Weber-Fechner law).

Later work in evaluating television impairments extended the ideas of comparison testing and introduced the comment-scale method of impairment evaluation. For example, one series of tests was based on the comparison of echo and random noise impairments with high quality, projected lantern slides of photographs impaired by defocusing the projection system by known amounts [10]. The comparisons made between defocused lantern slides and impaired television pictures involved the use of a unit called the *limen*, or *liminal unit*. (One liminal unit indicates a 75 percent observer vote preference for one picture condition over another.) The analysis of the preference votes approximately followed a normal distribution. It was also shown that the difference from one comment to another is approximately one limen where the comment scale used was the same as that given in Figure 23-2, with one minor exception—the present definition of comment 7 is "extremely objectionable," while that used in the earlier tests was "not usable." The results of these tests were plotted in various ways to determine the usefulness of the data for engineering purposes. One such plot is illustrated in Figure 23-3, in which echo attenuation is plotted as a function of the percent of observa-

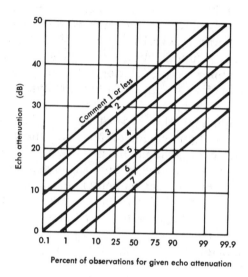

Figure 23-3. Illustrative plot of echo impairment characterization.

tions indicated for each value of attenuation. A separate plot is given for each comment number. Test data were smoothed for these plots.

As the emphasis shifted to comment scale testing, data analysis techniques have been improved, and data have been subjected to more and more critical evaluations. In early studies [11], data that had been smoothed by eye was often presented for engineering use. More recently, the smoothing process and methods of presentation have been derived by more systematic procedures and more analytic methods [12]. Further improvements in analysis are under study.

The analysis of subjective test results of telephone impairments has evolved in a somewhat similar manner. Some indication of how such analyses are currently made is illustrated by a report on the derivation of modern message circuit noise objectives [13]. One useful product of these tests and analyses is illustrated by Figure 23-4, which shows the variation of observer reaction to various amounts of noise on a telephone circuit, expressed in terms of comment scale testing.

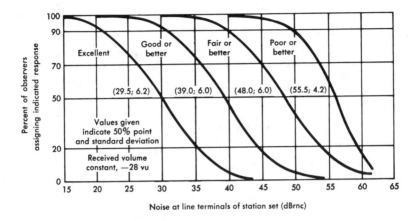

**Figure 23-4. Noise opinion curves.**

### REFERENCES

1. Fletcher, H. and J. C. Steinberg. "Articulation Testing Methods," *Bell System Tech. J.*, Vol. 8 (Oct. 1929).

2. French, N. R., C. W. Carter, Jr., and W. Koenig, Jr. "The Words and Sounds of Telephone Conversations," *Bell System Tech. J.*, Vol. 9 (Apr. 1930).

3. Castner, T. G. and C. W. Carter, Jr. "Developments in the Application of Articulation Testing," *Bell System Tech. J.*, Vol. 12 (July 1933).

4. French, N. R. and J. C. Steinberg. "Factors Governing the Intelligibility of Speech Sounds," *The Journal of the Acoustical Society of America*, Vol. 19 (Jan. 1947).

5. Martin, W. H. "Rating the Transmission Performance of Telephone Circuits," *Bell System Tech. J.*, Vol. 10 (Jan. 1931).

6. McKown, F. W. and J. W. Emling. "A System of Effective Transmission Data for Rating Telephone Circuits," *Bell System Tech. J.*, Vol. 12 (July 1933).

7. "IEEE Recommended Practice for Speech Quality Measurements" (New York: The Institute of Electrical and Electronics Engineers, Inc., 1969), Standard #297.

8. Riesz, R. R. and E. T. Klemmer. "Subjective Evaluation of Delay and Echo Suppressors in Telephone Communications," *Bell System Tech. J.*, Vol. 42 (Nov. 1963), pp. 2919-2941.

9. Mertz, P. "Perception of Television Random Noise," *Journal of the Society of Motion Picture Engineers*, Vol. 54 (Jan. 1950), pp. 8-34.

10. Mertz, P., A. D. Fowler, and H. N. Christopher. "Quality Rating of Television Images," *Proceedings of the Institute of Radio Engineers*, Vol. 38 (Nov. 1950), pp. 1269-1283.

11. Fowler, A. D. "Observer Reaction to Low-Frequency Interference in Television Pictures," *Proceedings of the Institute of Radio Engineers*, Vol. 39 (Oct. 1951), pp. 1332-1336.

12. Lessman, A. M. "The Subjective Effects of Echoes in 525-Line Monochrome and NTSC Color Television and the Resulting Echo Time Weighting," *Journal of the Society of Motion Picture and Television Engineers*, Vol. 81 (Dec. 1972), pp. 907-916.

13. Lewinski, D. A. "A New Objective for Message Circuit Noise," *Bell System Tech. J.*, Vol. 43 (Mar. 1964), pp. 719-740.

14. Guilford, J. P. *Psychometric Methods* (New York: McGraw-Hill Book Company, Inc., 1954).

Chapter 24

# Grade of Service

Transmission grade of service is a measure of the expected percentage of telephone users who rate the quality of telephone connections excellent, good, fair, poor, or unsatisfactory when the connections include the effects of a given class of transmission impairments. It combines the distribution of customer opinions with the distribution of plant performance parameters to obtain the expected percentage of customer opinions in a given category or categories. While the term is usually applied to overall communication service, the grade-of-service concept can be applied in theory to one aspect of communications such as transmission; to one specific impairment such as noise, loss, or echo; or to various combinations of these impairments. It is usually expressed in terms such as *a noise grade of service of 95 percent good or better* or *a noise/loss grade of service of 3 percent poor or worse*. Among the impairments that tend to degrade speech signal transmission performance, noise, loss, and echo are predominant.

Transmission management of the telecommunications network involves the establishment of transmission objectives, the measurement of transmission performance, and the measurement of customer opinions of the quality of service rendered. The grade-of-service concept is a useful tool for fulfilling these responsibilities. While the concept is usually used directly in establishing objectives, it is also sometimes used in inverse applications to determine what performance must be achieved to meet established grade-of-service objectives.

## 24-1 A GRAPHIC DERIVATION OF GRADE OF SERVICE

The transmission grade of service is usually determined by mathematical derivation from opinion and performance survey data. A simplified graphical analysis is offered first in order to illustrate the

551

basic concept as an aid to an understanding and interpretation of the mathematical derivation. This analysis is, of course, hypothetical and uses values of opinion and performance parameters that represent a typical problem of loss grade of service.

## Customer Opinion Distributions

The distribution of customer opinions would normally be determined by a subjective test program in which the distributions of connection losses and talker volumes are inherently combined. For a loss grade-of-service analysis, the combined distribution would then have to be interpreted in terms of loss only. Assume that the results of the subjective tests are finally expressed as the percentage of observers that rate connections with each of seven loss values (0 through 6 dB, inclusive) according to the standard rating scale — excellent, good, fair, poor, unsatisfactory. For this impairment, the unsatisfactory category is increased by those observations that the received volume is considered too loud when the connection loss is low.

The distribution of good ratings is plotted as a histogram in Figure 24-1. In the analysis it might be found that observers rated more than one value of loss in a given category, and the total number of observations in that category might then exceed 100 percent. To avoid this ambiguity, assume that for each observer only the highest value of loss rated in a given category is used; thus, the ordinate in Figure 24-1 is really in terms of the threshold of good ratings.

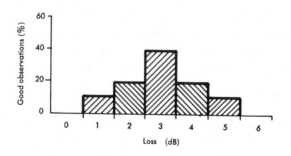

Figure 24-1. Histogram of good ratings of loss.

This manner of displaying the results of the subjective tests may now be extended to show how the connection is rated by all the observers for each value of loss. Figure 24-2 illustrates the cumulative opinions of good or better and too loud, and fair or worse. The cumulative opinions of good or better never reach 100 percent because the low losses result in some opinions of too loud which must be subtracted from the total.

### Relation of Connection Losses to Subjective Test Results

To determine the grade of service, the losses of the connections being evaluated must be measured and plotted. Figure 24-3, a histogram of the distribution of such losses, represents the probability of encountering a given connection loss. Similarly, Figure 24-1 represents the probability that a connection loss of a given value is rated good by a random sample of observers. These distributions may now be combined as a product of the probabilities at each loss value to obtain the probability of a connection of a given loss rated good. The resulting probabilities are plotted as the distribution shown in Figure 24-4. Thus, of all possible combinations of con-

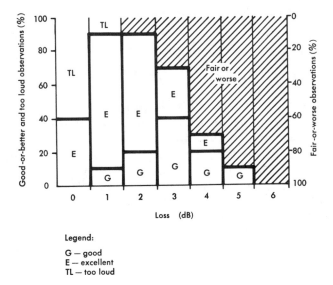

Figure 24-2. Overall ratings of loss.

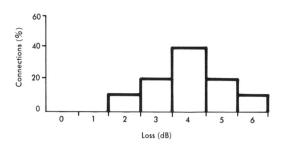

Figure 24-3. Histogram of connection losses.

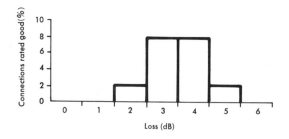

Figure 24-4. Histogram of connections rated good.

nections and observers, just 20 percent (the sum of the entries in the histogram in Figure 24-4) would be rated good.

Another, somewhat more detailed process may be used to combine the distributions of Figures 24-1 and 24-3. This approach involves finding the probability that the difference between the two is a certain number of dB for every possible combination of opinion and connection loss. The process is illustrated in Figure 24-5 where the values of $Z$ are the differences in loss values from Figure 24-1 and Figure 24-3. The total probability of the loss difference, $(Y-X)=Z$, is the sum of the products of all the ordinates at loss values with differences equal to $Z$.

Coincidence of the good threshold and connection loss values (i.e., 5,5; 4,4; 3,3; 2,2) is shown at $Z = 0$. As in Figure 24-4, the percentage of connections rated good by observers is again 20 per-

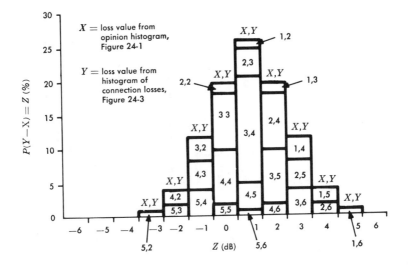

Figure 24-5. Histogram of dB differences between loss and loss rated good.

cent. When trunk losses are lower than the good threshold of each observer, the connection is rated good, excellent or too loud. When trunk losses are higher than the good threshold, the connection is rated fair or worse.

This general approach can be pursued further to extract from the result those combinations of connection losses and observer opinions that would result in a net judgment of too loud. Notice, however, that in the example chosen, the only connections for which too loud ratings are given are those having 0 or 1 dB of loss (Figure 24-2). The assumed distribution of connection losses (Figure 24-3) shows that there are no connections in the study having less than 2 dB of loss. Therefore, the resultant distribution is not affected by the too loud category of opinions.

If the distribution of the parameter $Z$, shown in Figure 24-5, were plotted as a cumulative distribution histogram, Figure 24-6 would result. Since there are no too loud ratings, Figure 24-6 shows that 37 percent ($Z = 0$ dB) of the connections are rated good or better. The remainder, 63 percent, are rated fair or worse. It may be said, then, that the example demonstrates a 37 percent good-or-better loss grade of service.

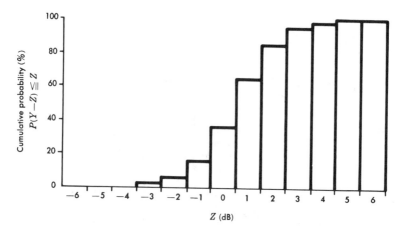

Figure 24-6. Histogram of cumulative distribution of $Z$.

For actual distributions of observer opinions and losses (and other impairments such as noise and echo which must be evaluated simultaneously), such a simplified approach using histograms and direct multiplication and summing of probabilities is impractical because the distributions are, or at least approach, continuous distributions. However, as pointed out in Chapter 9, distributions of these parameters are normal or may safely be assumed normal. In addition, the results of subjective tests are usually represented by normal distribution functions such as those illustrated in Chapter 23 (see Figure 23-4). Therefore, the combination of impairment and opinion distributions may be treated mathematically by using representative values for their means and standard deviations. As covered in Chapter 9, the combined distribution can be determined for two independent distributions by subtracting their mean values and adding their variances.

## 24-2 MATHEMATICAL DERIVATION OF GRADE OF SERVICE

As previously mentioned, grade of service can be computed from mathematical expressions of measured performance parameters and of user opinions of performance. While manipulation of grade-of-service relationships has seldom been necessary in the operating companies, the concept is becoming more widely applied as more use is made of computer programs, such as GOSCAL (grade-of-service calculation). Such programs will be made available for field applications. The mathematical grade-of-service relationships may be used to determine the expected grade of service for certain combinations of performance parameter measurements and user opinions.

To describe grade of service mathematically,* let $R$ denote the rating assigned by an individual selected at random whose telephone connection is subject to an impairment designated $X$. Also, for convenience, assign numerical values, designated $c_j$, to the 5-category comment scale as follows:

$$\text{Excellent} = 5$$
$$\text{Good} = 4$$
$$\text{Fair} = 3$$
$$\text{Poor} = 2$$
$$\text{Unsatisfactory} = 1.$$

Next, suppose that connections are made and rated by each of a random sample of $k$ customers; their corresponding ratings are designated $R_1, R_2, \ldots, R_k$. Then, the fraction who rate their connections in any category, $c_j$, is

$$(1/k) \sum_{i=1}^{k} \delta(R_i - c_j)$$

where

$$\delta(R_i - c_j) = 1 \text{ when } R_i = c_j,$$

and

$$\delta(R_i - c_j) = 0 \text{ when } R_i \neq c_j.$$

Thus, the expected fraction is

$$E\left\{ (1/k) \sum_{i=1}^{k} \delta(R_i - c_j) \right\} = (1/k) \sum_{i=1}^{k} P\{R_i = c_j\}$$
$$= P\{R = c_j\}$$

---

* Mr. N. A. Marlow, Bell Telephone Laboratories, Holmdel, N. J., provided this mathematical treatment of the grade-of-service relationships in an unpublished memorandum.

The corresponding grade of service for category $c_j$, i.e., the expected percentage of customers who would place their connections in category $c_j$, is then $100P\{R=c_j\}$ percent. The discussion to this point is general in that the impairment, $X$, either can be fixed or can vary randomly according to a given distribution.

To obtain an expression for the probability $P\{R=c_j\}$, which explicitly reflects both the variability of customer rating for a given impairment value and the variability of impairment values, let $P\{R=c_j \mid X=x\}$ denote the conditional probability that the rating is $c_j$, given a fixed value $x$ of the impairment $X$. If $f_X(x)$ denotes the density of $X$, it follows from the law of total probability that

$$P\{R=c_j\} = \int_{-\infty}^{\infty} P\{R=c_j \mid X=x\} f_X(x)\, dx \quad . \tag{24-1}$$

By writing $g_j(x) = P\{R=c_j \mid X=x\}$ and fixing a particular value of $x$, it follows that as $c_j$ varies, $g_j(x)$ is a conditional (discrete) distribution of customer opinion which can be estimated by subjective testing in a controlled laboratory environment.

To illustrate a particular application of Equation (24-1), suppose that the rating category under consideration is good or better; i.e., $R = 4$ or $5$. The corresponding grade of service is

$$100P\{R=4 \text{ or } 5\} = 100 \int_{-\infty}^{\infty} P\{R=4 \text{ or } 5 \mid X=x\} f_X(x)\, dx \tag{24-2}$$

where

$$P\{R=4 \text{ or } 5 \mid X=x\} = P\{R=4 \mid X=x\} + P\{R=5 \mid X=x\}$$

$$= g_4(x) + g_5(x)$$

Suppose also that $X$ is normally distributed with mean value, $\mu$, and standard deviation, $\sigma$. Then, by substitution in Equation (24-2), the grade of service becomes

$$100P\{R=4 \text{ or } 5\} = \frac{100}{\sigma\sqrt{2\pi}} \int_{-\infty}^{\infty} P\{R=4 \text{ or } 5 \mid X=x\} e^{-(x-\mu)^2/2\sigma^2}\, dx \quad . \tag{24-3}$$

For a category of good or better, the function $P\{R=4 \text{ or } 5 \mid X=x\}$ varies with the value, $x$, of the impairment $X$ and can often be adequately represented in practice by the function

$$g_4(x) + g_5(x) = P\{R=4 \text{ or } 5 \mid X=x\} = F_z\left(\frac{a-x}{b}\right) \qquad (24\text{-}4)$$

where $a$ and $b$ are constants and $F_z(z)$ is the standard normal integral, as given in Equation (9-25), evaluated at $z = \frac{a-x}{b}$. Here, it is written

$$F_z\left(\frac{a-x}{b}\right) = \frac{1}{\sqrt{2\pi}} \int_{-\infty}^{\frac{a-x}{b}} e^{-y^2/2} dy \quad .$$

It has been assumed that as $x$ increases in value, the expected fraction of ratings which fall into the good-or-better categories decreases. Equation (24-4) does not imply that customer opinion is normally distributed; as mentioned earlier, opinion for a given impairment level has a discrete distribution, and Equation (24-4) simply expresses the analytical fact that a complementary normal distribution function is used to describe the function $g_4(x) + g_5(x)$. By combining Equations (24-3) and (24-4) the grade of service can be written

$$100\, P\{R=4 \text{ or } 5\} = \frac{100}{\sigma\sqrt{2\pi}} \int_{-\infty}^{\infty} F_z\left(\frac{a-x}{b}\right) e^{-(x-\mu)^2/2\sigma^2} dx. \qquad (24\text{-}5)$$

The last integral can be evaluated in closed form with the result

$$100P\{R=4 \text{ or } 5\} = 100\, F_z\left(\frac{a-\mu}{\sqrt{b^2+\sigma^2}}\right) \quad . \qquad (24\text{-}6)$$

In the above illustration, it was assumed that the conditional probability of a good-or-better rating, $P\{R=4 \text{ or } 5 \mid X=x\}$, decreases monotonically as the impairment value $x$ increases. This might occur, for example, when $X$ represents noise. On the other hand, if $X$ represents received volume, then the function $P\{R=4 \text{ or } 5 \mid X=x\}$ would tend to increase up to a point, as $x$ increases, and then decrease because both low and high volumes tend to be objectionable.

When the conditional probability $P\{R=4$ or $5 \mid X=x\}$ varies monotonically with the value of the impairment, $x$, it is possible to derive the grade of service by a different process from that just described. In particular, suppose that for a customer selected at random, it is possible to define a unique threshold, $T$, such that if the value of the impairment $X$ is less than $T$, then a good-or-better rating would result. This would occur, for example, if customers consistently give poorer ratings as the magnitude of the impairment increases; the threshold would then be the value of impairment where the transition occurs from a fair to a good rating.* Next, suppose as before that a random sample of $k$ customers with thresholds $T_1, \ldots, T_k$ establish connections having corresponding impairments $X_1, \ldots, X_k$. For the $i^{th}$ subscriber, a good-or-better rating occurs if and only if $X_i \leqq T_i$ so that the fraction who would rate their calls good or better is

$$(1/k) \sum_{i=1}^{k} V(T_i - X_i)$$

where

and

$$V(T_i - X_i) = 1, \; T_i - X_i \geqq 0,$$

$$V(T_i - X_i) = 0, \; T_i - X_i < 0.$$

Thus, the expected fraction is

$$E\left\{ (1/k) \sum_{i=1}^{k} V(T_i - X_i) \right\} = (1/k) \sum_{i=1}^{k} P(X_i \leqq T_i)$$

$$= P(X \leqq T) \quad ,$$

and the corresponding grade of service for the good-or-better category is

$$100P\{X \leqq T\}$$

---

*In some cases it may not be possible to identify a unique transition because of subjective rating inconsistencies. For example, as the impairment value is increased, the ratings may show a reversal such as is illustrated by the sequence 5,4,4,3,4,3,3,2,2,1. This represents a major drawback in the approach based on the threshold concept.

To determine the probability, $P\{X \leq T\}$, assume that an individual's threshold, $T$, is statistically independent of the impairment, $X$. Then

$$P\{T \geq x \mid X=x\} = P\{T \geq x\} \quad,$$

and it follows from the law of total probability that

$$P\{T \geq X\} = \int_{-\infty}^{\infty} P\{T \geq x \mid X=x\} \, f_X(x) \, dx$$

$$= \int_{-\infty}^{\infty} P\{T \geq x\} \, f_X(x) \, dx \qquad (24\text{-}7)$$

where $f_X(x)$ is the density of the impairment $X$. On the other hand, if $X = x$, a good-or-better rating occurs if and only if $T \geq x$; hence

$$P\{T \geq x\} = P\{T \geq x \mid X=x\}$$

$$= P\{R=4 \text{ or } 5 \mid X=x\} \quad. \qquad (24\text{-}8)$$

Thus, the distribution of thresholds among individuals is given by the function

$$P(T < x) = 1 - P\{R=4 \text{ or } 5 \mid X=x\} \quad, \qquad (24\text{-}9)$$

and substitution into Equation (24-8) gives the grade of service.

To illustrate, suppose again that $X$ is normally distributed with mean, $\mu$, and standard deviation, $\sigma$, and assume that

$$P\{R=4 \text{ or } 5 \mid X=x\} = F_Z\left(\frac{a-x}{b}\right)$$

where

$$F_Z\left(\frac{a-x}{b}\right) = \frac{1}{\sqrt{2\pi}} \int_{-\infty}^{\frac{a-x}{b}} e^{-y^2/2} dy \quad.$$

From Equation (24-9),

$$P\{T < x\} = 1 - F_z\left(\frac{a-x}{b}\right)$$

$$= F_z\left(\frac{x-a}{b}\right).$$

Thus, $T$ is normally distributed with a mean $a$ and standard deviation $b$. To determine the good-or-better grade of service, write

$$P\{X < T\} = P\{X - T < 0\} \tag{24-10}$$

Since $X$ and $T$ are both normally distributed and independent, it follows that $X - T = S$ is normal with mean, $\mu - a$, and standard deviation, $\sigma_s = \sqrt{\sigma^2 + b^2}$. Thus, if $d$ is any real number,

$$P\{X - T < d\} = F_s\left(\frac{d - (\mu - a)}{\sqrt{\sigma^2 + b^2}}\right) \tag{24-11}$$

where $F_s(s)$ is the normal distribution function.

In particular, setting $d = 0$,

$$P\{X < T\} = F_s\left(\frac{a - \mu}{\sqrt{\sigma^2 + b^2}}\right) \tag{24-12}$$

Consider now a simple but useful application of these concepts to the determination of a noise grade of service. The function $f(S)$, is plotted in Figure 24-7 as a normal probability density function representing the combined distribution of noise and customer opinions. Its median value is shown as $(\mu - a)$. The 0-db value is by definition the value at which the stimulus just meets the given rating, *good*. The 0-db point is related, for a particular set of conditions, to the median value by a value proportional to the standard deviation, $K\sigma_s$, in the figure. Noise is more disturbing as its magnitude increases, and so it is necessary to express the integral $f(S)$ so that

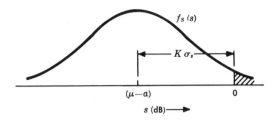

Figure 24-7. Normal probability density function for noise grade-of-service determination.

the area to the left of the 0-dB point in Figure (24-7) represents the good-or-better grade of service. That is,

$$\text{Grade of service} \atop \text{good or better} = 1 - \int_0^\infty f_S(s)\ ds \quad .$$

$$= 1 - \frac{1}{\sigma_S \sqrt{2\pi}} \int_0^\infty e^{-s^2/2\sigma_S^2}\ ds \quad . \quad (24\text{-}13)$$

The process of combining the noise distribution and opinion distribution functions to determine the grade of service can be conveniently carried out by using arithmetic probability paper. Suppose, for example, that the distribution of noise for a group of trunks, as measured and translated to the station set terminals, is found to have a mean value of 36 dBrnc and a standard deviation of 4 dB. Also, suppose that subjective tests have been conducted, that the results show that the median value of noise rated good or better at the station set is 39 dBrnc, and that opinions vary normally with a standard deviation of 6 dB (see Figure 23-4). The two functions are plotted in Figure 24-8.

To determine the good-or-better noise grade of service provided by these trunks, the two functions of Figure 24-8 may be combined and plotted as in Figure 24-9. Here, the mean value of the combined distribution is the difference between the means of the two distributions of Figure 24-8 ($36 - 39 = -3$ dB) and the standard deviation

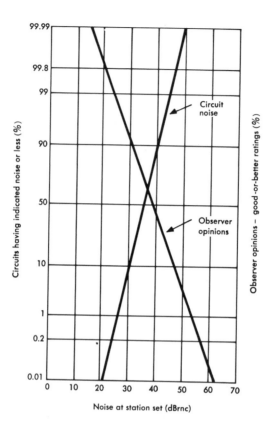

Figure 24-8. Illustrative probability density functions for noise and noise ratings.

of the combined distribution is the square root of the sum of the squares of the two standard deviations ($\sqrt{4^2 + 6^2} = 7.2$ dB). The resulting distribution crosses the 0-dB value, the value of noise relative to that rated good or better, at the 62 percent point on the ordinate. Thus, these trunks yield a noise grade of service of 62 percent good or better.

To determine a probability density function needed to achieve a given grade of service, such as 95 percent, the inverse process must be used. The resulting function might then be used to set equipment performance criteria. The process would be an iterative one in which the density function of Figure 24-7 would first be assumed. The 0-dB

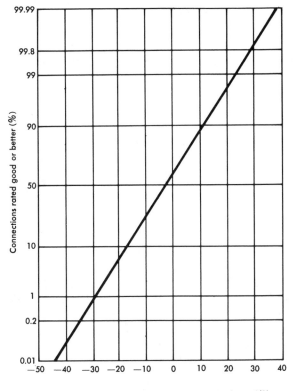

**Figure 24-9.** Combined probability density function for grade-of-service rating of illustrative trunk group.

point would be determined by referring to Figure 9-15 to determine the value of $K\sigma_S$ for which the area to the left of the 0-dB point is 95 percent of the total $(1.65\ \sigma_S)$. Achievable performance curves would then be studied to be sure they would fit the assumed conditions. By continuing the measurement of performance, it is possible to verify that the 95 percent grade-of-service criterion is still being satisfied.

The preceding discussion has been confined to the grade-of-service evaluation of a single impairment; in reality a number of impairments always coexist. This type of evaluation is sometimes made with the assumption that other impairments are held constant or that the impairment of interest is dominant. To establish overall transmission grade-of-service evaluations, all important impairments must be considered. Many evaluations of telephone circuits now are based on composite grades of service combining the effects of more than one impairment. The loss/noise grade of service is an example [1, 2]. Techniques are also now available to extend the concept to loss/noise and echo grade of service. For such evaluations, equations like (24-5) must be evaluated as multiple integrals. Computer programs such as GOSCAL are useful as an aid in evaluating these integrals.

## 24-3  USES OF THE GRADE-OF-SERVICE CONCEPT

Alternate solutions to many transmission engineering problems can be compared by using the grade-of-service concept. The information obtained by such analyses often provides the basis for making engineering compromises in establishing or allocating objectives, identifying weak spots in performance, evaluating improvements, showing the effects of time on services rendered, or conducting a subjective test program or a field performance survey. Grade of service is also used to evaluate combinations of service parameters and to optimize performance with respect to such combinations so that no single parameter is allowed to dominate.

### Engineering Compromises

Solutions to engineering problems almost always involve compromise when requirements conflict. One of the more obvious conflicts is achievement versus cost; performance can nearly always be improved if costs are ignored.

Grade-of-service relationships can often be used to determine what compromises are most appropriate and most economical. As an example, consider a study to determine the allocation of noise objectives to long-haul and short-haul carrier systems. The allocations were derived to satisfy customer-to-customer grade-of-service opinions of 95 percent good or better for 1000- to 4000-mile connections made up of combinations of long-haul and short-haul trunks having various assumed noise density functions. The resulting noise values were translated through a representative loop loss distribution to equivalent values at the station set for which grade-of-service opinions had been established by subjective testing. Throughout the study, the noise values were determined for each assumed trunk length from measured data. The standard deviation was assumed constant at 4 dB, independent of trunk length. One overall result of the study, shown in Figure 24-10, was a curve showing the locus of points for which combinations of long-haul and short-haul trunk noise satisfied a constant criterion of 95 percent good or better. Other curves, not shown, were determined for the other percentages, 97, 93, etc.

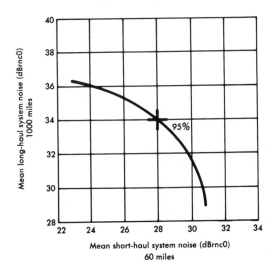

Figure 24-10. Noise allocation study results.

The choice of objectives, shown to be 34 dBrnc0 for long-haul and 28 dBrnc0 for short-haul systems, was made on the basis of considerations of expected noise performance of systems under development at the time of the study and the economics of reducing the noise in those systems. Figure 24-10 shows that a 2-dB relaxation of the long-haul objective to 36 dBrnc0 would require a 4-dB more stringent short-haul objective in order to maintain the 95 percent good-or-better grade of service. Also, if the short-haul objective were relaxed by 2 dB to 30 dBrnc0, the long-haul objective would have to be made more stringent by about 2.5 dB.

Another example of how grade of service might be used to solve an engineering problem involves the design of a group of new circuits needed to provide service to some remote location. If the most inexpensive design were applied, circuit losses would be significantly higher than the objective mean value. The reduction of these losses to meet the objective would necessitate the use of gain devices whose cost would make the project appear economically unattractive. There might even be a third alternative in which losses might be reduced to values somewhat short of the objective at only a modest cost increase. The three alternatives could be analyzed to determine how the losses in each case would affect the overall grade of service. The results might be evaluated economically by making a present worth of annual cost analysis, and then the final decision could be made as a compromise in which the reduction in grade of service could be evaluated in terms of cost to the company. Any of the alternatives might be a reasonable solution to the problem, provided the overall grade of service remained within the satisfactory range.

## Performance Evaluation

Two aspects of performance can be studied conveniently by using grade-of-service concepts. These are the identification of performance weak spots in a geographical area and the evaluation of performance improvements that have been introduced (evaluated by measurement) or proposed (evaluated by analysis).

Consider as an example of weak spot identification a situation in which transmission performance in an administrative unit (for example, a district or division) is shown to be deteriorating by a succession of dropping indices or an increasing number of customer complaints. Usually, the source of trouble can be identified by direct analysis of the index data, but in some cases it may be necessary to make measurments and then to compute the grade of service for loss, noise, and echo (return loss). The results might show that the falling index and/or the basis of complaints is a result of deteriorating echo performance. Further investigation might reveal that balance measurements have not been made according to schedule in just one central office, and with concurrent rearrangements in that office, the resulting impedance mismatches are the cause of performance deterioration.

Performance improvements that are introduced in the system or in some part of the system (a central office, a district, a division) may be evaluated by making a grade-of-service comparison using data obtained by measurements taken before and after the improvement is introduced.* This type of evaluation might include an analysis of the expected improvement, based on predicted performance before implementation, and a comparison of the measured results with those predicted.

## Time Effects on Grade of Service

Grade-of-service evaluations have their greatest value when they are made periodically. Variations and trends can be observed by comparing results of succeeding evaluations. Such comparisons are most informative in situations where poor maintenance practices, aging plant, or uncontrolled rearrangements lead to the introduction of increasing impairments, though the trends can often be observed and the cause of deterioration identified by direct analysis of indices.

Improvements in overall plant performance are also best evaluated by making periodic grade-of-service evaluations. Design changes cannot generally be made in the field instantaneously, and the periodic monitoring of performance can be used to verify that an improvement program is having the desired effect.

Similarly, the introduction of a more stringent transmission objective cannot be expected to result in a rapid improvement in overall performance. It is often uneconomical or technically unfeasible to introduce the necessary improvements in existing systems, apparatus, or equipment. The introduction of new systems designed to meet the new objectives also takes time; years often pass before customer satisfaction (as measured by appropriate performance indices) reflects the results of introducing such new systems.

Finally, communication system customers appear always to expect improved service with time [3]. The response of communication networks to increased public demands is necessarily a matter of considerable time.

Sometimes a grade-of-service evaluation shows the need for a new subjective test program or for a field survey of performance. When customer satisfaction with transmission performance appears to de-

---

*Caution must be observed when making a grade-of-service analysis on a small data base. The percentages obtained are valid on an absolute basis only for large universes. However, observations of the *change* in good-or-better or poor-or-worse percentages with changes in transmission parameters for one or more subcomponents are valid even for small data bases.

teriorate with time (as indicated by survey results, for example) and no performance degradation can be identified, the cause may be a result of changing public opinion regarding acceptable performance; and new subjective tests may be needed to establish the new bases of opinion.

With the introduction of new services and systems, there is also the possibility that unexpected changes in performance are the cause of deteriorating customer satisfaction. A survey might show the deterioration of customer satisfaction, and the change in plant performance might be reflected in the results of transmission measurements. When this occurs, a new survey of plant performance may be needed.

## 24-4 LOSS-NOISE-ECHO GRADE OF SERVICE

Grade of service is being used increasingly to establish transmission objectives, to relate performance to objectives, and to evaluate given performance parameters through the use of performance indices. In a number of these applications, grade of service provides a link between the subjective evaluation of an impairment and the establishment of an objective.

As discussed previously, grade of service can be applied to one specific impairment (such as loss, noise, or echo) or to combinations of these impairments. One important combination of impairments is reflected in the combined loss-noise grade of service recently developed. Earlier work had resulted in a model based on received volume; this was followed by work on a model to incorporate both received volume and idle circuit noise. These models were replaced as a result of more recent work on loss and noise. Finally, effects of talker echo, obtained in combination with loss and noise, resulted in models of the subjective effects of loss, noise, and talker echo on telephone connections [4]. These models have provided a basis for performance evaluation in terms of combined loss-noise-echo grade of service.

In the analysis of subjective test results, it was recognized that different tests yielded somewhat different results even when the same impairments were tested. This complicated the combining of results from different tests into a composite model of subjective opinion and led to the concept of a general transmission rating scale, referred to as the *R-scale,* which assigned a single numerical value to any specific impairment.

In this concept of the R-scale, it was recognized that subjective test results can be affected by various factors such as the subject group, type of test, and range of conditions included in the test. This could cause difficulties in trying to establish unique relationships between impairments and subjective opinion. The introduction of the R-scale tended to reduce this difficulty by looking at the problem in two parts. First, the transmission rating as a function of the impairment was anchored for two specific impairment conditions which tended to lessen the dependence on individual tests. Second, the subjective opinions could still be displayed for the individual test base from the R-scale results plus the reverse transformation.

## Connection Loudness Loss and Noise Model

The loss-noise grade of service has been recognized for some time as a valuable element in the evaluation of transmission performance. Therefore, loss and noise were first treated together as a step toward the larger result of a loss-noise-echo grade of service. This loss-noise work led to the establishment of the anchor points for the R-scale shown in Figure 24-11. These two points were selected to be well separated in quality. One point is typical of a short intertoll connection and the other represents an extreme condition of loss and noise which should rarely occur even on long intertoll connections between long loops.

| LOSS, $L_E$ (dB) | NOISE (dBrnc) | TRANSMISSION RATING |
|---|---|---|
| 15 | 25 | 80 |
| 30 | 40 | 40 |

Figure 24-11. Transmission rating anchor points.

Results from a number of different subjective tests were used in deriving the loss-noise subjective opinion model. The loss values from these tests were expressed in terms of loudness loss values which represent the acoustic-to-acoustic transfer efficiency of overall telephone connections in terms of the Electro-Acoustic Rating System (EARS) method [5]. Noise values used in the model are expressed at the line terminals of a telephone set having a reference receiving efficiency of 26 dB based on the EARS method. This efficiency value is approximately the receiving efficiency of a loop made up of a 500-type telephone set, a short line facility, and a standard central office battery feeding bridge.

The R-scale representation of subjective opinion for connection loudness loss and noise is denoted $R_{LN}$ and is given in Equation (24-14). Curves generated from the equation for $R_{LN}$ are plotted in Figure 24-12.

$$R_{LN} = 147.76 - 2.257\sqrt{(L_e - 7.2)^2 + 1} - 2.009N_F$$
$$+ 0.02037\, L_e N_F \qquad\qquad (24\text{-}14)$$

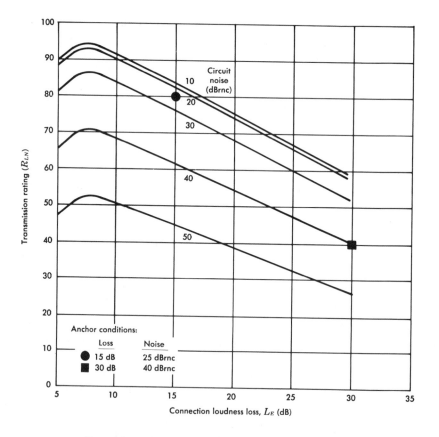

Figure 24-12. Transmission rating for loss and noise.

In Equation (24-14), $L_e$ is the acoustic-to-acoustic connection loudness loss (in dB) of an overall telephone connection and $N_F$ is the power addition of the circuit noise, $N$, and 27.37 dBrnc where $N$ is expressed in dBrnc at the line terminals of a telephone set with a reference receiving efficiency of 26 dB based on the EARS method.

## Talker Echo Model

The R-scale representation of subjective opinion for talker echo, denoted $R_E$, is given in Equation (24-15).

$$R_E = 95.01 - 53.45 \log \left[ \frac{1+D}{\sqrt{1 + (D/480)^2}} \right] + 2.277E. \qquad (24\text{-}15)$$

In this equation, $D$ is the round-trip echo path delay in milliseconds and $E$ is the round-trip acoustic-to-acoustic loudness loss in dB of the echo path. Curves generated from Equation (24-15) are plotted in Figure 24-13 for a range of echo path delay and echo path loss values.

## Loss-Noise-Echo Model

The R-scale representation of subjective opinion for the combined impairments of loss, noise, and talker echo is denoted $R_{LNE}$; it is related to $R_{LN}$ and $R_E$ as shown in Equation (24-16) where $R_{LN}$ and $R_E$ are given in Equations (24-14) and (24-15) respectively.

$$R_{LNE} = \frac{R_{LN} + R_E}{2} - \sqrt{\left(\frac{R_{LN} - R_E}{2}\right)^2 + (10)^2}. \qquad (24\text{-}16)$$

Transmission rating for loss, noise, and talker echo, ($R_{LNE}$) is shown plotted in Figure 24-14 as a function of echo path loudness loss for a range of round-trip echo path delay values with a connection loudness loss value of 15 dB and a circuit noise value of 30 dBrnc. The asymptotic limits of the curves at large values of echo path loud-

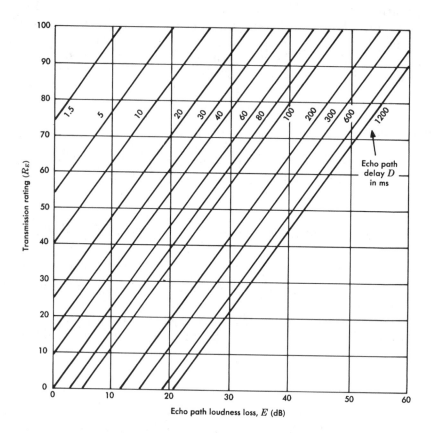

Figure 24-13. Transmission rating for talker echo.

ness loss vary with connection loudness loss and circuit noise in accordance with the curves of Figure 24-12.

## Subjective Opinion Models

The transmission rating scale was selected so that most telephone connections have positive ratings between 40 and 100 with higher ratings denoting better quality. One important feature of the transmission rating is that ratings can be computed without reference to any particular subjective test. As users become more familiar with the R-scale, the test-independent measure of subjective quality may replace the presently used measures such as percent good or better, percent poor or worse, etc., which are test dependent.

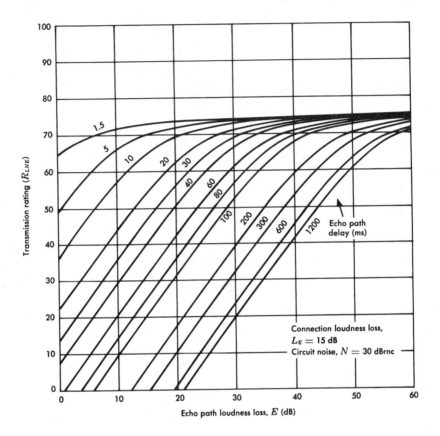

Figure 24-14. Transmission rating for loss, noise, and talker echo.

For present use, relationships have been established between the R-scale and the subjective opinions for specific subjective tests. For example, contours of constant percent good or better and poor or worse are shown in Figure 24-15 for a specific set of loss-noise subjective tests performed in 1965. The results of these tests have been widely used and thus represent a subjective test data base that is well known [6].

### REFERENCES

1. Lewinski, D. A. "A New Objective for Message Circuit Noise," *Bell System Tech. J.*, Vol. 43 (Mar. 1964), pp. 719-740.

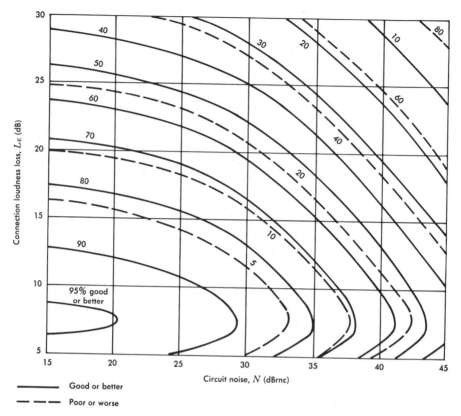

Figure 24-15. Constant percent contours.

2. Sen, T. K. "Subjective Effects of Noise and Loss in Telephone Transmission," *I.E.E.E. Transactions on Communication Technology*, Vol. COM-19 (Dec. 1971), pp. 1229-1233.

3. MacAdam, W. K. "A Basis for Transmission Performance Objectives in a Telephone Communication System," *A.I.E.E. Transactions, Part 1, Communications and Electronics*, Vol. 77 (1958) pp. 205-209.

4. Cavanaugh, J. R., R. W. Hatch, and J. L. Sullivan. "Models for the Subjective Effects of Loss, Noise and Talker Echo on Telephone Connections," *Bell System Tech. J.*, Vol. 55 (Nov. 1976), pp. 1319-1371.

5. Sullivan, J. L. "A Laboratory System for Measuring Loudness Loss of Telephone Connections," *Bell System Tech. J.*, Vol. 50 (Oct. 1971) pp. 2663-2739.

6. Spang, T. C. "Loss-Noise Echo Study of the Direct Distance Dialing Network," *Bell System Tech. J.*, Vol. 55, (Jan. 1976), pp. 1-36.

Chapter 25

# Determination and Application of Objectives

Transmission objectives must be processed in a number of ways to make them useful in system design and operation. They can be considered as goals that are established and used as criteria for the achievement of a quality of service that is economical and ultimately satisfactory to nearly all customers. The processes to which objectives are submitted produce a set of requirements, i.e., performance parameters that *must* be satisfied if the objectives are to be met.

Ambiguity between the terms *objectives* and *requirements* may be noted. The distinction may be a matter of definition, point of view, or terminology. When an expression such as "a 95 percent good-or-better grade-of-service objective" is used, there can hardly be any doubt that an objective is being discussed. If it is stated that "the transistor must have a single-frequency fundamental-to-third harmonic ratio for a milliwatt output," there is little doubt that a requirement is being stated. In between, there are many uncertainties and shades of meaning. The processing of requirement and objective data, therefore, is often similar and involves what shall here be called *determination, interpretation, allocation,* and *translation.*

These subjects are all broad in nature. It is not the intent here to discuss them in detail with respect to the many possible applications. Rather, general considerations of the processing are reviewed and several examples are given. It must be recognized, also, that the processes are generally reversible. The measurement of performance of devices, circuits, and parts of systems can often be extrapolated to determine or estimate the overall performance of a transmission system.

## 25-1 DETERMINATION

The determination of transmission objectives most often begins with a subjective test program designed to establish the relationship

577

between an impairment (or combination of impairments) and observer opinions of the effects of various amounts of the impairment. The data representing the test results are then combined with performance parameters so that the combination can be expressed in terms of grade of service. Finally, the grade-of-service relationships are analyzed through engineering economy studies to determine relative costs of furnishing various grades of service. On the basis of these studies, a value for the objective can be selected that represents a reasonable compromise between providing customer satisfaction and the cost of providing the service.

Another way of processing data in order to form an objective is to derive an objective *index* for a given impairment; an index is a single number, based on a scale of 1 to 100, which can be used as a broad measure of plant performance. Indices are used mainly as tools for transmission management; they are particularly valuable in showing trends of performance for large geographical or administrative units of the plant. The objective indices are always expressed according to the following scale:

|         |                  |
|---------|------------------|
| 99 - 100 | Excellent        |
| 96 - 98  | Fully satisfactory |
| 90 - 95  | Fair to mediocre  |
| Below 90 | Unsatisfactory    |

Requirements that must be met in order to satisfy the established objectives are usually derived from the objectives. The dependency of the one upon the other distinguishes requirements from objectives. An *objective* is a goal; a *requirement* must be met to satisfy the goal.

## 25-2  INTERPRETATION

In whatever form it may be stated, an objective is subject to a great deal of interpretation to make it applicable to a particular set of circumstances. The interpretive treatment of objectives may call for identification of the following: (1) the static characteristics of the impairment (for example, a pure single frequency versus one having high sideband content or intelligible crosstalk versus nonintelligible crosstalk); (2) the probabilistic phenomena which might cover the probability of occurrence or the statistical properties of a varying impairment; (3) the simultaneous effects of multiple impairments of the same or different types; (4) the way of expressing the impairment and the objective for various types of signals; (5) the establishment of a requirement in terms of a *limit*. The term *limit* implies that some form of corrective action must be implemented if the requirement is exceeded.

## Static Impairment Characteristics

The interpretation of an objective often depends on some characteristic of the impairment. For example, the transmission of television signals in the presence of speech signals introduced new types of interference to telephone transmission. Some of these interferences may be regarded as single-frequency interferences for which objectives have long been established. However, since they result from line scan frequency components of the television signal, there is sideband energy at multiples of 30 and 60 Hertz on both sides of each line scan frequency multiple. These sidebands produce a subjective effect that makes the single-frequency interference sound distorted. The objective for this distorted single-frequency interference is a matter of interpretation and had to be established by subjective tests designed to compare the interfering effect of the distorted tone with that of a pure single frequency. The distorted tone was found to be less interfering than the pure single frequency, so that 2 dB more interference power can be tolerated. Thus, the single-frequency interference objective may be interpreted for application to television tone interferences so that if the single-frequency objective is $x$ dBrnc0, the television tone objective is $x + 2$ dBrnc0.

Another example of interpretation concerns crosstalk objectives, usually expressed in terms of minimum allowable crosstalk coupling loss derived from crosstalk indices. In some cases, the nature of the coupling path results in nonintelligible crosstalk due to some phenomenon such as frequency inversion. In the past, the objectives applied to nonintelligible crosstalk were the same as those applied to intelligible crosstalk. The reason for this interpretation was that when nonintelligible crosstalk is heard, the syllabic character of the interference is quite recognizable, and the listener finds it as annoying as if it were intelligible. More recently, this type of crosstalk has been treated as noise, with the objective made about 3 dB more stringent than that for random noise.

## Probabilistic Characteristics

A statement of objectives for impairments having probabilistic characteristics must include appropriately qualifying phrases to account for the variability. Several examples may be cited.

In digital signal transmission, one of the most common ways to express transmission impairment is in terms of error rate. When the effect of impulse noise is evaluated, performance must be related to error rate in such a way that the rate of signal transmission, the amplitude distribution of the interference, the probability of occurrence of the interference, and certain characteristics of the transmitted signal must all be considered. Simplistic statements of an error rate objective are often inadequate; block error rates, or the expected percentage of unimpaired transmitted blocks of information, are sometimes more meaningful because error detection codes and the basic coding of the signal often permit retransmission of impaired blocks. A burst of errors might completely ruin a single block of information, causing an apparently excessive error rate. The result, however, might be the retransmission of that single block of information, one of many perfect blocks. Thus, an error rate objective simply expressed as an objective of $10^{-6}$ must be further interpreted to account for signals and interferences having various parameters.

Impulse noise may also impair television signals. Here again, any expression of an impulse noise objective must take into account the statistical variation of impulse amplitudes and the probability of occurrence.

Interference, though continuously present, may also vary in amplitude and/or in frequency and in the subjective effect of one or the other. The previously cited example, comparing the interfering effects of television signal components and single frequencies, involved subjective tests that included amplitude variations (due to changes in picture content) and frequency variations (due to frequency drift in power sources) in addition to the sideband distortion components discussed. The interpretation of the final objective had to recognize that the mean amplitude value and the nominal frequency were represented.

Very low-frequency interference in television signals sometimes has the same subjective effect as a flickering of the picture. Careful interpretation of subjective test results must be made to insure that both bar pattern and flicker effects are covered by the objective.

## Multiple Impairments

Another situation that involves interpretation of objectives occurs when multiple impairments are simultaneously present. When they are of different types, a portion of the objective must be allocated to each impairment. If the impairments are of the same type, the objectives are usually established or interpreted in terms of the combined effect.

One illustration of the way multiple impairments may be treated is the handling of multiple sources of intelligible crosstalk. The performance and objectives for intelligible crosstalk are usually expressed in terms of the crosstalk index. As discussed in Chapter 17, the crosstalk index is derived by mathematical relationships which include the probability of hearing intelligible crosstalk. If the sources are independent, the number of sources is included as a parameter in the derivation. If the crosstalk paths can cause simultaneous exposures to the same source, the coupling is increased by $10 \log N$ or $20 \log N$ ($N$ is the number of paths), as appropriate, and the crosstalk is treated as if it were from a single source.

Sometimes there are multiple interferences of a random nature whose combined interfering effect is best evaluated by summing the powers of the individual contributors. In multichannel analog transmission systems, for example, the combined effect of thermal noise and interchannel modulation noise is determined by adding the powers of the two contributors. The objective must be interpreted as applying to the sum of the interferences.

## Objectives, Requirements, and Limits

There is a multitude of expressions used for performance, objectives, and requirements in telecommunications. These expressions are all subject to interpretation according to the nature of the impairment, the characteristics of the channel and the signal, or the manner in which these characteristics interact. Depending on circumstances, digital signal transmission performance or objectives may be expressed in terms of signal-to-noise ratio, noise impairment, error rate, or percentage of eye closure. Television impairments are usually expressed in terms of signal-to-interference ratios where the signal is measured in peak-to-peak volts; however, the interference may be expressed in rms, peak, or peak-to-peak volts. Random noise and echoes are weighted by frequency and/or time delay. Telephone objectives and requirements are often expressed in terms of absolute values of interference as measured at specified transmission level points.

All these expressions must be thoroughly understood since often it is necessary to interpret one in terms of another or to derive one from another. Part of the interpretation process requires a thorough understanding of the transmission level point concept. When the concept is properly applied, telephone system noise in dBrnc0, for example, can easily be interpreted as a signal-to-noise ratio for data signal transmission analysis.

Objectives have been defined as desired goals. Requirements are performance parameters that must be met if objectives are to be satisfied. Limits are performance parameters that, when exceeded, indicate a need for some form of corrective action and, in some cases, removal of a circuit from service until the corrective action is completed.

When a new transmission system is being developed, *design objectives* are applied to guide the generation of *design requirements* that must be met. The design requirements are maximum or minimum values that must be met in the controlled development environment if the design objectives are to be met.

When a system is installed in the field, *engineering objectives* are those that are recommended for the application or layout of the system in the field environment. After installation and during the useful life of a transmission system, trunks and other transmission channels are connected through the system. When these circuits are connected and before they are released for service, they are subjected to a series of tests specified on the circuit order to ensure that all the equipment is properly aligned and to verify that the circuit meets its engineering objectives. The minimum and maximum values established for these tests are sometimes called circuit order requirements.

*Maintenance requirements* are intended to reflect performance that is practical to obtain in the field environment using existing equipment and operating procedures. *Maintenance limits* are those within which performance is satisfactory; when the limits are exceeded, maintenance action is required. A maintenance limit may be established at some value which, if exceeded, requires that a circuit be taken out of service. Such a limit is sometimes called a *turn-down limit*.

## 25-3 ALLOCATION

Objectives are usually established and applied in a format that expresses the performance goal for overall systems or broad categories of service. To be useful in development, design, or operation, these broad objectives must be appropriately allocated to a variety of impairments, to portions of the plant or parts of systems.

### Allocation Assuming Power Addition

The allocation process requires the exercise of considerable judgment and a knowledge of many system performance parameters and cost relationships. The process is seldom arbitrary, but in the absence of data indicating otherwise, it is common to assume that different types of impairments add according to their powers and that like impairments from different parts of a system also add by power. It does not necessarily follow that the several impairments are given equal weight. If economic factors or system parameters are significant, the impairments may be allocated different proportions of an overall objective.

An overall objective for a single impairment may also be allocated to different parts of the plant or to different elements of connections according to the assumption of power addition. Allocations of noise to a hypothetical layout of video circuits may be used to illustrate this technique. Figure 25-1 shows the principal elements in a 4000-mile layout. The elements include a toll transmission circuit or circuits, local transmission circuits, carrier switching centers, video television operating centers (TOCs), and intraoffice trunks. The number of each of the various elements must be assumed or determined from engineering studies of the service needs.

If the layout is known and power addition is assumed, the allocation of the objective is a straightforward process of dividing the power of the interference that just meets the objective among the

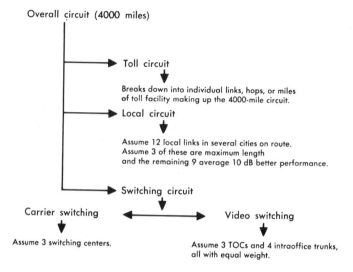

Figure 25-1. Assumed configuration of the overall circuit for objective allocations.

elements that may make up a 4000-mile overall connection. This breakdown is illustrated by Figure 25-2. Allocation such as that illustrated cannot be rigidly followed because some degradations are present only in certain sections of the overall connection and are completely absent in other sections. Thus, it is often necessary or desirable to reallocate larger portions of the objective to the troublesome sections.

## Cost Effects

When allocation to various parts of the plant or system is considered, the relative costs of achieving the allocated objective must be carefully weighed. The relationships between allocations, the characteristics of different parts of the plant, and the relative extent to which the parts of the plant are used must be considered. Two illustrations, taken from a study of message circuit noise allocations, show how these parameters interact [1].

The study, which culminated in grade-of-service calculations for telephone connections over all distances, showed that the overall noise grade of service was about 97 percent good or better. Yet, for

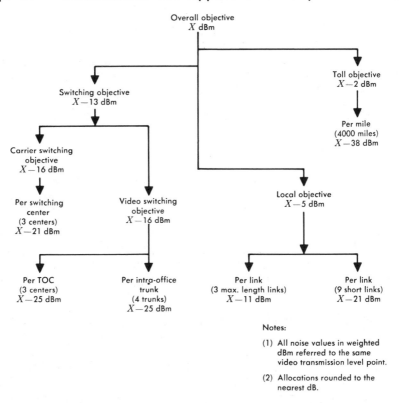

**Figure 25-2.** Noise objectives for a 4000-mile circuit allocated on the assumption of power law addition.

longer connections the grade of service was calculated to be below 90 percent due to the fact that noise increases at a rate of about 3 dB for each doubling of the distance. The relatively poor performance on long connections had little effect on the overall grade of service because the number of calls decreases rapidly with distance.

The conclusion from this part of the study was that noise objectives for longer systems should be made more stringent by 3 to 4 dB to improve the grade of service on long connections. It was recognized, however, that the achievement of 3 to 4 dB improvement in noise on existing systems would not be economically feasible, so the more stringent objectives are applied only to newly designed systems.

The second observation made in the study was that noise from short- and long-haul trunks influenced the overall noise at the station set to the extent that a reduction of loop noise below 20 dBrnc could not be justified. Regardless of money or effort expended to reduce this noise contribution, the grade of service would not be significantly improved. Thus, loop noise in excess of 20 dBrnc would be observed as a degradation in grade of service. These results are illustrated in Figure 25-3. As a result of these studies, an overall loop noise objective of 20 dBrnc was adopted.

### Allocation for Digital Transmission

Allocation problems for digital signal transmission are the same in principle as those encountered in analog signal transmission. How-

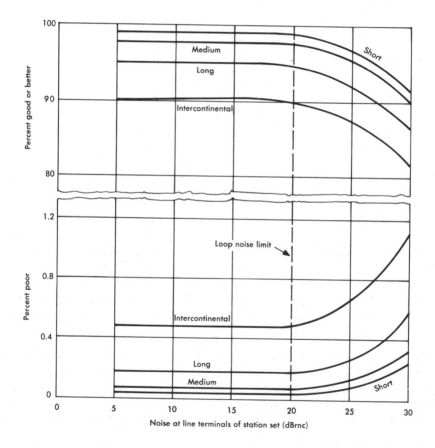

Figure 25-3. Grade of service for connection objective as a function of loop noise.

ever, the problems differ in detail because of the discrete nature of the signal, the regenerative processes used in digital transmission system repeaters, and the properties of time division signal multiplexing. Digital signal impairments include errors, jitter, and misframes.

In a regenerative repeater system, it is assumed that transmission line noise impairments are noncumulative because, in theory, signal pulses are perfectly reconstructed at each repeater. However, margin must be provided to guard against pulse distortion to the point where errors may occur since, at that point, performance deteriorates very rapidly.

Because of the precipitous nature of performance degradation, digital system components, such as repeaters and multiplex equipment, normally operate at an error rate that is near zero. A high system error rate is often due to a single repeater operating with excessive pulse distortion or with insufficient margin. Therefore, it is common practice to assign the same error rate objective to parts of a system as that assigned to the whole system. The probability of exceeding this error rate is then allocated among the parts.

Certain impairments are cumulative and careful attention must be given to these when objectives are being allocated. Timing problems (jitter) are cumulative along a repeatered line. Therefore, the total objective must be allocated so that relatively large margins are maintained.

Misframes occur when the demultiplexing terminal loses synchronism with the incoming bit stream. Communication on all channels is interrupted until synchronism is recovered. The effect is usually noted at all levels of the digital hierarchy below that at which it occurs. Thus, misframe impairments must be allocated among the various multiplex levels.

Another class of impairments that may accumulate has its source in the terminal equipment used to convert analog signals to a digital format. The coding process produces noise that is called quantizing noise, an impairment that increases with the number of times the signal is so processed. Care must be taken to allocate objectives realistically in relation to laws of accumulation that may pertain to given situations.

## 25-4  TRANSLATION

The translation of objectives from one set of parameters to another (for example, from a broad objective to a specific requirement) or from one transmission level point to another frequently requires a knowledge of how systems operate or how they can be organized to adapt them to an assumed process of translation. Three examples of system operating parameters are the laws of addition of intermodulation products in a series of analog cable system repeaters, the effects of frogging on system performance, and the relationships among various noise contributors in an optimized system design.

### Objectives to Requirements

Requirements are generally derived from objectives by a process that may be regarded as translation. Many examples could be used to illustrate the process. Consider two that relate to the generation of intermodulation noise in analog systems.

In an analog cable system proposed for use up to 4000 miles, a system design objective is established as a goal to satisfy overall noise grade-of-service objectives. A number of interpretation and allocation processes are carried out so that the translation is finally carried to a point such that it is possible to specify the permissible magnitudes of second-harmonic and third-harmonic products for a 0 dBm0 signal in the 800 miles between frogging points. These new values may now be regarded as system design requirements that must be met in the 800-mile link if the initial overall noise objective is to be satisfied.

In a microwave radio system, a noise objective consistent with overall Bell System noise grade-of-service objectives is similarly established for 4000-mile transmission. Through interpretation and allocation, this objective is processed so that it is possible to determine the maximum permissible value of interchannel modulation noise generated in one radio repeater. This value might be regarded as a requirement itself, but further translation may also be applied so that the requirement can be expressed in terms of return-loss values for the waveguide sections of the radio repeater. (In FM systems reflections due to low return losses produce intermodulation among the signal components.) Thus, the overall noise objective for the system is translated into return-loss requirements for the individual repeater waveguide sections.

## Transmission Level Point Translations

Objectives and requirements are often expressed in reference to a specific transmission level point, generally 0 TLP. Sometimes it is necessary to express the same objective in reference to some other TLP. This translation process is usually straightforward but care must be taken to make the translation properly.

Consider the previous discussion of the translation of an overall noise grade-of-service objective to the second and third harmonic requirements for an analog cable system. To illustrate the further translation to a different TLP, assume that the 800-mile requirements are a maximum second harmonic of $-40$ dBm0 and a maximum third harmonic of $-50$ dBm0 for a 0 dBm0 fundamental signal. (These values are illustrative only and are not necessarily representative.) Now, suppose that a convenient point of measurement of system performance is at a $-10$ dB TLP. A 0 dBm0 signal translates to a $-10$ dBm signal at the $-10$ dB TLP, the second-harmonic requirement translates to $-40 -10 = -50$ dBm, and the third-order requirement translates to $-50 -10 = -60$ dBm.

Now suppose that these 800-mile requirements must be expressed in terms that are consistent with the magnitude of a test signal specified as $-16$ dBm0. At the 0 TLP, the second-harmonic requirement is then $-40 -2(16) = -72$ dBm0; translated to the $-10$ dB TLP, the requirement is $-72 -10 = -82$ dBm. Similarly, the third-harmonic requirement at 0 TLP is $-50 -3(16) = -98$ dBm0 or $-98 -10 = -108$ dBm at the $-10$ dB TLP. (Recall that the second and third harmonics vary in dB by 2:1 and 3:1, respectively, relative to the fundamental.)

The processes are similar when objectives or requirements are translated between transmission level points in video systems. However, the situation is further complicated by the fact that television objectives are usually expressed in volts or in dB relative to volts (rms, peak, or peak-to-peak) and the expression of the objectives and their translation must therefore take into consideration the impedances at the points of interest as well as the appropriate voltages.

## Indices

Objectives and requirements are used in the field as standards against which measured performance can be compared. One aspect of this process involves transmission management. The vastness of the plant and the many parameters to be evaluated have led to the concept of transmission indices. These are single-number evaluations of a large sector of plant which are derived from grade-of-service objectives and expressed in such a way that performance measurements can easily be translated into indices. Such indices are used extensively as a transmission management tool.

## System Parameter Effects

Three analog cable system parameters were previously mentioned for their significant impact on the translation of objectives or requirements. These include the laws of addition of intermodulation products, the effects of frogging, and the interrelationships among noise contributors in an optimized system design. Many other parameters could be mentioned, but these illustrate the importance of system effects on translation processes.

**Laws of Addition.** It can be shown that second-order and many third-order intermodulation products tend to accumulate in successive repeaters of an analog cable system by a law of power addition [2]. Also, some types of third-order products (usually termed $2A-B$ and $A+B-C$ products) tend to add systematically (by voltage) in successive repeaters. The translation of objectives must take these facts into account.

The hypothetical set of objectives for second and third harmonics, previously suggested as illustrative, can be used again here; for a 0 dBm0 fundamental signal, the second- and third-harmonic requirements were −40 dBm0 and −50 dBm0. Assume that the system under consideration requires 250 tandem repeaters in an 800-mile link. The translation of the harmonic distortion requirements to per-repeater requirements can be accomplished by taking these laws of addition into account. Thus, a per-repeater second-harmonic requirement may be obtained from the 800-mile requirement by subtracting the power of the other repeater contributions. For a 0 dBm0 fundamental, then, the per-repeater requirement for second harmonic is

$$-40 -10 \log 250 = -40 -24 = -64 \text{ dBm0.}$$

For the third harmonic, the per-repeater requirement is

$$-50 -20 \log 250 = -50 -48 = -98 \text{ dBm0.}$$

These values may be further adjusted for the TLP appropriate to the repeater and for the magnitude of the test signal.

A peculiarity in computing the third-harmonic requirement must be explained. Third-harmonic intermodulation products accumulate by a power law. The requirement is derived on the basis of voltage addition because the dominant ($2A-B$ and $A+B-C$) third-order products add by voltage and the same nonlinear coefficient applies to both third harmonics and the dominant products. Thus, it is a convenient fiction to assume voltage addition for third harmonics. The requirement must ultimately be translated into a per-repeater modulation coefficient requirement and, beyond that, into requirements on device linearity and amplifier feedback.

**Frogging.** Frequency frogging of telephone channels is specified in long analog cable systems to accomplish two principal goals: (1) to break up the systematic addition of third-order modulation products and (2) to break up systematic departures from ideal (flat) transmission in the system amplitude/frequency response. Both of these advantages are sought in order to ease the requirements on linearity and equalization of repeaters in short sections of line. The effectiveness of frogging can be seen by examining the translation from 4000-mile to 800-mile objectives and then examining how the resulting linearity requirements would have been made more stringent if the frogging advantage could not be realized.

The effect of in-phase addition on the third-harmonic requirement for a single repeater in an 800-mile link was previously noted. The $-50$ dBm0 requirement for an 800-mile link (from which the single-repeater requirement was derived) was, by inference, derived by translation from the 4000-mile objective and included the assumption of power addition between 800-mile links. If it had been necessary to assume in-phase or voltage addition between 800-mile links, the 800-mile requirement would have been $-43 -20 \log 5 = -57$ dBm0 for the third harmonic of a 0 dBm0 fundamental. The assumption of frogging and power addition of 800-mile links eased the 800-mile and per-repeater requirements by $57 -50 = 7$ dB.

**Signal-to-Noise Optimization.** Transmission system design analysis has shown that if the predominant source of intermodulation noise in an analog cable system is a result of second-order nonlinearity, the optimum signal-to-noise performance is obtained when signal amplitudes are adjusted so that the intermodulation noise is equal to the thermal noise in the system [2]. If the predominant intermodulation noise is due to third-order nonlinearity, the optimum performance is obtained when signal amplitudes are adjusted so that the intermodulation noise is 3 dB less than the thermal noise.

These relatively simple theoretical relationships are made complex by many of the detailed design parameters that enter into an actual computation and the resulting iterative process that must take place during design and development. The result of the design process is that one type of intermodulation phenomenon tends to dominate the other. The translation process, from objectives to device and circuit requirements, must then be applied rigidly to the dominant parameter. The extent to which the requirements on parameters of less importance can be relaxed depends on the amount of margin exhibited.

Some parameters can be safely ignored in spite of their potentially harmful effects on performance. In the analysis of intermodulation phenomena, for example, the second- and third-order nonlinearity terms of an input/output function are so dominant that terms higher than third order are usually ignored. During design and development the higher order terms must be checked, but they are usually found to be insignificant. Exceptions are found occasionally in the design of nonlinear modulator circuits used in multiplex equipment.

### REFERENCES

1. Lewinski, D. A. "A New Objective for Message Circuit Noise," *Bell System Tech. J.*, Vol. 43 (Mar. 1964), pp. 719-740.

2. Technical Staff of Bell Telephone Laboratories. *Transmission Systems for Communications*, Fourth Edition (Winston-Salem, N.C.: Western Electric Company, Inc., 1970), Chapter 13.

Chapter 26

# Transmission Objectives

Transmission objectives, derived from grade-of-service analyses of the results of subjective tests and performance measurements, are stated in terms of design, performance, or maintenance objectives. Requirements, derived from the objectives, are also given in these same terms and additionally in terms of maintenance limits. These limits define points at which performance is so poor and service is so adversely affected that circuits must be repaired or, in the extreme, taken out of service until repairs or adjustments can be made.

The objectives are continuously studied and modified to adapt to the changing environment caused by the introduction of new equipment and systems, new technology, new services, and by changes in customer opinions. The objectives are often established as overall values applicable to terminal-to-terminal connections. They are then allocated in various ways to appropriate portions of the plant, to various impairments, or to a particular type of service.

This chapter discusses well-established objectives, with numerical values given wherever possible. Where objectives have not become standard, only general discussion is included to indicate the nature of transmission problems involved. Space does not permit a comprehensive listing of objectives for all types of signals, systems, or services. Furthermore, the changing nature of transmission objectives might make the material obsolete within a short time.

## 26-1 VOICE-FREQUENCY CHANNEL OBJECTIVES

Transmission objectives for voice-frequency channels in the switched message network were initially established to satisfy the needs of speech transmission. As new types of signals and services have evolved and as new technology has been applied to all parts of the system, existing objectives have been modified and new objectives have been developed where necessary.

## Bandwidth

The bandwidth of telephone loops and trunks has evolved without a specific and consistent set of objectives. Initially, deficiencies in the bandwidth of such circuits were masked by station set limitations. The necessity for establishing an acceptable channel bandwidth allocation and carrier separation was recognized when AM carrier systems were introduced. At that time, the single sideband (SSB) mode of transmission was established; also, the 4-kHz spacing of carriers was deemed adequate in view of practical bandwidth limitations and in view of articulation tests conducted for the purpose of establishing bandwidth requirements for intelligibility and naturalness of speech.

As new systems have been introduced and as design technology has improved, efforts have continued to make the effective bandwidth of network channels as wide as is economically feasible within the constraints of (1) the 4-kHz carrier separation and achievable filter designs and (2) the unavoidably low singing return losses of loaded cable facilities near the high-frequency cutoff region.

More recently, subjective tests have shown that the preferred bandwidth for voice communications is approximately from 200 to 3200 Hz. As a result of these tests and technological advances, efforts are being made to establish standard design objectives for the bandwidth of each major portion of the network so that overall connections can meet the bandwidth objective. The natural increase in attenuation with higher frequencies for nonloaded cable and the sharp high-frequency cutoff of loaded facilities cause both types of facilities to exert considerable influence over the effective bandwidth of loops and VF trunks in an overall connection. Economics must be considered for making objectives for each portion of the network as stringent as possible, yet balanced with other portions. When finally established, the design objective may be expected to be close to the preferred bandwidth.

While this discussion centers about the transmission of speech signals, it should be pointed out that as new voiceband data services are introduced, the continued pressure to transmit at higher data rates creates additional demand for wider bandwidths.

## Frequency Response Characteristic Distortion

Formal message network objectives for inband amplitude and phase distortion are not generally available (except for the case of conditioned data loops). However, the same factors that tend to make the effective telephone channel bandwidth as wide as possible also work to make the inband response as uniform as possible. The difficulties have been to express the objective values in a generally acceptable manner and to allocate channel impairment requirements optimally among the many contributors.

**Amplitude/Frequency Distortion.** The only message network design objective for this impairment that has general acceptance applies to loops for DATAPHONE® service or to data access arrangement (DAA) loops for speeds of 300 bits per second or higher. The objective is that the loss at 2800 Hz shall be no more than 3 dB greater than the loss at 1000 Hz [1].

Other portions of the network for which amplitude/frequency response objectives are being studied include trunks, carrier- and cable-derived facilities, and central office equipment including the transmission paths through switching machines. Maintenance objectives are also under study for these parts of the network. Some installation requirements for allowable slope are available in terms of 400-Hz and 2800-Hz deviations from 1000-Hz loss values. The limits depend on the type of facility used.

**Phase/Frequency Distortion.** Message network phase/frequency distortion objectives, usually expressed in terms of envelope delay distortion, are applied to loops and other special-service circuits conditioned for data transmission [1]. If a loop is conditioned for data transmission at a rate of 300 bits per second or higher, a performance objective applies of no more than 100 microseconds of differential delay between any two frequencies over the band from 1000 to 2400 Hz.

Phase/frequency distortion objectives are under study for application to other parts of the plant. These studies are likely to result in an objective for signal peak-to-average ratio (P/AR); the P/AR meter method of expressing impairments is discussed in Chapters 18 and 21. This approach has considerable merit in its simplicity but has the undesirable attribute of evaluating all impairments simultaneously [2]. Thus, it can be used for expressing delay distortion objectives only where delay distortion is known to be the predominant impairment.

## Network Loss Design Plans

Echo and loss result from different impairing mechanisms and have different subjective effects on listeners. However, they are closely related in that a practical way of reducing echo effects is to increase loss. Since a loss increase introduces impairment (reduced received volume), a compromise must be sought that maintains circuit losses at satisfactory values, yet reduces echo effects to acceptable values.

Ideally, it might seem to be desirable to operate telephone circuits with no echo and no loss between end offices, but such ideal designs are impractical and can be achieved only in the laboratory or on selected circuits under well-controlled operating conditions. In a large complex network involving switching, the nature of the variables makes such a mode of operation uneconomical and impractical. Echo-free transmission implies high return loss at interfaces such as four-wire to two-wire conversion points. Lossless transmission implies stable operation of electronic circuits in the face of highly variable terminating impedances.

The transmission objectives for loss and echo in the toll portion of the network have been established on the basis of the *via net loss* concept [3, 4]. This concept and the resultant objectives, while still applicable to the network, are subject to continuous study, and details change from time to time. For example, echo suppressor application rules were recently revised. In addition, with the introduction of time-division switching arrangements in the toll portion of the network, a fixed-loss design is being considered for replacing or supplementing the VNL plan. Such a new plan is required where digital switching and digital transmission techniques are combined and where they must be integrated with the existing analog network.

**Via Net Loss Design Plan.** A significant factor related to echo impairment is the propagation time involved in a connection, *echo path delay*, which can be predicted from known parameters of the types of facilities used in making up a connection. The loss and propagation time of the echo path can be predicted only statistically because of the variations in facility properties and in return loss at the distant end office. Similarly, reaction to talker echo (the third controlling parameter in judging echo performance) can be predicted only statistically because of the variation in customer tolerance.

The required one-way overall connection loss for satisfactory echo is plotted in Figure 26-1 as a function of the round-trip echo delay and of the number of trunks in the connection. Also shown is a linear approximation for one trunk that proved useful in the evolving concept of VNL operation. This approximation was derived empirically by considering (1) the need for increased loss at low delays to prevent singing or near-singing; (2) the need to control noise, crosstalk, and transmission system loading; (3) the compromise between sufficient loss to control the effect of echo and the degradation introduced by echo suppressors; and (4) the analytical advantage of having the loss expressed as a linear function of echo path delay.

Linear approximations for more than one trunk may be derived by adding 0.4 dB for each additional trunk to the loss required for a single trunk. This loss is approximately the difference between loss curves at 45 milliseconds delay. The linear approximations may be drawn from the equation,

$$OCL = 0.102D + 0.4N + 4.0 \qquad \text{dB} \qquad (26\text{-}1)$$

where $OCL$ is the overall connection loss, $D$ is the echo path delay in milliseconds, and $N$ is the number of trunks in the connection. Note that 0.102 is the slope of the dashed line of Figure 26-1 and that $(0.4N$ and $4.0)$ dB is the zero-delay intercept of this line for various values of $N$. Equation (26-1) is used for connections involving round-trip delays up to 45 milliseconds. For delays in excess of 45 milliseconds, one of the trunks is equipped with an echo suppressor.

The final step in the VNL design process is to assign trunk losses so that each type of trunk in a connection operates at the lowest practicable loss consistent with its length and the type of facility used. In Equation (26-1), 2 dB of the constant is assigned to each toll connecting trunk, and the remainder is assigned to each trunk in the connection, including toll connecting trunks. The amount added to each trunk is in proportion to the echo path delay of the trunk and is defined as *via net loss*.* It is

$$\text{VNL} = 0.102D + 0.4 \qquad \text{dB} \qquad (26\text{-}2)$$

where $D$ is the echo path delay in milliseconds.

* The recent change in the method of defining toll connecting trunk loss, which involved the assignment of an additional 0.5 dB of loss (for central office equipment) to the toll connecting trunk, resulted in a design value of VNL + 2.5 dB. This added loss was previously assigned to the loop; thus, the customer-to-customer loss has not changed.

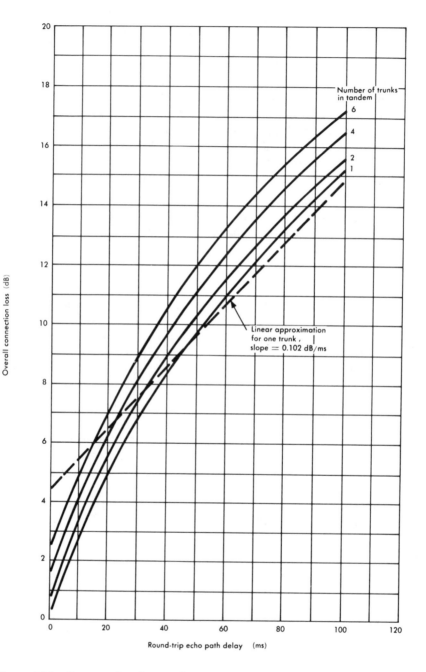

Figure 26-1. One-way loss for satisfactory echo in 99 percent of connections.

Since the echo path delay is directly related to the length of the circuit, Equation (26-2) is usually written in terms of length and a via net loss factor (VNLF).

$$VNL = VNLF \ (d) + 0.4 \qquad dB \qquad (26\text{-}3)$$

where $d$ is the distance in miles. The VNLF is

$$VNLF = \frac{2 \times 0.102}{v} \qquad (26\text{-}4)$$

where $v$ is the velocity of propagation in miles per millisecond. The velocity of propagation used in Equation (26-4) must allow for the delay in an average number of terminals as well as for the delay of the medium. Accepted values of VNLF are given in Figure 26-2 for commonly used facilities.

| TYPE OF FACILITY | VIA NET LOSS FACTOR, dB/MILE | |
|---|---|---|
| | TWO-WIRE CIRCUITS | FOUR-WIRE CIRCUITS |
| Toll cable (quadded low-capacity) | | |
| 19H172-62, 16H172-63 | 0.04 | 0.020 |
| 19B88-50 | 0.04 | 0.020 |
| 19H88-50 | 0.03 | 0.014 |
| 19H44-25, 16H44-25 | 0.02 | 0.010 |
| VF open wire | 0.01 | — |
| Carrier (cable, open wire, microwave radio) | — | 0.0015 |
| VF local cable (loaded or nonloaded) | 0.04 | 0.017 |

**Figure 26-2. Via net loss factors.**

**Fixed Loss Design Plan.** With the introduction of No. 4 ESS and with the expected evolution from analog to digital methods of transmission, a new loss plan for the toll portion of the network is needed because of the difficulty of associating circuit loss with digital signal processing. The introduction of controlled toll trunk losses in an all-digital network would require (1) the conversion of digital signals to their analog equivalents, insertion of the required losses, and reconversion to the digital format or (2) changing the encoded signal amplitude by some digital process. Either method would be costly and would introduce impairments.

A *fixed loss plan* is now being introduced for use in an all-digital toll network. This plan specifies a 6-dB trunk loss between class 5 offices, regardless of connection mileage, to provide a good compromise between loss/noise and echo performance over a wide range of connection lengths and loop losses. Under this plan, each toll connecting trunk is allocated a loss of 3 dB; all-digital intertoll trunks (digital facilities interconnecting digital toll switches) are operated at a loss of 0 dB.

Connections in an all-digital toll network will have lower trunk losses than similar connections in the present analog network. In addition, noise will be reduced because of the use of digital facilities and the elimination of channel banks at intermediate toll offices. Loss/noise and echo grade-of-service studies have shown that full implementation of an all-digital toll network will result in a significant improvement in transmission quality; the large improvement in loss/noise grade of service due to lower loss and noise will more than offset the slight degradation in echo grade of service due to lower loss. These conclusions are based on the assumed application of echo suppressors on trunks longer than 1850 miles and a 4-dB increase in the requirements on terminal balance for two-wire toll connecting trunks.

The fixed loss plan strictly applies only to an all-digital network. It will take many years for the present predominantly analog network to be converted to a predominantly digital network. Therefore, an operating plan to cover the transition period must be provided in order to assure satisfactory performance when analog and digital switching machines are interconnected.

The plan being implemented is designed to make the combined analog-digital network conform closely to the characteristics of the present analog network. The fixed loss plan will be implemented as portions of the network are converted to all-digital facilities. Following are the principal characteristics and constraints of the combined network:

(1) The expected measured loss and the inserted connection loss of each trunk must be the same in both directions of transmission.

(2) The −2 dB TLP at the outgoing side of analog toll switches and the 0 dB TLP at class 5 offices are retained and a −3 dB TLP is established for digital toll offices.

(3) The −16 dB and +7 dB TLPs at carrier system input and output are retained.

(4) Existing test and lineup procedures for digital channel banks are retained.

(5) Combination intertoll trunks, those terminating in digital terminals at a digital (No. 4 ESS) switching machine at one end and in D-type channel banks at an analog switching machine at the other end, are designed to have 1-dB inserted connection loss.

(6) Analog intertoll trunks are designed according to the via net loss plan.

## Echo Objectives

One way to express the network echo design objective is that trunk loss designs should be such that talker echo (the dominant impairment) is satisfactorily low on more than 99 percent of all telephone connections which encounter the maximum delay likely to be experienced. This way of expressing the overall objective recognizes: (1) that echo performance can be controlled by controlling trunk losses, (2) that the control of echo on *connections* implies further control of echo on the trunks used to form customer-to-customer connections, and (3) that echo impairment is a function of the amount of delay in the connection.

The parameters of echo amplitude and echo delay contribute significantly to echo performance in the network. Echo amplitude depends on the impedance relations at circuit interfaces and the losses of the involved circuits. The phenomenon is treated in terms of return loss and trunk losses that combine to produce echo path loss. On long circuits, minimum echo delay can be obtained by using carrier facilities where the velocity of propagation is much higher than in voice-frequency facilities. Carrier facilities are used for economic reasons (larger circuit cross sections and lower unit costs) as well as for echo delay control.

The most serious source of echo is low return loss found at class 5 offices where connections are made between loops and toll connecting trunks. While reasonable control of toll connecting trunk impedance can be exercised, the impedances of the randomly connected loops vary widely due to varying lengths and circuit make-up.

The distribution of echo return losses (ERLs) at class 5 offices, calculated from loop survey data, has a mean value of 11 dB and a standard deviation of 3 dB [5]. The distribution of singing return loss (SRL) has a mean value of 6 dB and a standard deviation of 2 dB. These return loss distributions are used in the overall process of establishing echo and loss objectives for other parts of the network.

Return loss objectives are specified for connection points in the toll portion of the network; generally the echo and singing return loss objectives for these points are more stringent than for the local portion since toll trunk parameters are more controllable. The following objectives are typical, but not all-inclusive.

(1) For four-wire trunks terminating at two-wire switches in class 1, 2, or 3 switching offices, the ERL objective is 27 dB (minimum 21 dB); the SRL objective is 20 dB (minimum 14 dB).

(2) For the interface between four-wire intertoll trunks and most two-wire toll connecting trunks at class 1, 2, 3, or 4 switching offices, the ERL objective is 18 dB (minimum 13 dB). For four-wire toll connecting trunks, the objective is 22 dB (minimum 16 dB). For both two-wire and four-wire trunks, the SRL objective is 10 dB (minimum 6 dB).

These return losses are measured against standard terminations, the values of which depend on the type of switching office involved. The measurement process and the complexity of impedance adjustment procedures that permit these return loss objectives to be met have led to the expression of objectives in terms of through balance and terminal balance requirements applied for many types of trunks at various types of toll switching offices.

Echo return loss objectives have less influence in the design of local trunks than in the design of toll trunks because echo problems are negligible for short trunk lengths. The objectives for singing return loss on these trunks are not firmly established, but the return losses are usually held to about 10 dB.

## Loss Objectives

The echo path loss involved in determining echo amplitude is made up of the return loss and twice the circuit loss between the speaker and the point of reflection. These circuit losses must be well controlled; they must be low enough to satisfy the requirements on talker volume and to avoid excessive contrast in volume from call to call yet they must be high enough to attenuate echoes to tolerable values.

**Volume.** The basic problem in telephone transmission is to provide a satisfactory signal amplitude at the receiver. The received signal amplitude is a function of many interacting parameters, starting with the transmitted signal amplitude. The latter depends on telephone speaking habits, station set efficiency of conversion from acoustic to electric signal energy, sidetone circuit design of the station set, and losses in the circuits between the transmitter and the receiver.

Received volume differs from many other quality parameters in that its effects are double-ended; volume can either be too low, causing difficulty in understanding the received message; or it can be too high, causing listener discomfort. Subjective tests have been made to determine listener reactions to different volumes. The results of one series of such tests, plotted in Figure 26-3, clearly show the double-ended nature of this parameter. Volumes to the left of the two left-hand curves are judged to be too low to satisfy listeners while volumes to the right of the right-hand curve are too high to satisfy listeners. Each of the curves, which divide regions of volume rated poor, fair, good, etc., is approximately normal with a standard

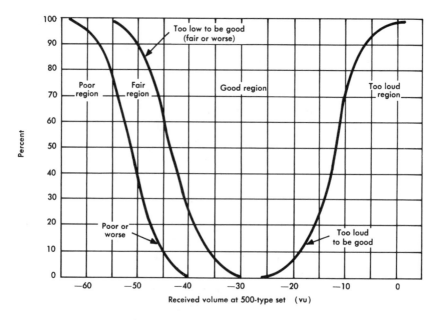

Figure 26-3. Judgment of received volume from subjective tests.

deviation of about 5 dB. The curves show a fairly wide range (from about −43 vu to about −12 vu at the median values) over which received volumes are rated good.

Data of the type shown in Figure 26-3 have been used to help establish allowable circuit losses in end-to-end customer connections. The total loss allowance is allocated to the various parts of the plant in accordance with the results of economic studies and with a satisfactory noise-loss-echo grade of service established by subjective testing.

**Loss Allocations.** Transmission objectives for loop loss have been derived on the basis of satisfying an overall loss/noise grade-of-service objective [3, 6]. Control of loop loss is accomplished by the application of carefully specified rules in the design and layout that produce a satisfactory distribution of losses. The three sets of rules are parts of the *resistance, unigauge,* and *long route design* plans. These three design plans permit straightforward application of the rules to the installation of new cables, of inductive loading, and of electronic equipment so that overall loss objectives are met because the objectives are built-in, integral parts of the plans. When the plans are properly applied, the resulting distribution of loop losses has a maximum value of about 9 dB (including the effects of bridged taps). The mean value and the standard deviation of the loss distribution depend on the geographical area served and on the concentration of customers within the area. For the Bell System, an average 1000-Hz value of 3.8 dB and a standard deviation of 2.3 dB are typical; these values are used in the determination of network loss objectives and grade of service.

The above values of loop losses were determined on the basis of measurements made in 1960 and 1964 [5]. Subsequent studies of losses and grade-of-service objectives resulted in some tightening of the objectives, particularly in the long route design plan.

Since a numerical loss objective (other than the maximum) is not expressed for individual loops, special treatment must be applied (1) when a loop is assigned to data transmission or another special service need and (2) when transmission complaints still exist after it has been verified that the loops involved have been installed according to appropriate design procedures.

Loss objectives for transmission circuits through switching machines have not been firmly established, but a loss of less than 1 dB is generally allowed for these circuits. Losses of various types of trunks constitute the remaining major allocation to parts of the plant, and for purposes of network administration, trunk losses are now defined in such a way as to include average switching system loss. Many of the loss values are given in terms of via net loss which varies according to the length and type of facility.

Losses allocated to trunks depend on the position of the trunk type in the switching hierarchy and the probability of encountering tandem connections of such trunks in an end-to-end telephone connection. In the toll portion of the network, interregional intertoll trunks are designed on the basis of maximum round-trip echo delay that can occur on connections involving the interregional trunks. If the delay can exceed 45 milliseconds, the interregional trunks are equipped with echo suppressors and the trunks are operated at 0 to 0.5 dB loss. (Losses high enough to satisfy echo requirements would generally be too high to satisfy volume and contrast objectives.) If the round-trip echo delays are less than 45 milliseconds, the interregional trunks are operated at VNL, with a maximum of 2.9 dB.

High-usage intertoll trunk groups are operated at via net loss where the value of loss is VNL $\leq$ 2.9 dB, equivalent to a maximum trunk length of about 1850 miles on carrier facilities. If echo requirements call for a loss greater than 2.9 dB, the trunks are operated at 0 dB loss and are equipped with echo suppressors unless they are in a final routing chain. To avoid having more than one echo suppressor in a connection, echo suppressors are generally permitted only in final groups between regional centers. Secondary intertoll trunks are operated as close to 0 dB as possible, with a maximum of 0.5 dB. Final intertoll trunk groups are operated at via net loss, but at a maximum of 1.4 dB loss.

Toll connecting trunks are usually operated at VNL + 2.5 dB loss with a maximum loss of 4.0 dB. An alternate design allows a trunk to have 3.0 dB to 4.0 dB loss provided it contains less than 15 miles of VF cable facilities or less than 200 miles of carrier facilities. On long end-office trunks (usually interregional) between class 4 and class 5 offices where echo requirements indicate the need for loss greater than 4 dB, an echo suppressor may be added and the loss set at 3 dB.

In the local portion of the network, direct trunks are designed to a nominal loss of 3 dB with a maximum of 5 dB. Tandem trunks are operated at a nominal loss of 3 dB and a maximum of 4 dB, and intertandem trunks are operated at via net loss. Loss values are assigned similarly to all service and miscellaneous trunks used in the network. Long interregional direct trunks (between class 5 offices) may be operated without echo suppressors at VNL + 6 dB loss (maximum 8.9 dB) over distances of up to 4000 miles.

### Loss Maintenance Limits

In order to maintain network performance, 1000-Hz measurements of trunk losses are made periodically in accordance with maintenance programs described in Volume 3. The objectives are set in accordance with indices which have been derived in relation to grade-of-service objectives. The percentage of measurements showing deviations from design values in excess of 0.7 or 1.7 dB (the larger deviations carry heavier weighting) determines the index for the group of trunks under study. If the index is 96 or higher, performance is satisfactory and no action is necessary. If the index is below 96, investigation and corrective action are indicated. If the loss of any trunk deviates from its design value by 3.7 dB or more, it must be removed from service.

### Message Circuit Noise

Message circuit noise is defined as the short-term average noise measured by means of a 3A noise measuring set or its equivalent [7]. Objectives for message circuit noise, allocated to various parts of the network, are based on subjective tests in which noise was evaluated by telephone listeners in the presence of speech signals held at a constant volume. Noise and volume were expressed in dBrnc and vu, respectively, at the line terminals of the station set; observers were asked to rate the performance in the usual manner (excellent, good, fair, poor, or unsatisfactory) for a wide range of noise values. The results of these tests are shown in Figure 26-4.

**Loop Noise Objectives.** The message circuit noise objective applied to loops is that noise measured at the line terminals of the station set shall not exceed 20 dBrnc.* Noise at or below this value has little effect on grade of service, but noise in excess of 20 dBrnc deteriorates grade of service appreciably.

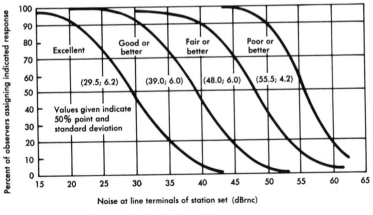

Figure 26-4. Noise opinion curves.

* Most loops have measured noise well below this value. The average is about 0 dBrnc.

In recognition of the special circumstances relating to long routes (those in excess of 1300 ohms controlled by resistance design and those extended by the unigauge design, both discussed in Volume 3), the noise objective is made somewhat more lenient. For long routes the noise objective is administered at 30 dBrnc. For routes on which the limit of 30 dBrnc is exceeded, special treatment (shielding, separation from power lines, balancing, etc.) must be employed according to circumstances.

**Trunk Noise Objectives.** The performance objectives for trunk noise have been allocated to allow for the tendency of noise to accumulate with distance and the smaller number of calls of very long distances compared with intermediate and short distances. To give weighting to these two factors, trunk noise objectives have been selected to achieve a 99 percent good-or-better grade of service for short toll connections (0 to 180 miles, airline distance), 97 percent good or better for medium length toll connections (180 to 720 miles), and 95 percent good or better for long toll connections (over 720 miles). Consistent with these overall objectives, allocations have been made for short-haul carrier facilities (for use on trunks less than 250 miles long), and long-haul carrier facilities (for use on trunks over 250 miles long). These allocations, which recognize the inherent variability of performance in the field environment, are expressed in terms of mean values and standard deviations. For short-haul carrier, the mean value of the objective is 28 dBrnc0 at 60 route miles and for long-haul carrier, 34 dBrnc0 at 1000 route miles. The standard deviation is $\sigma = 4$ dB in each case. These allocations allow for a 3 dB increase in noise for each doubling of the distance. This increase is typical of analog carrier system performance but is not usually experienced in pulse-type carrier systems. Design objectives for carrier systems are based on these performance objectives but are normally expressed in terms of worst channel noise in a nominal environment. The current design objective for 4000-mile coaxial cable systems, including multiplex equipment, is 40 dBrnc0. The design objective for microwave radio systems is approximately 1 dB higher [3]. Generally, where these message circuit noise objectives are met for speech transmission, objectives for voiceband data transmission are also met.

## Impulse Noise

Impulse noise is any burst of noise that produces a voltage in excess of about 12 dB above the rms noise as measured by a 3-type noise measuring set with C-message weighting; in a speech channel, these bursts are usually less than 5 but may very rarely be as long as 45 or 50 milliseconds in duration. The ratio of the voltage excess to the rms noise voltage is nominally at least 12 dB for a 3-kHz bandwidth; it may be as great as 40 dB in some systems, particularly microwave radio. Impulse noise is usually viewed as superimposed on background message circuit noise [8]. Objectives are dominated by requirements for digital data signal transmission, and circuits that are satisfactory for data are generally satisfactory for speech signal transmission.

Impulse noise objectives are usually established on the basis of the number of counts obtained on a 6-type impulse noise counter during a prescribed measurement interval and may be expressed for loops, trunks, or customer-to-customer connections. The application of acceptable objectives to station sets and central office equipment is under study.

The objective for any loop or single voice channel is that there should be no more than 15 impulse noise counts in 15 minutes at a given threshold. For a sampled trunk group, there should be a maximum of 5 counts in 5 minutes at a given threshold. Sampling plans are specified and the noise thresholds are set at different values for loops, for VF trunks, and for compandored and noncompandored carrier trunks. The threshold values are also weighted to take into account the expected increase of noise with distance in carrier systems. The trunk impulse noise thresholds are shown in Figure 26-5. The loop impulse noise threshold is 50 dBrnC referred to the local central office.

### Intelligible Crosstalk

Intelligible crosstalk objectives are generally expressed in terms of the crosstalk index, a measure of the probability of receiving intelligible crosstalk. The derivation of the crosstalk index, its relationship to the impairing effects of intelligible crosstalk, and the use of generalized crosstalk index charts are presented in Chapter 17.

Objectives have been established for most types of trunks. A maximum crosstalk index of 1 is used for intertoll and secondary intertoll trunks. An index of 0.5 is applied to toll connecting, direct, tandem, and intertandem trunks. No index objective has yet been established for loops.

| TRUNK LENGTH (MILES) | VF TRUNKS, dBrnc0 | COMPANDORED* CARRIER AND MIXED COMPANDORED-NONCOMPANDORED, dBrnc0 | NONCOMPANDORED CARRIER, dBrnc0 |
|:---:|:---:|:---:|:---:|
| 0-60 | 54 | 68 | 58 |
| 60-125 | 54 | 68 | 58 |
| 125-250 | 54 | 68 | 59 |
| 250-500 | | 68 | 59 |
| 500-1000 | | 68 | 59 |
| 1000-2000 | | 68 | 61 |
| over 2000 | | 68 | 64 |

\* Compandored trunks, including those with D-type channel banks, are measured with a −10 dBm0 tone transmitted from the far end and filtered out ahead of the measuring set by a C-notched filter or equivalent. The C-notched filter is a C-message weighting network with a narrowband suppression section to provide at least 30 dB of attenuation at the tone frequency.

Figure 26-5. Impulse noise thresholds for trunks.

Crosstalk objectives for central office equipment are usually expressed in terms of equal level coupling loss. In four-wire offices, the objective for minimum coupling loss between the two sides of one circuit is 65 dB. The coupling objective for different circuits is 80 dB in two-wire and four-wire offices.

## Single-Frequency Interference

Well documented and generally accepted transmission objectives for single-frequency interferences are not now available. When new systems have been designed, design objectives have been applied in a generally conservative manner. The factors that have made it difficult to derive acceptable objectives include the frequency and amplitude of the interference, the stability or variability of frequency and amplitude, the harmonic content of the interference, the presence or absence of masking message circuit noise or other interferences, the possible presence of other single frequencies, and the constancy or intermittency of the interference. As a rule of thumb, single-frequency interferences must be well below other noise in the circuit.

The most conservative estimate, one that makes single-frequency noise inaudible to nearly everyone, is that the interference should be 30 dB below message circuit noise. More lenient estimates have led to design objectives of 10 to 12 dB below message circuit noise. These objectives apply to speech signal transmission and, when met, usually result in satisfactory transmission of other voiceband signals.

### Frequency Offset

Frequency offset objectives are set primarily to satisfy the needs of program signal transmission. While the determination of the threshold for frequency offset is as critical to speech transmission as it is to music transmission, subjective tests have shown that listeners are more tolerant of offset in speech signals than in music signals. The overall performance objective for offset is a maximum value of $\pm$ 2 Hz; the maintenance objective is $\pm$ 5 Hz.

### Overload

Overload of broadband or single-channel electronic systems produces signal impairments in the form of noise and distortion. The objective for overload is expressed as a degradation of the grade of service in an individual channel. While objectives have not been firmly established, a reduction of about 1 percent in good-or-better and an increase of about 0.1 percent in poor-or-worse grades of service appear to be reasonable performance objectives for the overload phenomenon. These criteria, when applied to D-type channel banks used with T-type carrier systems, have resulted in the objective that these banks transmit a $+3$ dBm0 sine wave signal without causing overload impairment.

A signal transmitted at higher amplitude than the design value may cause intelligible crosstalk or single-frequency tone interference as a result of intermodulation or other crosstalk paths. This impairment is not considered as overload unless it is so extreme that the entire system is affected.

### Miscellaneous Impairments

A number of miscellaneous impairments are recognized as having potentially serious degrading effects on voice-frequency channel transmission; they include phase and gain hits, phase and gain jitter, incidental frequency modulation, and dropouts. Formal objectives for these types of impairments have not been established.

## Telephone Station Sets

The transmission performance of station sets is controlled primarily by design, and there are no specific transmission performance or maintenance objectives. The great majority of sets in service are the 500-type, which were developed to meet a set of stringent design objectives [9]. There are no transmission options or adjustments on these sets. Therefore, where troubles can be identified with the station set or where trouble complaints cannot be identified with other parts of the local connection, the transmitter, the receiver, or the entire set may be replaced and returned to the manufacturer.

A unique consideration is involved in operator and auxiliary services wherein the operator headset (receiver and microphone) must be regarded as the station set. One of the more stringent objectives that must be met by these circuits is that pertaining to sidetone. In this case, sidetone is a design parameter of the access circuits rather than the headset circuits. The objectives are commonly expressed in terms of the acoustic sidetone path loss, which is defined as the ratio in dB of the loudness-weighted acoustic sound pressure produced by the receiver for a given loudness-weighted acoustic sound pressure input to the transmitter (or microphone). The objective for this loss is 12 dB, an optimum determined by subjective tests; values as low as 8 dB and as high as 16 dB are considered tolerable.

## 26-2 WIDEBAND DIGITAL SIGNAL TRANSMISSION OBJECTIVES

As in the case of transmission objectives for voice-frequency channels, the expanding use of existing channels for new types of signals and services has made it necessary to refine and redefine channel transmission objectives. Similarly, the adaptation of analog systems and portions of analog systems for wideband digital signal transmission has led to new objectives for wideband channel applications. Frequency bands that were originally provided only as parts of the voice transmission network are being adapted for wideband digital signal transmission, and as a result, transmission objectives for the wider bands and new signals are in process of refinement and redefinition. In addition, digital transmission systems are being developed and introduced into the network, thereby requiring that objectives be established for their design and operation.

The transmission objectives to be established and the manner of adapting systems and signals for compatibility depend on the signal format, the sensitivity of the signal to various impairments, and the characteristics of the system or channel involved. The parameters involved include load capacity, bandwidth, signal-to-noise performance, jitter, error rates, and the rate of digital transmission.

The wide range of bandwidths, signal formats, impairments, services, and digital systems makes it difficult to present a complete set of wideband digital transmission objectives. Therefore, this discussion is limited to a number of examples of objectives that have been established for specific signal formats and to the approach used in several digital system designs. In most cases, the determination of the objective ultimately rests on subjective judgment of the required grade of service.

There are two types of wideband digital signals commonly transmitted on analog systems: the 1A Radio Digital System (1A-RDS) signal, a 1.544 Mb/s signal transmitted at baseband (0 through 500 kHz) over microwave radio systems in a multilevel signal format containing seven discrete levels and a family of binary digital data signals that may be transmitted at 19.2 kb/s, 50.0 kb/s, or 230.4 kb/s in the half-group, group, or supergroup bands, respectively, of the L-multiplex (FDM) equipment [10, 11]. Transmission objectives for these signals and for digital transmission systems are evolving as the technology advances.

### Performance Evaluation

Transmission objectives for wideband digital signals are expressed variously in terms of error rate, noise impairment, and eye diagram parameters. In addition, objectives must be expressed for signal power when digital signals are to be transmitted on analog systems.

**Error Rate.** A commonly used design objective for wideband digital signal transmission, one that has not been sanctioned for general application, is an error rate of $10^{-6}$; i.e., the terminal-to-terminal error rate shall not exceed one error in $10^6$ bits. Error rate counters, or violation counters as they are sometimes more properly called, are used with many systems to determine error performance for the complete end-to-end connection or for some link in the connection. Violations of a predetermined code format are counted and compared with the objective which must be expressed in the same terms. The objective must be that value allocated to the particular link under surveillance.

**Noise Impairment.** The expression of an objective in terms of noise impairment is used to equate the degradation of channel performance by various impairments to an equivalent degradation due to Gaussian noise. This equivalence can be explained in another way. A certain error rate can be expected from a given channel whose characteristics are ideal in all respects except for the presence of Gaussian noise. The noise impairment due to the introduction of some other degradation, such as delay distortion, is measured by the improvement in Gaussian noise (improved signal-to-noise ratio) that would be required for the same channel performance as in the channel impaired only by the original value of Gaussian noise.

Two goals are met by expressing objectives in terms of noise impairment. First, objectives can be allocated to a variety of impairments in an orderly manner that lends itself readily to changes necessary to meet specific conditions. Second, a straightforward method is provided for determining how good the channel signal-to-noise ratio must be to meet a specified error rate objective. Both advantages are especially desirable for studies of digital signal transmission on analog channels.

**Eye Diagram Closure.** When a random stream of digital pulses is properly impressed on an oscilloscope, the successive pulses can be made to form a pattern, called an eye diagram. As the pulse stream is impaired by channel imperfections (such as noise, gain and delay distortion, and crosstalk), the opening in the eye (or eyes for multi-level signals) is reduced by predictable amounts. Thus, the eye pattern may be used as a measure for performance, and transmission objectives can be expressed in terms of the percentage of eye closure.

This manner of stating objectives has not proved to be useful in operating and maintaining systems, but it has found considerable use in system design where measurements are made under laboratory conditions [12, 13]. The approach has been used to compare performance and objectives; it has also been used as a means of allocating objectives among a number of different impairments, each being allowed a certain percentage of eye closure in the horizontal (timing) or vertical (amplitude) dimensions or in both.

**Signal Power.** When a signal is impressed upon a transmission channel, the channel must be capable of transmitting the signal satisfactorily; in addition, the signal cannot be allowed to degrade other signals that may share the same transmission system. Overload performance is one criterion that must be satisfied in both respects.

The impressed signal amplitude must be limited so that the signal itself is not degraded by the overload characteristics of the channel. The degradation would fall between two extremes, one in the form of peak clipping that might be relatively innocuous and the other in the form of excessive distortion that would render the signal useless. The limiting value depends in each case on the characteristics of the channel or system to be used.

Simultaneous transmission of digital and other kinds of signals on analog facilities further requires that the load imposed by the digital signals does not seriously impair the other signals. The usual criteria for the loading objective are (1) that the average power in the digital signal shall not exceed the average power allotted to the displaced speech channels (−16 dBm0 per 4-kHz band), and (2) that any single-frequency component of the digital signal shall not exceed −14 dBm0. The latter criterion is sometimes relaxed if the component is not a multiple of 4 kHz or if the amplitude variability results in a low probabiilty of its exceeding −14 dBm0.

## Design Applications

Since most transmission objectives for wideband digital signals have not yet been formally accepted or generally applied, it is best to illustrate for specific cases the ways objectives evolve, are derived, and are applied.

**Bit Rate and Bandwidth.** In the design of a new digital transmission system or the adaptation of analog facilities to the transmission of digital signals, the first consideration is the overall system design problem of relating available bandwidth to the desired transmission rate. First-order effects on the design include: (1) the achievable signal-to-noise ratio of the proposed facility, (2) the desirability of designing a synchronous system that permits regeneration, (3) the cost involved in terminal and signal regeneration equipment, (4) the feasibility and cost of equalizing the medium, and (5) the transmission objectives that must be satisfied if the service needs are to be met. While the concern here is primarily with the objectives, all of these effects interact in ways that make discussion of objectives meaningless unless the interactions are explored as well.

The need for digital signal transmission over the analog microwave radio network evolved partly from the Digital Data System (DDS) development program. The feasibility of transmitting a DS-1 signal on a TD-type radio system was established but this possibility was deemed undesirable because the DS-1 signal carries significant energy at frequencies up to 1.544 MHz. A substantial number of telephone channels would thus have to be dropped to accommodate the digital signal. It was also shown that the upper half of the DS-1 spectrum might be filtered or the signal might be coded as a 3-level, class IV, partial response signal with spectral nulls at 0 and 772 kHz. The former approach was more theoretical than practical; the latter still appeared too costly because about 120 message channels would have to be dropped to provide a roll-off band.

A 7-level, class IV, partial response signal with a 15 percent roll-off band was chosen and is now used in the 1A Radio Digital System (1A-RDS) which provides a digital facility for DDS. The signal has spectral nulls at 0 and 386 kHz and extends only to 444 kHz, well below the 564-kHz multiplex low-end frequency. Thus, no message channels are displaced.

**Performance Objectives.** Objectives for 1A-RDS were derived from those established for DDS. They were based on a level of performance which was judged would provide a high-quality service at the customer sub-rates of 56 kb/s and below. The basic criterion was stated in terms of percentage of error-free seconds. Allowances were included for known sources of hits, such as those caused by protection switching initiated by maintenance activities and fading. A sub-set of objectives covers the number of errored-seconds that occur in shorter periods of time and the number and length of error bursts.

**Designs Based on Noise Impairments.** In setting objectives for transmitting wideband digital data signals in the half-group, group, and supergroup bands of the L multiplex equipment, a major concern was the equalization of gain and delay distortion in those bands. The objectives for these services were derived initially from the basic goal of achieving an error rate of $10^{-6}$ or better (between terminals) 95 percent of the time. Portions of this objective were then allocated to various well-defined impairments (random and impulse noise, for example), and the remainder was allocated to misequalization, data set and terminal limitations, net loss variations, and jitter.

These allocations first involved the derivation of a required signal-to-noise ratio of 12.7 dB. After noise impairments had been assigned to each of the principal sources of degradation anticipated, it was concluded that an overall signal-to-noise ratio (Gaussian noise) of 22 dB would be required to meet the service objective; this signal-to-noise ratio was used as a design objective.

## 26-3　VIDEO TRANSMISSION OBJECTIVES

The Bell System transmits three types of video signals that might be reviewed in detail in terms of applicable transmission objectives: broadcast television signals, closed circuit television signals, and PICTUREPHONE signals. Only broadcast television signals are covered, however, since closed circuit television and PICTUREPHONE objectives are not well established. Generally, closed circuit television objectives tend to be somewhat more lenient than those for broadcast quality signals. Thus, it is usually safe to use broadcast objectives; if there appears to be serious difficulty in meeting them, the case must be considered separately. For PICTUREPHONE, only some early design objectives have been used in preliminary studies and experimental work [14].

The objectives to be discussed are, for the most part, expressed in terms of overall 4000-mile objectives. These, of course, must be allocated to different parts of the plant in accordance with some logical procedure, as outlined in Chapter 25. Most of the objectives given are design objectives, and each must be interpreted carefully and applied judiciously when operational variations and limits are considered for use as performance and maintenance objectives.

### Random Noise

The degree of noise impairment to television signals is a complex function of the distribution of noise power versus frequency and the characteristics of the impaired signal (for example, whether it is a monochrome or a color signal). When the noise is at a high enough amplitude, it may appear as fine, closely packed dots in rapid, random motion. When observed in monochrome signal transmission, the dots appear to have the characteristics of a swirling snowstorm; as a result, the impairment has commonly been referred to as "snow."

If the noise is concentrated at the lower video frequencies, the dots are relatively large or may appear as streaks in the picture. If the noise is concentrated at high frequencies, the dots are much finer and harder to see. Hence, equal powers of noise are judged to be more annoying at low than at high frequencies. When the noise is concentrated in relatively narrow bands, it produces fleeting herringbone patterns in the received pictures. If the band is made narrower, the pattern approaches that of a single-frequency interference. Thus, equal powers of noise tend to be more objectionable as the bandwidth of the noise is decreased.

These observations have led to the expression of random noise objectives in terms of a single weighted value applicable to monochrome or color signals. The weighting, which takes into account the more objectionable nature of low-frequency noise, makes possible the use of a single number as an objective; i.e., equal measured values mean equal subjective effects, regardless of the type of noise. The effect of narrowband noise is accounted for simply by weighting its effect with that of broadband noise on the basis of total power. Thus, if single-frequency interference is present in a channel, the random noise objective must be made more stringent by an amount that makes the power sum of random and single-frequency noises meet the random noise objective. In addition, the single-frequency objective must also be met.

The random noise weighting characteristic is shown in Figure 26-6. In spite of some differences in annoying effects in monochrome and color signal transmission, it is found that satisfactory results are obtained when this single weighting curve is used to evaluate noise on facilities used for both types of signals [15]. The objective generally applied is that the noise introduced by a 4000-mile system produce a signal-to-noise ratio of 53 dB or better. This ratio is expressed in terms of the peak-to-peak composite signal voltage (including synchronizing pulses) to the weighted rms noise voltage in the frequency range of 4 kHz to 4.2 MHz. The noise from zero to 4 kHz is treated separately.

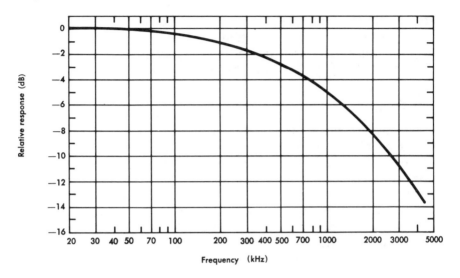

Figure 26-6. Monochrome and color random noise weighting for broadcast television signals.

## Low-Frequency Noise

Noise in the band from zero to 4 kHz is measured in a manner similar to that for random noise. It is treated separately because of the likely presence of power-frequency interference (hum in telephone circuits), which can cause bar pattern interference in the received picture. If hum is not present, the low-frequency random noise is simply added to the broadband random noise.

The objective for low-frequency interference is expressed in terms of the ratio of the peak-to-peak signal voltage to the rms interference voltage in the band from 0 to 4 kHz. The objective for a 4000-mile circuit is a 50 dB signal-to-noise ratio.

## Impulse Noise

The characteristics of impulse noise are not well-defined for evaluation as a television impairment. Generally, impulse noise is any interference that affects a small portion of the received picture for only a short interval of time.

The objectives for impulse noise are also not well-defined. A ratio of the peak-to-peak composite signal voltage to the peak impulse voltage of 20 dB is sometimes considered to be an acceptable objective. It is applied specifically to interferences that occur at a rate of about one per minute. No quantitative data are available for impulses of different durations or for other frequencies of occurrence.

## Single-Frequency Interference

A single-frequency interference usually appears on a television receiver as a discernible bar pattern that may be stationary or in motion. If the interference is an integral multiple of the nominal 60-Hz field frequency, it appears as a broad, stationary, horizontal pattern. If the interference differs slightly from a 60-Hz multiple, the bars travel up or down the picture. If the interference is weak, the impairment may more nearly resemble a flickering than a bar pattern, an impairment much more annoying than a stationary pattern. The effect depends on the flicker rate.

For frequencies at or near multiples of the line scanning frequency, the patterns are stationary or moving, vertical or diagonal bars. The bar structures become finer as the interfering frequency increases; the most critical frequencies are in the range of 100 to 300 kHz.

Similar phenomena are produced by single frequencies near the color carrier frequency. The high- and low-frequency characteristics must be determined as high or low frequencies relative to (i.e., displaced from) the color carrier frequency of 3.579545 MHz.

While there is a wide variation of subjective reaction to single-frequency interferences according to their frequency, stability, multiplicity, etc., the objective is usually stated as two simple numbers. First, the objective for a single interferer is taken as a signal-to-noise ratio of 69 dB where the signal amplitude is expressed in peak-to-peak volts (including the synchronizing pulse) and the interference is expressed as an rms voltage. The second expression for the interference is that the total weighted interference (including random noise) is to be 53 dB below the signal, the same value as that given previously for weighted random noise.

## Echo

Echo refers to a signal produced by reflection at one or more points in a transmission path or generated by transmission irregularities and having sufficient magnitude and time difference to be perceived as distinct from the signal received over the primary transmission path. Echoes may lead or lag the main signal and have characteristics that are described by four different picture impairments.

**Types of Picture Impairments.** A number of different picture impairments may occur as a result of reflections at points of discontinuity in the transmission path or as a result of transmission irregularities. All such picture impairments are subject (at least in theory) to control and reduction by some form of transmission equalization. These impairments are discussed in Chapter 18 but are mentioned again here to stress the facts that all are due to transmission discontinuities or departures from ideal transmission characteristics and may be dealt with in terms of echoes.

*Streaking and Smearing.* These are often considered separately but, for convenience, are considered here as one type of impairment. Both are described as unwanted lines or areas of brightness, usually observed to the right of a sharp brightness change in a picture, extending toward the right edge of the picture. Streaking extends undiminished to the right-hand edge; smearing diminishes substantially toward the edge of the picture. Both result from transmission irregularities at frequencies in the region of the field repetition rate (60 Hz), frequencies in the region of the line scanning frequency (15.75 kHz), and the first 10 to 15 harmonics of each.

*Ringing.* An oscillatory transient, called ringing, may occur in a signal at the output of a system as a result of a sudden amplitude change of the input signal. This results in closely spaced multiple repetitions of some picture elements whose reproduction requires frequency components approximating either the cutoff frequency of the system or the frequency of a sharp discontinuity within the passband. The ringing occurs at approximately the frequency of the discontinuity or of the band edge and is often accentuated by a rising gain characteristic preceding the discontinuity or band edge. Performance can be improved by extending delay equalization through the cutoff region.

*Overshoot.* This impairment is due to an excessive response to a sudden change in signal amplitude. It appears as a black outline to the right of white objects and as a white outline to the right of black objects. A sharp overshoot may be referred to as a spike; it is caused by excessive gain at high frequencies.

*Flat and Differentiated Echoes.* Echoes are complex phenomena whose interfering effects depend on echo amplitude, time separation from the main signal, the nature of the original signal, and the frequency characteristic of the echo source. If the echo essentially covers the entire transmitted band, it is referred to as a flat echo. If it has a sharp frequency characteristic, usually with stronger reflections at high frequencies, it is known as a differentiated echo. Differentiated echoes are generally less interfering than flat echoes. If the echo path accentuates the high frequency echo components at a rate of 6 dB per octave, the echo is less interfering than flat echo by about 15 dB.

**Echo Objective.** The echo objective for video transmission is a 40 dB signal-to-echo ratio. It is expressed in terms of a single, well-defined, long delayed (10$\mu$s or more) echo. In practice, many echoes are usually present, and each component echo must be weighted in accordance with a weighting function that represents the change in subjective effect with the time displacement of the echo. The weighted components are then combined on a power basis for comparison with the objective. A typical time-weighting function is shown in Figure 26-7. Recent analysis of subjective test data has shown that the function also varies according to picture content and the polarity of the echo [16].

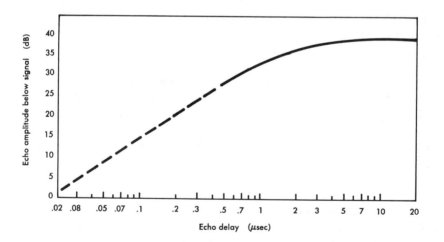

Figure 26-7. Single flat echo objectives (echo time weighting curve).

Any departure from flat amplitude response or linear phase response of a transmission channel can be expressed in terms of the Fourier components of the response functions. These components are expressed as cosinusoidal functions of the amplitude response and as sinusoidal functions of the phase response. The Fourier components can then be regarded as generating echoes which may be summed by power after the weighting function has been applied.

## Crosstalk

Video crosstalk occurs when an undesired signal interferes with a desired signal. The objectives for crosstalk are expressed in terms of dB of loss in the coupling path between the two signals at 4.2 MHz at equal transmission level points. When the coupling path is flat with frequency, the crosstalk is called flat crosstalk. When the crosstalk path loss decreases with frequency at a specified rate in dB per octave, the coupling is called $x$ dB differentiated crosstalk where $x$ is the rate of loss decrease.

Where crosstalk can be seen, the interference appears as an image of the unwanted picture moving erratically across the wanted picture. The motion occurs because of the lack of synchronization between independent signals. As the crosstalk image moves across the picture, it appears to be framed. The apparent framing is formed by the synchronizing pulses of the interfering signal. The framing tends to be more noticeable than any feature in the image. The side frames, which extend from the top to the bottom of the wanted picture, interfere with the total wanted picture. The effect is similar to a windshield wiper moving across the picture; the term "windshield wiper effect" is sometimes applied.

If the crosstalk is weak (high coupling loss), neither the frame nor the image is discernible. At such a near-threshold point, only a slight flicker can be seen as the frame moves across certain portions of the desired picture. The subjective effect is more dependent on flicker rate than on crosstalk magnitude.

If the coupling loss varies with frequency, resulting in differentiated crosstalk, the interfering image may appear to be in bas-relief. However, the synchronizing pulses are still the most prominent feature in the crosstalk image since they have the largest rate of change.

The overall objective for crosstalk coupling loss between equal level points is dependent on the nature of the coupling path. Some typical path characteristics that may be encountered in practice are illustrated in Figure 26-8. The applicable objectives, expressed in dB of loss at 4.2 MHz, are as follows:

| CROSSTALK PATH | OBJECTIVE, dB |
|---|---|
| Flat crosstalk | 58 |
| 6 dB/octave | 37 |
| 12 dB/octave | 21 |
| 24 dB/octave | 17.5 |

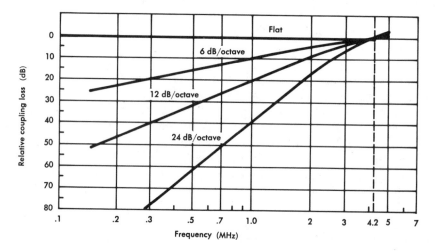

Figure 26-8. Coupling path loss characteristics.

## Differential Gain and Phase

These impairments, which have serious effects on color television signal transmission, are described in Chapters 18 and 21, respectively. The objective for differential gain, which may produce undesired changes in color saturation, is 1.4 dB. The objective for differential phase, which produces changes in color hue, is 5°.

## Audio/Video Delay

It is customary in the Bell System to transmit video and associated sound signals over separate transmission paths. If the difference in absolute delay between the two paths is excessive, an impairment results because the picture and sound are out of synchronism; the sound is heard before or after the producing action in the picture. The objective is that the delays in the two transmission paths differ by no more than 55 milliseconds.

## Luminance/Chrominance Delay

While the luminance and chrominance information in a video signal is transmitted over the same channel, the dominant components of one part of the signal are so far removed in frequency (over 3 MHz) from the other part that there can be a significant delay difference between the two. When this delay difference is excessive, the color portions of the signal are shifted relative to the luminance portions; i.e., there is a misregistration of color. Such an effect is most noticeable at sharp vertical edges of highly saturated color areas that are bounded by low-saturated color areas relatively free of detail [17].

The objective for the delay difference is 50 nanoseconds. It is expressed as the difference in delay between 3.6 MHz and frequencies below 200 kHz. Since the delay below 200 kHz tends to be constant, measurements are usually made at 200 kHz and 3.6 MHz to evaluate performance relative to the 50 ns objective.

### REFERENCES

1. Bell System Technical Reference PUB 41005, *Data Communications Using the Switched Telecommunications Network* (American Telephone and Telegraph Company, May 1971).

2. Campbell, L. W. "The PAR Meter — Characteristics of a New Voiceband Rating System," *IEEE Transactions on Communications Technology*, Vol. COM-18 (Apr. 1970), pp. 147-153.

3. Andrews, F. T., Jr. and R. W. Hatch. "National Telephone Network Transmission Planing in the American Telephone and Telegraph Company," *IEEE Transactions on Communications Technology*, Vol. COM-19 (June 1971), pp. 302-314.

4. Huntley, H. R. "Transmission Design of Intertoll Trunks," *Bell System Tech. J.*, Vol. 32 (Sept. 1953), pp. 1019-1036.

5. Gresh, P. A. "Physical and Transmission Characteristics of Customer Loop Plant," *Bell System Tech. J.*, Vol. 48 (Dec. 1969), pp. 3337-3385.

6. Lewinski, D. A. "A New Objective for Message Circuit Noise," *Bell System Tech. J.*, Vol. 43 (Mar. 1964), pp. 719-740.

7. Cochran, W. T. and D. A. Lewinski. "A New Measuring Set for Message Circuit Noise," *Bell System Tech. J.*, Vol. 39 (Jul. 1960), pp. 911-931.

8. Fennick, J. H. "Amplitude Distributions of Telephone Channel Noise and a Model for Impulse Noise," *Bell System Tech. J.*, Vol. 48 (Dec. 1969), pp. 3243-3263.

9. Inglis, A. H. and W. L. Tuffnell. "An Improved Telephone Set," *Bell System Tech. J.*, Vol. 51 (Apr. 1951), pp. 239-270.

10. Seastrand, K. L., and L. L. Sheets. "Digital Transmission over Analog Microwave Radio Systems," *IEEE International Conference on Communications*, June 19, 20, 21, 1972, Philadelphia.

11. Mahoney, J. J., Jr. "Transmission Plan for General Purpose Wideband Services," *IEEE Transactions on Communications Technology*, Vol. COM-14 (Oct. 1966), pp. 641-648.

12. Technical Staff of Bell Telephone Laboratories, *Transmission Systems for Communications*, Fourth Edition (Winston-Salem, N.C.: Western Electric Company, Inc., 1970) Chapter 27.

13. Mayo, J. S. "A Bipolar Repeater for Pulse Code Modulation Signals." *Bell System Tech. J.*, Vol. 41 (Jan. 1962), pp. 25-97.

14. Baird, J. A. et al — "The PICTUREPHONE System," *Bell System Tech. J.*, Vol. 50 (Feb. 1971), Special Issue.

15. Cavanaugh, J. R. "A Single Weighting Characteristic for Random Noise in Monochrome and NTSC Color Television," *Journal SMPTE*, Vol. 79 (Feb. 1970), pp. 105-109.

16. Lessman, A. M. "The Subjective Effects of Echoes in 525-Line Monochrome and NTSC Color Television and the Resulting Echo-Time Weighting," *Journal of the SMPTE*, Vol. 81 (Dec. 1972), pp. 907-916.

17. Lessman, A. M. "Subjective Effects of Delay Difference Between Luminance and Chrominance Information of the NTSC Color Television Signal," *Journal of the SMPTE*, Vol. 80 (Aug. 1971), pp. 620-624.

Chapter 27

# Economic Factors

The quality of service provided by a telecommunications network must be based on an appropriate balance between customer satisfaction and the cost of service. To make service objectives meet the criterion of reasonable cost, compromises must often be made among the objectives themselves or between objectives and system development or application parameters.

A number of compromises may be used to illustrate the process of adjusting designs, applications, and objectives for economic reasons. There are, of course, no unchanging and absolute relationships among these factors. Guidelines tend to change with time because new systems, new services, and changing customer opinions bring about changes in the objectives. Furthermore, economic relationships are significantly affected by local and national economic factors such as inflation.

## 27-1 OBJECTIVES

The derivation and application of transmission objectives often involve judgment as to what can be accomplished within reasonable cost constraints, the compromises that result from such judgments, the reconciliation of one set of objectives with another, and the existing economic, environmental, and human resources factors. Consider first the determination of transmission objectives and the economic factors involved.

### Determination of Objectives

The determination of telephone transmission objectives requires the use of subjective testing to establish the relationship between an impairment and observer opinions of its effect. The test results are then related to measured or derived performance parameters to obtain values for the grade of service that can be expected for

**627**

the combination of parameters involved. It is often possible at this point to determine the cost of achieving this grade of service and the effects of changing the objectives or the performance parameters. These changes can then be evaluated economically by comparing the results with the initial cost.

Qualitatively, the results are usually predictable. In nearly all cases, costs increase when objectives are made more stringent or when performance is improved. The characteristics of a cost/grade-of-service curve are obviously important, and the judgment that must be exercised in establishing the objectives is influenced by the nature of this curve.

In Figure 27-1, curve A shows a gradual increase in cost with improving grade of service and demonstrates many situations in which the simple prediction of increasing cost with improving grade of service is verified. Since the simple prediction does little to support engineering judgment, the establishment of the objective must be based on other criteria. On the other hand, curves B-B′ and B-B″ represent very different sets of circumstances.

Curve B-B′ shows that a relatively small increase in cost yields a substantial improvement in grade of service up to a good-or-better rating of 97 percent and that, regardless of cost, the grade of service cannot be increased beyond 98 percent. Thus, from the point of view

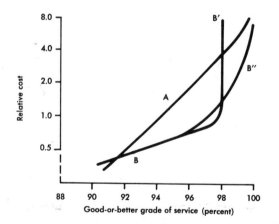

Figure 27-1. Typical cost curves.

of economic effects, any attempt to achieve higher than a 97 percent good-or-better grade of service would be wasteful.

Curve B-B″ shows another type of relationship in which costs increase somewhat faster above 95 percent good-or-better grade of service than below. The change is not nearly as abrupt as for Curve B-B′, and the achievement of a 96 or 97 percent grade of service (considered satisfactory) would be justified. The important point to note is that the derivation of cost curves such as those illustrated often provides strong support for determining objectives, either in terms of grade of service or more directly in terms of transmission objectives. Such curves may also be used to judge the cost of making desired improvements in performance.

### Allocation of Objectives

As described in Chapter 25, objectives may be allocated to different sources of an impairment, the total objective for one type of impairment may be allocated to different parts of the plant (e.g., local or toll), or the total objective may be allocated to various parts of a transmission system. Each method of allocation is either directly dependent on or indirectly tempered by economic factors.

An illustration of how economic factors can affect the allocation of an objective to different sources of the impairment is seen in the design of analog submarine cable transmission systems. In most analog cable systems for land application, signal amplitudes are adjusted to produce optimum signal-to-noise performance. In submarine cable system design, the cost of cable repairs enters into the problem of noise allocation to various sources. If cable laying, aging, or other phenomena cause unanticipated misalignment of signal amplitudes in the positive (overload) direction, the increase in intermodulation noise might necessitate installation of additional equalizers in the cable, a costly operation. To guard against this possibility, submarine systems are usually operated at low signal amplitudes. The results are high margin against overload and the allocation of most of the message circuit noise objective to thermal noise. Intermodulation is seldom a controlling source of message circuit noise in submarine cable systems.

An illustration was given in Chapter 25 of how economic factors influence allocation of the objective for one type of impairment to various parts of the plant. It was pointed out that the 20 dBrnc maximum noise allocated to loops is such that no amount of expenditure in the loop plant could possibly improve the noise grade of service unless both loop and trunk objectives are made more stringent. At present, the noise resulting from trunks (carried predominantly on carrier systems in the intertoll portion of the network) controls the grade of service.

Finally, economic factors may affect the allocation of an impairment to different parts of a system. For example, in long analog cable transmission systems, the design objective for message circuit noise for a 4000-mile system is 40 dBrnc0. A possible allocation of this objective might be 37 dBrnc0 to the line repeaters and 37 dBrnc0 to terminal multiplex equipment. However, the difficulty and cost of achieving high quality performance in line repeaters (used in large numbers compared to the number of terminals) is recognized by a higher allocation to the line equipment. In most systems, the line repeaters are allocated 39.4 dBrnc0 and the terminal equipment 31.2 dBrnc0. This allocation, when further translated to individual units (repeaters or terminal equipment), still results in a per-unit allocation that is more stringent for a repeater than for the terminal equipment. However, the economic balance is such that a further allocation of the objective to the repeater (which already has been allocated about 87 percent of the total) would not result in significantly lower overall costs.

## Economic Objectives

At certain times and under certain circumstances economic objectives may supersede all others. In times of economic stress, the desirability of improving performance or increasing route capacity may have to be subordinated to the necessity of reducing capital and operating expenditures. Such circumstances, undesirable as they may seem, must be recognized and improvements or expansion must be deferred.

In addition to the effects of economic stress, other less dramatic effects must be considered. Among the most significant of these is the availability of capital funds versus anticipated revenue. Sometimes it is necessary to keep outmoded equipment in service by paying for its maintenance from operating funds even though the results of engineering economy studies have demonstrated the desirability of replacing the old equipment with new. When capital funds are in short supply, it is impossible to update equipment in the desired manner. This type of situation may be disclosed by engineering economy studies which compare initial capital outlays and estimated operating costs to available capital funds and anticipated revenue.

## 27-2　DESIGN COMPROMISES

Most of the compromises that must be made between objectives and cost are made during the development and design of new systems and new equipment. These compromises are made at every stage of development and design; the type of system to be developed, the features to be provided, the choice of circuits and physical designs, and the selection of components all relate to the balance between objectives (grade of service) and cost.

### Circuit Devices

Devices used in electronic circuits include such elements as resistors, capacitors, inductors, transformers, transistors, and diodes. Each device selected for the circuit under design must obviously meet the requirements imposed by its function in the circuit. It must be of the correct value, capable of dissipating a certain amount of power, characterized by input/output relationships that are adequately linear, sufficiently reliable, etc. Even with these constraints, there is often a wide choice within which circuit needs can be met. Making that choice with good judgment involves consideration of costs and their relationship to the circuit requirements. Two significant factors are the manufacturing costs of the devices used and the ingenuity of the designer in utilizing a device to serve more than one function.

The benefits of mass production are evident in the reduced cost of devices. Also, economic benefits are usually effected when a device can be made to serve multiple functions, as do many of the devices in telephone station sets. The quantity of sets manufactured annually is so high that even a fraction of a cent saved in one device yields a significant manufacturing cost saving. As a result, a great deal of effort is devoted to design and redesign of the station sets and of each device used. In such applied cost reduction studies, careful attention must always be given to every aspect of the design, including the environmental conditions that are found in the operating plant (heat, humidity, voltage, handling, etc.) as well as the circuit requirements.

## Circuits

As discussed here, circuits are packaged entities of interconnected electronic devices that provide some specific function such as modulation, multiplexing, or amplification. A circuit may include electronic networks, filters, and equalizers, often referred to as *apparatus*.

The design of circuits has progressed rapidly in recent years from point-to-point connection of devices through printed wiring techniques to a gamut of thin film, thick film, and integrated circuit arrangements that have evolved with the development of solid state technology. With the wide choice of circuit arrangements available, careful attention must again be paid to economic factors. If large numbers of identical circuits are to be built and close control of circuit performance is required, integrated circuits are likely to be a good first choice. Sometimes, the added expense of integrated circuits in small quantities is justified because the reproducibility of integrated circuit performance is high.

An interesting illustration of circuit selection based on economic factors hinges on the selection of devices involved in the design of narrowband elimination or bandpass filters. In cases where the total available band for achieving prescribed characteristics is wide, the design may employ electronic devices, but if the efficiency of bandwidth utilization must be high and the available bandwidths are small, piezoelectric crystals may be needed to achieve the desired characteristics. The cost of the resistors, capacitors, and inductors used in an electrical filter design is, in most cases, much lower than the cost of crystals and the necessary additional devices. The choice depends on available bandwidth and the stringency of the requirements.

## Physical Design

The physical design of equipment and facilities is greatly influenced by the costs of maintenance and operation as well as by the costs of manufacture and installation. Recent trends in physical design have been influenced by the decision to adopt new standards in building design and by the recognition that both transmission and operation could be improved by integrated designs of equipment bays [1, 2]. These integrated designs, sometimes called unitized bays, include many more combinations of transmission, signalling, and switching system interface equipment than were formerly provided in a single bay. Some of these combinations have been made possible by the development of miniature devices and some by improved techniques of bay wiring and functional circuit interconnection. The new designs result in a significant reduction in office wiring, the elimination of a number of cross-connect frames, reduced congestion of cable racks and cross-connect frames, and a reduction in the number of jack fields.

The most significant feature of the new building design standards is the reduction of ceiling height and the concomitant standardization of 7-foot equipment bay heights. The packaging of electronic circuits must now be consistent with the 7-foot bay standard, but in order to serve existing buildings with reasonable efficiency, bays are also designed to old standards. The necessity for designing equipment for several bay heights has led to a number of design compromises that will eventually be unnecessary. As buildings of new design become predominant, bay designs for the older buildings will no longer be economically justifiable.

Unitized bay designs have led to a set of design compromises different from those relating to building design. Facility terminals, as the new designs are called, contain all voice-frequency terminal equipment needed for a specific facility. In general, transmission performance improves, but there may be some minor limitations on the features that can be provided and the flexibility of equipment use.

The advent of solid-state technology has also led to situations in which the solutions to design problems have resulted in various compromises. A transistor dissipates less power than an electron tube, but transistors are so much smaller that many more can be packaged into a given volume than electron tubes. The result is higher heat dissipation per unit of volume for transistor circuits, so temperature control has become a problem in packaging solid-state devices. Since the higher density of components has lead to higher weight per unit of volume in many designs, floor loading must be reconsidered. Thus, physical designs have interacted with circuit and system designs to bring about new adjustments in objectives and design features. The process of adjustment and compromise is continuous and parallels the development of all aspects of new technology.

## Systems

The design of systems follows the same pattern of compromise as has been outlined for components, circuits, and physical designs. System features and design criteria must be considered in respect to feasibility and cost. Reliability, maintainability, restoration of service, automatic versus manual testing, remote control and telemetry, and many other operational features must all be weighed carefully in terms of service and cost.

The balance among system alternatives and cost factors plays an important role in determining whether to develop a new system. An example, illustrated by Figure 27-2, involves the cost of a carrier system relative to the cost of copper pairs for voice-frequency transmission. Costs for carrier and voice-frequency transmission are normalized to a value of unity at the point where the two costs are equal. As illustrated, the cost of carrier transmission has a base, $A$, representing the fixed cost of the terminal equipment. To this base cost is added the line cost (medium and electronics), which increases approximately linearly with distance. The cost of voice-frequency transmission increases linearly from a base of zero except for discontinuities, designated $B$, introduced by possible gauge changes and the periodic need for VF repeaters. The slope of the VF facility cost curve is directly affected by the total cable cost and the number of pairs per circuit.

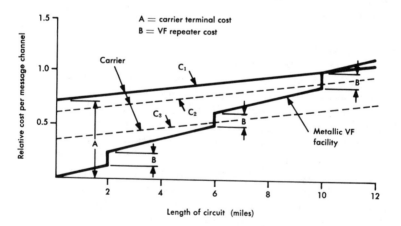

Figure 27-2. Comparison of costs for very short circuits.

It can be seen that the cost/distance curve for the carrier system is already less steep than for voice-frequency transmission. Further, it is evident that even if this slope is significantly reduced, the cost of a circuit is not materially affected because of terminal costs. The conclusion is that only a significant reduction of the terminal cost, $A$, can be expected to improve the position of carrier transmission relative to that of voice-frequency transmission. Curve $C_2$ shows the effect of a terminal cost reduction of about ten percent, a reduction that has no effect on the relative markets for the two transmission modes because the crossover point of the cost curve is still at ten miles. However, with a different set of curves and crossover points, a ten percent reduction might be very significant and lead to a different conclusion.

Curve $C_3$ shows the effect of a terminal cost reduction of about 50 percent. This may well provide encouragement for the development of new carrier terminals if the cost reduction appears to be possible, because the crossover point of the carrier and voice-frequency transmission cost curve is now at six miles. In addition, it would be necessary to show that there are large numbers of circuits in the range of six to ten miles and that there could be a high expectancy of achieving the 50 percent cost reduction by terminal redesign.

Many studies of the type described have been made to guide the development of T-type and N-type carrier systems. The curves of Figure 27-2 are representative but are not based on any specific study results. Many other details must be included in a transmission system development study, such as the gauge of wire and the loss to which the circuits are designed.

A second example of cost factors in transmission design is shown in Figure 27-3, which illustrates the effect of electronic equipment costs on the total line costs in long-haul analog cable transmission systems. The channel capacity or bandwidth of a number of systems is shown relative to an arbitrarily selected bandwidth taken as unity and to an arbitrarily selected unit of line cost. The line costs for a number of systems are then plotted in terms of electronic and nonelectronic components. Nonelectronic cost components include the cost of cable, installation, and right-of-way.

Examination of the curves of Figure 27-3 shows that as the normalized bandwidth increases beyond a value of 3, the cost of

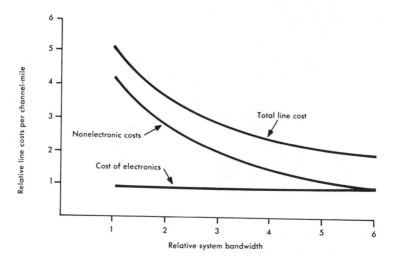

Figure 27-3. Electronic and nonelectronic line costs, analog cable systems.

electronic equipment increases gradually. More important, the cost of nonelectronic components and electronic components is about equal for a normalized bandwidth of 6. Thus, it may be expected that any additional increase in bandwidth would show that the total cost starts to increase, and the development of new systems would be financially questionable. Other means of transmission should probably be considered.

## 27-3  APPLICATION COMPROMISES

Economic factors influence transmission problems in the field much the same as in the laboratory environment. In the field, the questions that arise involve long-range and short-range planning activities; the optimum selection of equipment, transmission media, and systems; and the allocation of funds to satisfy necessary operating functions.

### Components

In the field environment, the word components refers to units of equipment such as filters, equalizers, amplifiers, or balancing networks. In designing and laying out loops and trunks, the proper selection of such components plays a significant part in achieving performance consistent with established objectives and, at the same time, in satisfying economic constraints.

In addition to the decisions that must be made to satisfy technical objectives, the choice of equipment components must often include consideration of general trade equipment components that can be purchased outside the Bell System. Many suppliers have introduced equipment in the market that meets the needs of telephone operating companies. Thus, a knowledge of outside suppliers' equipment, its performance capabilities and cost is imperative.

## Systems

In an operating company the planning function typically includes engineering studies that involve system choices in which economic factors are important. Many economic studies must be made since the choice of system depends heavily on cost relationships. Cost curves similar to Figure 27-2 provide a good basis for solving system applications problems as well as design and development problems. By combining cost curves and a projected distribution of circuit lengths, it is often a simple process to decide the most economical choice of facilities. However, where system capacities are large, analysis may be quite complicated.

In very heavily populated areas, where large circuit cross-sections are needed, there may be more than the usual choices of systems. In most cases, the choice is primarily made from several alternate plans for increasing the amount of paired copper cable, and the use of T- and N-type carrier systems. In larger metropolitan areas, the choices may be increased by the possibility of using microwave radio systems or digital coaxial cable systems. Several such systems are now available for metropolitan area applications.

Efficient use of maintenance personnel is another economic factor in choosing a system. A number of equipment types are now available to make measurements and surveillance tests automatically and even to control certain operational functions remotely. In any planned installation or expansion program, the cost of using such equipment must be compared with the more conventional manual and on-site maintenance methods.

### REFERENCES

1. Pferd, W. "NEBS: Equipment Buildings of the Future," *Bell Laboratories Record* (Dec. 1973), pp. 359-364.
2. Giguere, W. J. and F. G. Merrill. "Getting It All Together with Unitized Terminals," *Bell Laboratories Record* (Jan. 1973), pp. 13-18.

Chapter 28

# International Telecommunications

The evolution of telegraph communications, the invention of the telephone, the development of radio communication, and the expansion of the total communication network have led to one of the world's most highly developed telecommunication systems on the North American continent. Whatever the principal reasons for this tremendous, but isolated growth in the United States and Canada (invention, corporate organization, common language, few and open international frontiers, etc.), little need was initially evident for coordination with other nations or for the establishment of international objectives or standards of performance. Thus, standards which in some cases, are quite similar to those now applied internationally, yet in other cases are quite different, have evolved independently in the United States and Canada.

When transatlantic telegraph and then radio-telephone communications were established, the need for some form of intercontinental coordination was recognized for the first time. The need for continuing and expanding coordination has been more evident as submarine cable and satellite communications have become realities. Today, the U. S. Government, a number of American common carriers, and American industrial and scientific organizations (including manufacturers) are members and active participants in international telecommunications organizations. The increasingly intense use of the radio spectrum, the enlarged international market and the increasing use of international direct distance dialing are bringing about increased standardization of international communications. This trend is supported, in part, by the proliferation of telecommunication equipment manufacturers, whose international marketing efforts are naturally enhanced when their equipment satisfies international recommendations and standards. International transmission planning permits the use of various internal trans-

638

mission plans nationally, but international standards are met at international switching offices and on international circuits.

## 28-1  THE INTERNATIONAL TELECOMMUNICATION UNION

While international coordination of telecommunications did not directly affect the development of the North American network, the need for coordination and cooperation among European nations was evident as long ago as 1865. At that time, the International Telegraph Union was formed by a convention in Paris attended by twenty delegations from different European countries. Today there are nearly 150 members of the International Telecommunication Union (ITU), the modern descendent of the International Telegraph Union. The ITU engages in worldwide activities and is one of the specialized agencies of the United Nations, which recognizes it as the sole specialized agency competent for telecommunications [1, 2, 3].

Three international technical advisory committees were formed to deal with various aspects of international telecommunications. These committees were an outgrowth of the International Committee for Long Distance Telephony*, commonly referred to as the CCI, inaurated in 1924. The three committees were called (1) the International Consultative Committee for Telegraph (CCIT), the International Telephone Consultative Committee (CCIF), and the International Radio Consultative Committee (CCIR). By 1939, when work was interrupted by World War II, these committees had succeeded in solving most of the traffic, operating, and radio coordination problems to the satisfaction of all Administrations [4]. The American Telephone and Telegraph Company participated in the activities of these early committees, first as an observer and then, after 1929, as a fully qualified member.

At present, the relationships between the ITU and the U. N. (and several other specialized agencies of the U. N.) provide for cross-representation between organizations, but the ITU has retained its independence. The ITU acts essentially as a technical advisory and

---

* The original name of this organization was the *Comité Consultatif International des Communications Téléphoniques à Grande Distance*.

administrative body. In 1956, two of its committees, the CCIF and the CCIT merged into the International Telegraph and Telephone Consultative Committee (CCITT).

The ITU and its principal organs maintain an active interest in all aspects of international telecommunications, including the studies of a wide variety of technical problems and the coordination of international traffic and operating procedures. In addition, a special working party of the CCITT and the CCIR (two of the principal organs of the ITU) have produced a handbook to assist the administrations and private operating agencies in an appreciation of the technical and economic problems involved in the planning of transmission systems [5]. Other special working parties have prepared handbooks on national automatic networks and on local networks. These handbooks contain information on current practices in countries that have highly developed telecommunication facilities and networks and are intended to help other countries fill their telecommunications needs.

International cooperation regarding satellite communications has been fostered by studies and recommendations of the CCIR. With the advent of launch vehicles capable of placing substantial radio communication equipment into earth orbit, the CCIR began studies of the commercial feasibility of international communication satellites which culminated in recommended criteria for this mode of communication. The radio frequencies required by early satellites were located in a frequency band bounded by excessive rain absorption above 10 GHz and high galactic noise below about 1.0 GHz. In order to share the available useful frequencies with existing microwave radio systems, equitable criteria had to be devised and international agreement obtained. In 1963, an Extraordinary Administrative Radio Conference (of the ITU), to which the CCIR is the consultative technical organization, adopted initial sharing criteria (e.g., signal powers, frequencies, and allowable interference) for satellites, earth stations, and the terrestrial systems affected. Their conclusions became part of the international radio regulations upon treaty ratification by the various countries involved, including the United States. The initial criteria prevailed until 1971 when a World Administrative Radio Conference adopted new criteria and allocated additional frequencies above the earlier 10 GHz maximum. This action was based on recommendations resulting from CCIR studies.

## Organizational Structure of the ITU

The ITU consists of four permanent organs: (1) a General Secretariat directed by the Secretary-General and a Deputy Secretary-General, (2) the International Frequency Registration Board (IFRB), (3) the International Radio Consultative Committee, and (4) the International Telegraph and Telephone Consultative Committee. The General Secretariat provides liaison between Administrations and private operating agencies throughout the world and is entrusted with the administrative and financial services of the ITU; it also has a Technical Cooperation Department whose experts work in various countries to provide technical assistance where needed [3, 6].

**International Frequency Registration Board.** The IFRB acts as the recipient of information from the Administrations to record the frequency assignments of certain types of radio stations. It also acts, wherever possible, to predict potential interference and to adjudicate complaints of radio interference between Administrations by suggesting solutions to real or incipient problems.

**International Radio Consultative Committee.** This committee studies technical questions relating to radio transmission and operations and issues recommendations based on technical reports resulting from their studies. The committee is made up of representatives from all members of the ITU Administrations and recognized private operating agencies. When authorized, industrial and scientific organizations may participate on a consultative basis. The plenary assembly assigns work to various study groups and working parties whose reports are received at plenary assemblies held by the CCIR approximately every three years. The reports of the study results submitted and of the resulting actions taken by the plenary assemblies are published by the ITU.

**International Telegraph and Telephone Consultative Committee.** The CCITT, like the CCIR, is made up of representatives from all members of the ITU, some recognized private operating agencies, and industrial and scientific organizations. The CCITT studies technical, operating, and tariff questions connected with international telecommunications. The study groups and working parties present the results of their studies to plenary assemblies of the CCITT, held about every three years. These reports, together with other actions of the plenary sessions of the CCITT, are published by the ITU in volumes whose colors are chosen to be distinctively related to a particular plenary session.*

## Study Groups and Working Parties

Most of the technical work of the CCIR and the CCITT is carried out by study groups, special study groups, and joint working parties that are assigned responsibility for specific types of problems or fields of investigation. These groups meet as required in order to consider their assigned questions and problems. Figure 28-1 shows a number of study groups of the CCITT, some of which operate jointly with the CCIR. Regular study groups are designated by Roman numerals and, in respect to their responsibilities, they fall within a functional division of one field of study. Special study groups, designated by letter, are involved in more than one field of study. They may be formed of members from the CCIR, the CCITT, or both. In Figure 28-1, arrows are used to indicate interactions between special study groups (SP. A, SP. C, and SP. D) and the functional study groups. Interactions within a field of study are not shown. Working parties are formed within a study group to study a particular problem or field of investigation; they may be permanent or they may exist only for the time necessary to complete an assignment. Study groups have the responsibility of responding to specific questions assigned by a plenary assembly of the CCIR or CCITT; these questions and subsequent recommendations are included in the official publications of the CCIR or CCITT. The study groups also prepare a list of questions and study programs for the following plenary period (the period between plenary assemblies). The questions are proposed by the members of the CCIR or CCITT. Formal approval by the plenary assembly is required for recommendations and questions to become official.

---

\* Examples are the Red Books, which cover the meetings of 1958 at Geneva and 1960 at New Delhi (Ist and IInd Plenary Assemblies); the Blue Books, which cover the meeting of 1964 at Geneva (IIIrd Plenary Assembly); the White Books, which cover the meeting of 1969 at Mar Del Plata (IVth Plenary Assembly); and the Green Books, which cover the meeting of 1972 at Geneva (Vth Plenary Assembly).

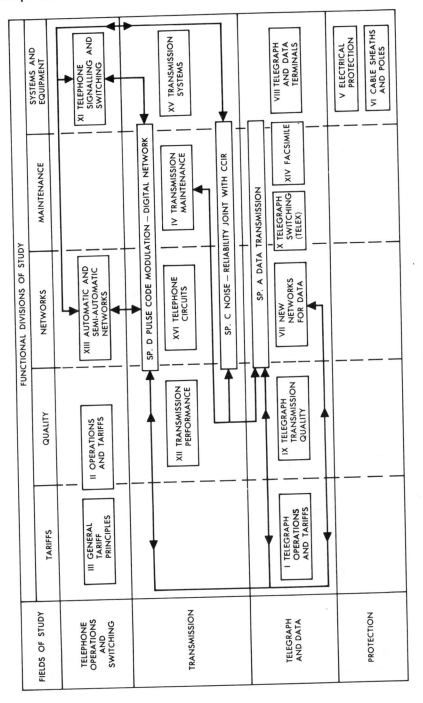

Figure 28-1. CCITT study groups.

### Characteristics of International Operations

The major differences between national and international operations of telecommunication networks involve geography, language, law, and politics. The ITU has performed admirably in solving the related problems by patient negotiation and by the dedicated services of the many persons representing its membership. The ITU operates on the basis of voluntary membership with the assumption that it is in the members' best interests to observe the conventions, regulations, and recommendations of the Union [7].

## 28-2 THE EVOLVING INTERNATIONAL NETWORK

The members of the ITU, through recommendations of the CCITT, have agreed on the goal of providing customer dialing of international calls on a worldwide basis. A general plan for achieving this goal has been agreed upon and is being implemented on a step-by-step basis. Operator and customer dialing of international calls is already a reality in many countries. The capability for operator and customer dialing requires all-number calling, already in effect or planned for most countries.

### World Numbering Plan

Worldwide operator and customer dialing requires a worldwide numbering plan. An appropriate plan, worked out by the CCITT, divides the world into eight zones, as shown in Figure 28-2. Each country is given a 1-, 2-, or 3-digit country code number, the first digit of which identifies the world numbering zone. In the multinational North American zone 1, which is already organized into a single integrated numbering plan, the single digit 1 is used as the country code of all the countries in zone 1. Another interesting detail is that the European zone has so many countries warranting two-digit codes that it has been assigned the initial digits 3 and 4.

For worldwide dialing, a customer or operator must first dial an international prefix. Ambiguity between national and international numbers employing the same initial digits is overcome by first dialing the international prefix. In the North American network, the prefixes for international dialing are 01 for calls requiring operator assistance and 011 for other calls. After the international prefix, the country code and the national number of the called station are dialed. The international numbering plan places the restriction that, after the

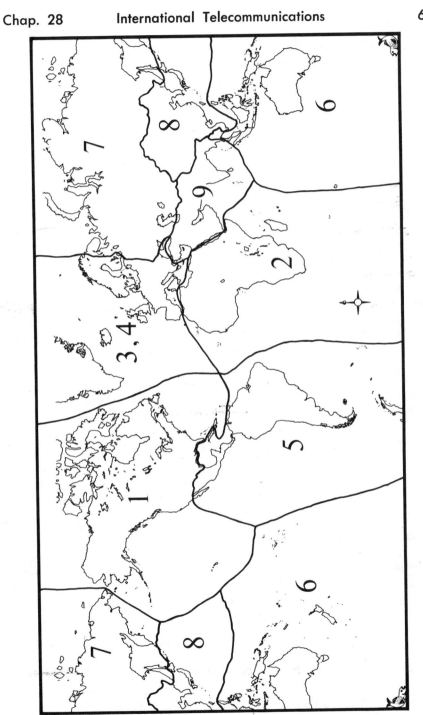

Figure 28-2. World numbering zones.

international prefix, there may be a maximum of twelve digits in the international telephone number.

Some of the capabilities implied by the numbering plan just described are yet to be implemented. In some cases, the digit capacity of registers in local and toll offices must be expanded. Call routing from an originating office to the appropriate international switching center must be provided.

## Signalling

When serious consideration was first given to direct dialing of transatlantic calls, it was immediately noted that the European and North American signalling systems were incompatible. Furthermore, none of the existing systems were compatible with the needs of the TASI system designed for use with the early submarine cable systems.

Early submarine cable operation was carried out by manual ringdown signalling. In 1960, agreement was reached by the British, French, and German Administrations with the Bell System on a specification for an intercontinental signalling system. The system, sometimes called the Atlantic system, was compatible with TASI, provided for two-way operation of circuits, and provided practical interfaces with the European and North American signalling systems. It used a modified version of the North American multifrequency pulsing for the transmission of address information and a new 2-frequency system for supervisory signals. In 1960, both of the latter frequencies were in standard use in Bell System signalling systems.

The CCITT was requested to study the Atlantic system for standardization as a recommended intercontinental signalling system. The system was accepted by the CCITT with some minor changes and was designated the CCITT Signalling System #5. Subsequently, the CCITT standardized a common channel signalling system (CCITT #6), designed to operate with stored program switching systems and capable of providing features not available in System #5.

## Traffic and Operating

While most international calls were formerly person-to-person, station-to-station calling is commonly used between many countries and is increasingly available for both incoming and outgoing U. S. international calls. Credit card and collect calls are also accepted for calls between many countries and are also more widely used in international telephony.

An international operator may sometimes have language difficulties or be unable to interpret a tone. To alleviate these problems, calling operators are able to ring forward and bring in an assistance operator in the terminating country. A language digit is prefixed to the called number by the switching machine and pulsed forward to prepare the distant equipment for receipt of a subsequent language-assistance signal.

## Routing Plan

A routing plan that is similar in many ways to the routing plan used in the North American network has been recommended by the CCITT. The hierarchical arrangement of the worldwide plan utilizes three levels of international switching centers (transit centers), designated CT1, CT2, and CT3; CT1 represents the highest rank in the hierarchy. The plan is shown in Figure 28-3.

High usage trunk groups between any pair of CT offices are established wherever they can be economically justified. Provisions are made for alternate routing of overflow traffic from high usage groups to alternate transversal trunk groups and then to the final group. An example is given in Figure 28-3. For a call from CT3 on the left to CT3 on the right, attempts would be made first to use the direct group between these offices. As implied by the arrows, attempts would be made to route the call to CT2 and then to CT1 on the right. Finally, the call would be offered to the final route.

According to the routing plan, CT1 offices are to be interconnected in pairs by circuit groups having low probability of blocking. In exceptional cases, two CT1 offices may be interconnected through an intermediate transit center of unspecified rank (CTx). This is done only where significant economies may be realized and only if transmission and other standards of service quality are met.

The CT1 offices are important in the world routing plan. Locations are chosen to satisfy national and international economic considerations as well as technical requirements for switching and trans-

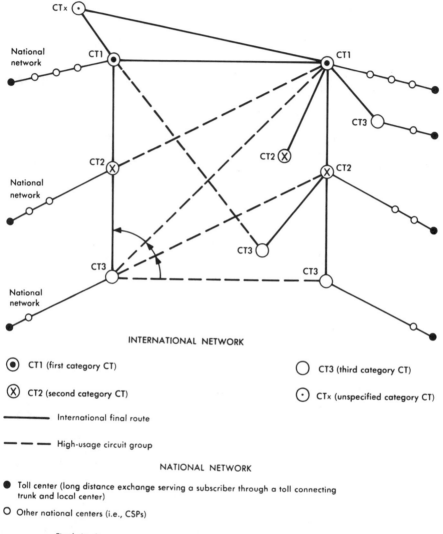

Figure 28-3. International routing plan.

mission. Each country in which a CT1 is located must be concerned with the costs of interconnecting that CT1 with all others by direct circuit groups. The CT1 offices are few in number and their recommended locations are strategically chosen on the basis of the in-transit flow of international transit traffic. These offices are concentration points for the traffic from a very large area. A number of CT1 offices have been designated or proposed by ITU members to implement the world routing plan.

The maximum recommended number of tandem-connected trunks that may be used for an international call is fixed by the CCITT at 12, with a maximum of six international circuits. In some cases and for only a small percentage of calls, the total number of tandem circuits may be as high as 14; even in these cases, the maximum number of international circuits is limited to six. An international call involving six international circuits would be one routed through transit centers in the following manner: CT3, CT2, CT1, CTx, CT1, CT2, CT3.

Final route engineering of a worldwide network for efficient handling of busy-hour traffic poses interesting problems caused by the concentration of traffic during a few hours of the day due to time zone differences. The CCITT has initiated a study of solutions to these problems by flexible routing and some form of network management.

## Transmission and Maintenance

The possibility of 12 or even 14 tandem-connected circuits increases the likelihood that international connections may be impaired more than domestic connections. In addition to the greater lengths and increased number of circuits involved, the variation in types of facilities also increases. These factors increase the probability of increased loss, loss variation, noise, distortion, and propagation delays. Unless very stringent controls are imposed, there is an increased likelihood of encountering multiple echo suppressors on an international connection. All of these factors make necessary a high degree of control over transmission design and maintenance.

The procedures involved in establishing and maintaining international circuits have been and continue to be the subjects of study by members of the CCITT. The resulting recommendations cover such aspects of international circuits as the types of facilities, switching systems, and signalling arrangements; detailed responsibilities for control, trouble locating, testing, and maintenance; and procedures for operating the international network. At the time international circuits are established between two countries, detailed agreements (largely based on current recommendations of the CCITT) are reached on all of the specific items involved in maintenance.

## 28-3  TRANSMISSION PARAMETERS AND OBJECTIVES

The CCIR and CCITT have defined a large number of transmission parameters and have established or recommended many transmission objectives for the international telecommunication network. These are thoroughly covered in documents published by the ITU [8, 9]. Space does not permit a comprehensive discussion here, but transmission level points, noise, and channel loading serve to illustrate the manner in which transmission problems are treated internationally.

Currently, in the reports published by ITU, transmission parameters are expressed in decibels. Some parameters are also expressed in decimal units of the international system of units. For example, noise and noise objectives are expressed in picowatts and picowatts per kilometer.

International practices and recommendations include the use of a *reference level,* a term that is analogous to the 0 TLP used in the Bell System. The reference level point is sometimes referred to as 0 dBr (dB relative level). For four-wire operation, the transmitting end of the circuit is defined as a −3.5 dBr point at the "virtual" switching point, a theoretical point whose exact location depends on national practice.

### Noise

The basic unit of noise measurement used in international practice is the picowatt (pW), i.e., $10^{-12}$ watt. It should be noted that for a 1000-Hz signal, this is the same reference as that used in the Bell System. In international maintenance practice, the standard test signal may be 800 or 1000 Hz. The picowatt may be expressed in decimal or logarithmic terms; the equivalent values are $1 \text{ pW} = 10^{-12} \text{ W} = 10^{-9} \text{mW} = -90 \text{ dBm}$.

Message circuit noise is measured, according to CCITT recommendations, by a noise measuring set called a *psophometer.* The set is equipped with a weighting network that has a characteristic somewhat similar to the C-weighting characteristic used in the Bell System. The two characteristics are shown for comparison in Figure 28-4. For general conversion purposes, it is usually sufficient to assume that the psophometric weighting of 3-kHz white noise decreases the average power by about 2.5 dB (compared with the 2.0-dB factor for C-message weighting). The term *psophometric voltage* refers to the rms weighted noise voltage and is usually expressed in millivolts.

The (rounded) conversion factor recommended by the CCITT for practical comparison purposes is that 0 dBm of white noise measured by a psophometer (1951 weighting) is equivalent to 90 dBrn measured on a 3A-type noise meter with C-message weighting. This conversion, which applies to white noise in the 300 to 3400 Hz band, is not valid for other noise shapes because of the differences between psophometric and C-message weighting [10].

The relationships between various CCITT and Bell System noise units are summarized in Figure 28-5. The data are particularly useful for conversion from one noise unit to another since an estimate of the frequency spectrum effects can be obtained by comparing the three conditions tabulated. The 1-kHz values are given for comparison of the various conditions used. The 1-kHz psophometric reading appears 1 dB high because the psophometric reference is 1 pW at 800 Hz. The 0- to 3-kHz band of white noise approximates the noise obtained from a message channel. The broadband white noise readings are proportional to the total area under the weighting curve and

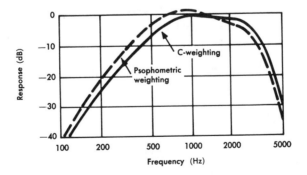

Figure 28-4. Comparison of noise weighting characteristics.

| NOISE UNIT | TOTAL POWER OF 0 dBm | | WHITE NOISE OF −4.8 dBm/kHz NOT BAND-LIMITED |
|---|---|---|---|
| | 1 kHz | 0 TO 3 kHz WHITE NOISE | |
| dBrnc | | 88.0 dBrnc | 88.4 dBrnc |
| dBrn 3 KHz FLAT | 90.0 dBrn | 88.8 dBrn | 90.3 dBrn |
| dBrn 15 KHZ FLAT | 90.0 dBrn | 90.0 dBrn | 97.3 dBrn |
| Psophometric voltage (600 ohms) | 870 mV | 582 mV | 604 mV |
| Psophometric emf | 1740 mV | 1164 mV | 1208 mV |
| pWp | $1.26 \times 10^9$ pWp | $5.62 \times 10^8$ pWp | $6.03 \times 10^8$ pWp |
| dBp | 91.0 dBp | 87.5 dBp | 87.8 dBp |

Figure 28-5. Comparison of noise measurements.

thus give significant information concerning the weighting function above 3 kHz. Similar data for other conditions or weightings can be obtained by integrating the appropriate weighting characteristic over the required frequency band.

## Channel Loading

To simplify calculations in carrier system design, the CCITT has adopted (Recommendation G.223) a conventional value, −15 dBm0, to represent the mean absolute power of speech and signalling currents [8]. When the −15 dBm0 value was established, it was based on a determination of expected channel signal loads. Analog system overload is discussed in Chapter 7 where *Definition 2* is the CCITT-recommended definition of overload.

The problem of loading has been under further study in recent years to determine whether the adopted value should be changed to reflect the transmission of new signal types. The approach that now appears most promising is that the amplitudes of data and other types of signals will be made compatible with −15 dBm0. One step has been to recommend a maximum sending reference equivalent (minimum loss) from the subscriber to the first international circuit. Also, several study groups have agreed that data and voice-frequency telegraph signals are to be reduced to −13 dBm0.

## REFERENCES

1. Michaelis, A. R. *From Semaphore to Satellite* (Geneva: International Telecommunication Union, 1965).

2. Leive, D. M. *International Telecommunications and International Law: The Regulation of the Radio Spectrum* (Dobbs Ferry, N. Y.: Oceana Publications, Inc., 1970).

3. Mili, M. "International Jurisdiction in Telecommunication Matters," *Telecommunication Journal*, Vol. 40 (April, 1973), pp. 174-182.

4. Timmis, A. C. *Compendium of the Technical Recommendations Issued by the CCIF, the CCIT and the CCIR of the International Telecommunication Union* (Liverpool: Automatic Telephone & Electric Co., Ltd., Aug. 1958), Chapters 1 and 2.

5. CCIR and CCITT. *Economic and Technical Aspects of the Choice of Transmission Systems* (Geneva, International Telecommunication Union, 1968).

6. Richards, D. L. *Telecommunication by Speech* (New York: Halsted Press Division, John Wiley and Sons, Inc., 1973), Appendix 2.

7. Cox, R. W., and H. K. Jacobson. *The Anatomy of Influence* (New Haven: Yale University Press, 1973), pp. 59-101.

8. CCITT. *Vth Plenary Assembly, Geneva, Green Book Volume III* (Geneva, International Telecommunications Union, 1973).

9. CCITT. *Vth Plenary Assembly, Geneva, Green Book Volume VI* (Geneva, International Telecommunications Union, 1973), Part II, Chapter VI.

10. CCITT. *Vth Plenary Assembly, Geneva, Green Book Volume V* (Geneva, International Telecommunications Union, 1973), Rec. P.53.

11. Mili, M. "International Jurisdiction in Telecommunication Matters," *Telecommunication Journal*, Vol. 40 (Sept. 1973), pp. 562-566.